# Specimen Science

**Basic Bioethics**
Arthur Caplan, editor

A list of the books in the series appears at the back of the book.

# Specimen Science

Ethics and Policy Implications

edited by Holly Fernandez Lynch, Barbara E. Bierer, I. Glenn Cohen, and Suzanne M. Rivera

The MIT Press
Cambridge, Massachusetts
London, England

© 2017 Massachusetts Institute of Technology

All rights reserved. No part of this book may be reproduced in any form by any electronic or mechanical means (including photocopying, recording, or information storage and retrieval) without permission in writing from the publisher.

Set in ITC Stone Sans Std and ITC Stone Serif Std by Toppan Best-set Premedia Limited. Printed and bound in the United States of America.

Library of Congress Cataloging-in-Publication Data

Names: Lynch, Holly Fernandez, editor. | Bierer, Barbara E., editor. | Cohen, I. Glenn, editor. | Rivera, Suzanne Marie, 1969- editor.
Title: Specimen science : ethics and policy implications / edited by Holly Fernandez Lynch, Barbara E. Bierer, I. Glenn Cohen, and Suzanne Rivera.
Description: Cambridge, MA : The MIT Press, [2017] | Series: Basic bioethics | Includes bibliographical references and index.
Identifiers: LCCN 2016041471 | ISBN 9780262036108 (hardcover : alk. paper)
Subjects: LCSH: Biological specimens--Moral and ethical aspects. | Bioethics--Legal issues.
Classification: LCC QH231 .S64 2017 | DDC 174.2--dc23
LC record available at https://lccn.loc.gov/2016041471

10  9  8  7  6  5  4  3  2  1

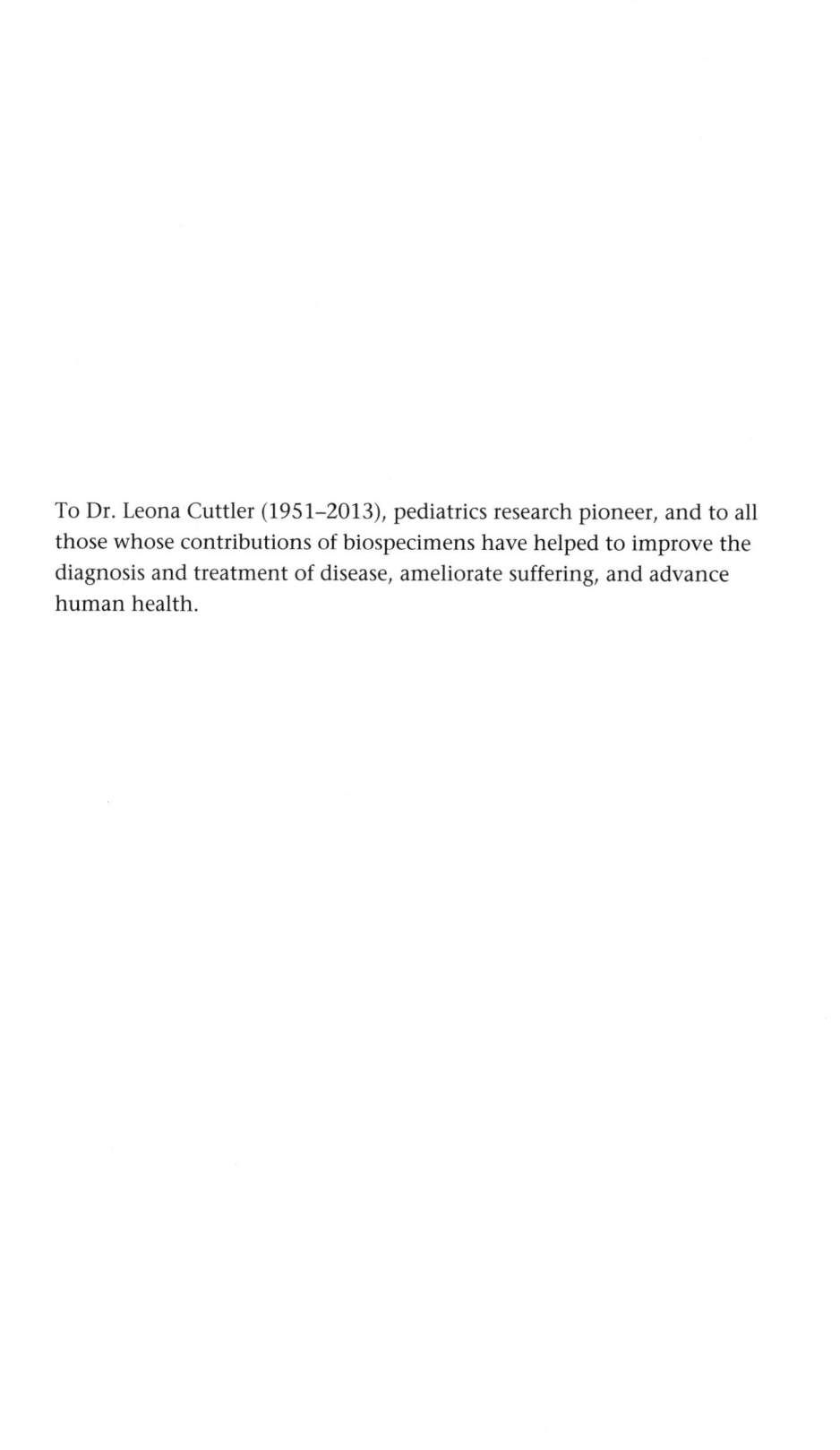

To Dr. Leona Cuttler (1951–2013), pediatrics research pioneer, and to all those whose contributions of biospecimens have helped to improve the diagnosis and treatment of disease, ameliorate suffering, and advance human health.

# Contents

Series Foreword  xi
Acknowledgments  xiii

**Introduction**  1
   Suzanne M. Rivera, Barbara E. Bierer, I. Glenn Cohen, and Holly Fernandez Lynch

**I   Background and Foundations**  19

**Introduction**  21
   Aaron S. Kesselheim

1  **Legal and Regulatory Issues in Biospecimen Research: National and International Perspectives**  25
   David Peloquin, Mark Barnes, and Barbara E. Bierer

2  **Property Rights and the Control of Human Biospecimens**  47
   Russell Korobkin

3  **Research with Biospecimens: Tensions, Tradeoffs, and Trust**  67
   Elisa A. Hurley, Kimberly Hensle Lowrance, and Avery Avrakotos

**II   Roots of the Debate: Autonomy, Justice, and Privacy**  85

**Introduction**  87
   Steven Joffe

4  **Research on Human Tissue Samples: Balancing Autonomy vs. Justice**  91
   David Korn and Rachel E. Sachs

5  **Biospecimens, Commercial Research, and the Elusive Public Benefit Standard**  107
   Barbara J. Evans and Eric M. Meslin

6   What Specimen Donors Want (and Considerations That May Sometimes Matter More)   125
    Suzanne M. Rivera and Heide Aungst

7   Assessing Risks to Privacy in Biospecimen Research   143
    Ellen Wright Clayton and Bradley A. Malin

III   Consent and Its Implications   159

Introduction   161
    P. Pearl O'Rourke

8   Broad Consent for Research on Biospecimens   167
    Christine Grady, Lisa Eckstein, Benjamin Berkman, Dan Brock, Sara Chandros Hull, Bernard Lo, Rebecca Pentz, Carol Weil, Benjamin S. Wilfond, and David Wendler

9   Evolving Consent: Insights from Researchers and Participants in the Age of Broad Consent and Data Sharing   185
    Nanibaa' A. Garrison

10   The Ethics of the Biospecimen Package Deal: Coercive? Undue? Just Wrong? Or Maybe Not?   201
    Ivor Pritchard and Julie Kaneshiro

IV   Special Populations and Contexts   219

Introduction   221
    Pamela Gavin

11   Biorepositories and Precision Medicine: Implications for Underserved and Vulnerable Populations   225
    Aaron J. Goldenberg and Suzanne M. Rivera

12   The Ethical Management of Residual Newborn Screening Bloodspots   243
    Jeffrey R. Botkin, Erin Rothwell, Rebecca A. Anderson, and Aaron J. Goldenberg

13   Informed Consent for Genetic Research on Rare Diseases: Insights from Empirical Research   257
    Sara Chandros Hull

14   Considerations for the Use of Biospecimens in Induced Pluripotent Stem (iPS) Cell Research   273
    Geoffrey Lomax and Heide Aungst

## V  Governance, Accountability, and Operational Considerations   291

**Introduction**   293
Barbara E. Bierer

**15**  Governance Issues for Biorepositories and Biospecimen Research   299
Karen J. Maschke

**16**  The Rise of Patient-Driven Research on Biospecimens and Data: The Second Revolution   317
Susan M. Wolf and Isaac S. Kohane

**17**  Informing the Public and Including It in Discussions about Biospecimens   335
Jane Perlmutter and Heide Aungst

**18**  Investigator's Commitment during the Consent Process for Biospecimen Research   353
Erin Rothwell and Erin Johnson

**19**  Biospecimen Repositories in the Era of Precision Medicine: Perspectives from a Biobanker "in the Trenches"   367
Quinn T. Ostrom and Jill S. Barnholtz-Sloan

**20**  Operationalizing Institutional Research Biospecimen Repositories: A Plan to Address Practical and Legal Considerations   383
Kate Gallin Heffernan, Emily Chi Fogler, Marylana Saadeh Helou, and Andrew P. Rusczek

Contributors   403
Index   409

# Series Foreword

Glenn McGee and I developed the Basic Bioethics series and collaborated as series co-editors from 1998 to 2008. In fall 2008 and spring 2009 the series was reconstituted, with a new editorial board, under my sole editorship. I am pleased to present the forty-eighth book in the series.

The Basic Bioethics series makes innovative works in bioethics available to a broad audience and introduces seminal scholarly manuscripts, state-of-the-art reference works, and textbooks. Topics engaged include the philosophy of medicine, advancing genetics and biotechnology, end-of-life care, health and social policy, and the empirical study of biomedical life. Interdisciplinary work is encouraged.

Arthur Caplan

**Basic Bioethics Series Editorial Board**
Joseph J. Fins
Rosamond Rhodes
Nadia N. Sawicki
Jan Helge Solbakk

# Acknowledgments

This volume was based on a public conference held at Harvard Law School in November 2015, the proceedings of which are available at http://petrieflom.law.harvard.edu/events/details/specimen-science-ethics-and-policy/.

We gratefully acknowledge the support of our various collaborators, including The Center for Child Health and Policy at Case Western Reserve University and University Hospitals Rainbow Babies & Children's Hospital; the Petrie-Flom Center for Health Law Policy, Biotechnology, and Bioethics at Harvard Law School; the Multi-Regional Clinical Trials Center of Brigham and Women's Hospital and Harvard; and Harvard Catalyst | The Harvard Clinical and Translational Science Center. Both the conference and this volume were supported by funding from the National Human Genome Research Institute.

We also wish to thank Cristine Hutchison-Jones and Justin Leahey from the Petrie-Flom Center, and our fantastic research assistants—Ethan Stevenson, Shailin Thomas, and José LaMarque—who helped us line edit and format the entire manuscript.

Last, but certainly not least, we sincerely thank all of our contributors for their tremendous efforts.

# Introduction

Suzanne M. Rivera, Barbara E. Bierer, I. Glenn Cohen, and Holly Fernandez Lynch

When a blood specimen is drawn from a vein in your arm, is that specimen still you? Do you "own" it? Are you entitled to any intellectual property interest in or profit from a product derived from it? What if its "value" is only understood by pooling information from 10,000 blood specimens? Should you be allowed to direct who may use your left-over blood specimen for research or specify under what circumstances such uses may occur? What about your hair that lands on the barbershop floor? Or a tumor that has been excised to prevent it from killing you? These are real questions at the center of a vigorous ethical and legal debate surrounding the use of human biospecimens for research—and they are questions that affect us all.

Existing case law on the collection, storage and use of biospecimens suggests that people do not own their bodily tissue once it has been removed from them. But there is an emerging line of inquiry that questions whether individuals ought to be asked explicitly at the time of collection for consent to use their blood or tissues for research. How far such an ethical requirement might extend remains a heated topic of debate. Should research use of excess tissues that otherwise would be discarded after a surgical procedure also require explicit informed consent? What if there is no identifying information attached to the tissues because researchers need only 100 lung tumor biopsies and don't need data about their donors? What about tumor specimens collected years ago, before their research uses became evident—should they be unavailable for research now because the original source cannot be found and asked for consent? And if excess tissues are used for research, and ultimately result in the development of a lucrative medical treatment, should the source of the tissues have a right to share in the profits?

These questions fall into a significantly contested area for the ethical conduct of human subjects research. Humans have been experimenting on one another since the beginning of time, but the voluntary nature of participation in such experiments only developed as an ethical—and eventually legal—requirement over the course of the last century (Presidential Commission for the Study of Bioethical Issues 2011). The most famous articulation of this principle is found in the Nuremberg Code, promulgated by American judges in 1947 at the trial of Nazi doctors for war crimes following their experimentation on concentration camp prisoners during World War II. The Nuremberg Code establishes that "[t]he voluntary consent of the human subject is absolutely essential," and subsequent US government policy and regulations adopted the same requirement, although not without exception. The regulations governing human subjects research in the United States demand voluntary consent for research participation in most instances, while allowing consent to be forgone, waived, or modified in others—in particular when the research itself is deemed not to involve human subjects at all, which has been the case for certain uses of biospecimens.

A requirement of voluntary consent is not particularly burdensome in the context of prospective research—namely, when biospecimens are collected in person from living individuals for use in a particular study. In these cases, you simply ask permission of the person standing before you, and they can either agree or decline to provide a research specimen. Indeed, in such cases, consent is both ethically mandatory and required by regulation. It becomes much more challenging, however—both ethically and practically—when one wishes to conduct research on biospecimens that *already* have been collected in the course of clinical care or earlier research. In these cases, researchers do not have (and may not be able to find) the biospecimen source immediately before them to ask for explicit consent for the research use being contemplated. Should the research be forgone? Is consent really necessary in this context, and what are the limits of acceptable use before consent becomes necessary? These questions have come to the fore in a series of foundational cases—some in courts of law and others in the court of public opinion—over the last 25 years.

In *Moore v. Regents of California*, a patient underwent a splenectomy for therapeutic reasons and had to travel numerous times by airplane to see the physician who performed the procedure for follow-up treatment (including further collection of specimen samples), only to find out later that his physician had utilized the excised tissue for research. The physician then developed a cell line that he commercialized for substantial revenue. The California Supreme Court determined that the patient may have a legal claim for breach of fiduciary duty by the physician and for lack of informed consent to the research use of the tissues. However, the Court rejected a claim of conversion—civil property theft—by the physician, reasoning that the patient had not sufficiently asserted, nor was there sufficient case law to support, an ongoing property interest in the "discarded" spleen tissue.

In another famous case, the Havasupai, an American Indian tribe living near the base of the Grand Canyon in Arizona, developed a relationship with researchers at the Arizona State University and agreed to participate in research intended to evaluate whether there was any genetic basis for the tribe's extremely high incidence of diabetes. The researchers collected blood samples for diabetes research, and subsequently provided aliquots—stripped of identifiers but nonetheless identified as belonging to members of the tribe—to researchers studying schizophrenia and theories about the migratory history of the Havasupai. The results were viewed as stigmatizing and in conflict with ancestral tribal beliefs. After an extended legal battle over the precise nature of the consent that had been provided, the case ended with a financial settlement and the return of collected biospecimens for burial in accordance with tribal traditions.

Finally, as is detailed in Rebecca Skloot's book *The Immortal Life of Henrietta Lacks* (Crown, 2010), in 1951, Henrietta Lacks, a poor black woman from Baltimore, sought treatment at Johns Hopkins Hospital for cervical cancer. Before administering radium for the first time, the attending doctor cut two dime-size samples of tissue, one cancerous and one healthy, from Henrietta Lacks' cervix. As was the custom of the day, no one specifically asked Lacks' permission for collection of the tissue or informed her that her specimens might be studied. The treating physician gave the tissue to Dr. George Gey, a scientist who had been trying to establish a continuously reproducing, or immortal, human cell line for use in cancer research. According to protocol, a lab assistant scribbled an abbreviation of Lacks' name, HeLa, on the sample tubes. HeLa cells succeeded where all other human samples had failed, and Gey gave away laboratory-grown cells to interested colleagues. Scientists grew HeLa cells in mass quantities to test the new polio vaccine among other uses, and soon a commercial enterprise was growing batches for large-scale use. More than half a century later, Lacks' tissue has yielded an estimated 50 million metric tons of HeLa cells, and more than 60,000 scientific and medical studies, and are in continued use today. If the specimens had been truly anonymized, Lacks' identity would not be known and there probably would have been no story to generate a best-selling book.

These cases and others have generated substantial discussion and disagreement in the bioethics and regulatory communities, and beyond. Did the researchers involved do anything wrong? Do the regulations adequately protect the rights and welfare of specimen sources? What, precisely, are specimen sources entitled to by way of consent, control, and compensation? Such questions form the backbone of the urgent need to evaluate the regulation and conduct of "specimen science," particularly as regulatory revisions finalized in January 2017 addressed research with specimens but left open a number of difficult ethical questions. This volume looks backward to these cases as the basis for, and to inform, the next generation of policy development in this area.

This volume was born out of a need to make sense of the shifting landscapes of science, technology, public opinion, and law regarding research with biological specimens. In 2010, Dr. Leona Cuttler, a distinguished physician scientist at Case Western Reserve University, assembled a group of colleagues from across the country to propose for NIH funding a project focused on elucidating the ethical and policy challenges associated with collaborative genetic research. Dr. Cuttler was motivated by a desire to prevent and cure the diseases of childhood she saw in her pediatric endocrinology practice, and she believed that scientists working together could get answers more quickly than they could working in isolation. One researcher with access to only her own patients' specimens might take ten years to amass enough data to test a hypothesis. But a team of researchers combining their samples could make headway in a fraction of the time.

Of course, working collaboratively on genetic research requires the sharing of biological specimens and their associated data, precisely as was the case with the Havasupai specimens, the HeLa cell line, and in *Moore*. And this, Dr. Cuttler knew, was a matter of considerable difficulty and controversy. The difficulty is borne out of our complex regulatory landscape and the ways in which research rules are interpreted and enforced at academic medical centers across the country and the world.

In the United States, each project that meets the federal definition of "research" with "human subjects"—as will be the case for certain types of research involving specimens and data—and that is not otherwise exempted from regulatory oversight must obtain ethical review and regulatory approval from an institutional review board (IRB) specifically constituted to examine the risks and benefits to the individual of the research proposed. However, Dr. Cuttler knew that there were growing differences of opinion about application of the regulations to research with specimens and data. Were they too strict in some areas, and too permissive in others, especially with regard to promoting socially beneficial research and protecting patient autonomy? How could federal regulations in place and unchanged for nearly three decades appropriately foster cutting edge genetic research of the type Dr. Cuttler's patients needed in order to cure their diseases? Would harmonizing practices across the nation's leading research centers streamline approval in appreciable ways?

She set out to ask these questions with her grant team using funding from the NIH's National Human Genome Research Institute, until her untimely passing in 2013. Dr. Cuttler's project lived on, however. In the pages that follow, philosophers, ethicists, legal scholars, regulators, patients, and others weigh in on these important questions.

This book is the result of a day-long symposium planned by members of Dr. Cuttler's study team in collaboration with colleagues from Harvard University. The presenters and participants are thought leaders from the fields of law, medicine, and philosophy, each of whom took a different approach to questions about the appropriateness of our regulatory oversight apparatus and how proposed changes to the regulations might promote or impede the ethical conduct of biospecimen research.

One common theme of the chapters in this volume is that the regulations governing federally funded research with human subjects, including those that govern the research use of human biological materials and data—better known as the "Common Rule" and untouched from 1991 to 2017—were out of step with current technological and scientific conditions. Advances in genetic research, for example, have far exceeded what could have been contemplated by those tasked with writing the rules nearly three decades ago. Because science has evolved, so too must the regulatory landscape, in order to simultaneously protect the rights and welfare of people who participate in research (as well as those who are studied without their knowledge) and to avoid inappropriate hindrance of medical advancement.

When the Common Rule first was promulgated, it was considered by most a fair assumption that a human biospecimen, free of any identifying information such as a name or a patient identification number, could not reasonably be re-identified by a scientist in a laboratory (or anyone else). In fact, for this reason, the use of existing (left over from clinical care or earlier research) human biospecimens (and associated data) traditionally has been deemed not to be human subjects research under the Common Rule if the identity of the subject may not readily be ascertained by the investigator, and to be exempt from most forms of regulatory oversight if the specimens or data are themselves identifiable but information is recorded by the investigator in a non-identifiable manner. This is because the framers of the Common Rule felt confident that these mechanisms of segregating specimens and data from identifiers were adequate to protect any interests of the individuals from which they were derived. Any risk, such as the remote risk of re-identification, was small in comparison to the promise of scientific and medical advances that depended upon the specimens.

Today, this is a matter of considerable controversy. On one hand, technological advances, such as whole-genome sequencing and increasing access to big data, have made it possible for a scientist with access to both sophisticated equipment and a reference key to take an otherwise unidentified human biospecimen and re-identify it using the unique data gleaned from a person's DNA. Although such equipment is expensive and not widely available, and there is no universal reference key of everyone's DNA, the potential for unauthorized re-identification has caused some to argue that there is a pervasive and unavoidable risk of "informational harm" to specimen sources that requires a tightening of the rules. Those who want more restrictions are concerned about protecting individual liberties and most importantly privacy, and want to maximize the principle of autonomy.

On the other hand are those who advocate for fewer restrictions on and greater sharing of biospecimens and data to maximize their utility in solving important problems. This camp is a mix of scientists, ethicists, and disease-oriented patient advocates, each of whom views the risks of unauthorized re-identification as relatively small and better addressed through penalties for violations rather than tighter restrictions that could impede important scientific progress. People who want fewer restrictions on use of biospecimens typically believe that a more communitarian approach to research will increase beneficence and justice.

Despite the lack of consensus around these issues, millions of human biospecimens are collected and stored each year. Most are obtained during routine clinical encounters with patients for diagnostic and treatment purposes, while a minority are collected with explicit research intent following informed consent by a donor. Regardless of the circumstances, it is not possible to know with certainty at the time of collection all the possible future uses for which a specimen could be of value. Thus, the only way to guarantee full informed consent would be to re-approach donors *each time* a new use is imagined (which notably is not even possible once a specimen has been de-identified). Science is evolving more quickly than our regulatory procedures can accommodate. A permissive stance would be to allow unforeseen future uses without obtaining new, specific consent *unless* there is an obvious risk of harm, on the premise that such uses might contribute to important medical discoveries. A conservative approach would be to discard biospecimens immediately following the original use for which they were collected to avoid even the appearance of violating donor's expectations. The current regulatory approach—maintained following revisions finalized in 2017, described more fully below and in chapter 1 of this book—requires consent from the specimen source only for the active collection of specimens and data for research use, and for the secondary use of only *identified* specimens and data. In light of technological advancement and changing patient attitudes, however, the challenge is how best to balance the importance of protecting individual interests against the value of promoting the good of the larger human community.

The federal agencies that oversee human subjects research made an attempt to address this challenge by proposing a major revision to the research rules in 2015 via a Notice of Proposed Rule Making (NPRM) to amend the Common Rule (following a 2011 Advance Notice of Proposed Rule Making [ANPRM]). The NPRM proposed an approach on one side of the spectrum of possibilities and generated substantial dialogue regarding the best path forward. Although its most controversial proposals regarding biospecimen research were ultimately not adopted in the final rule issued in January 2017, many of the chapters that follow touch upon the NPRM to illustrate the significance of the debate and options that may become more relevant given questions left open in the final rule, especially regarding identifiability and broad consent.

In particular, the NPRM proposed to expand the definition of the term "human subject" to include a living individual about whom the investigator "obtains, uses, studies, or analyzes biospecimens" even when those biospecimens have been stripped of all the traditional pieces of identifying information. The presumption underlying this proposal was that all genetic material can potentially serve as a unique personal identifier—if not now, then in the relatively near future. The proposal also was based on the notion that individuals have greater autonomy interests in controlling their biospecimens than previously had been acknowledged. The consequence of this regulatory redefinition would be that secondary research with de-identified biospecimens would no longer be permissible without some form of consent, a dramatic change from the status quo. Under the NPRM's proposal, that consent could be "broad," applying to future unspecified use, rather than specific for each study, but the conditions for waiver of consent for biospecimen research would have been dramatically tightened, such that waiver would become "rare."

Just as there is no consensus about how to best balance autonomy and privacy concerns with the desire to advance scientific frontiers, so too there was great disagreement about the wisdom of the proposed rules. An earlier attempt to solicit feedback from the public on the ANPRM resulted in four years of hand-wringing before a substantially similar set of proposals was released as an NPRM, the last step before issuance of a final rule. This, despite the fact that the government received over 1,100 comments reflecting considerable differences of opinion about the appropriateness of the suggested changes.

Not surprisingly, nearly twice as many public comments were submitted in response to the NPRM (2,186 in total, many of which were from organizations), and they, too, showed a clear lack of consensus. According to a systematic analysis of these comments released in May 2016 led by the Council on Government Relations (COGR), with support from the Association of Public and Land-grant Universities (APLU), there was significant opposition to most major NPRM proposals regarding the regulation of biospecimen research. More specifically, the comments submitted by patients and the research community (e.g., researchers, universities, medical centers, and industry) overwhelmingly opposed the proposed changes involving non-identified biospecimens on the grounds that they would have reduced the availability of specimens for research, negatively impacting medical advances. Influential advisory groups, including the Presidential Commission for the Study of Bioethical Issues and the Department of Health and Human Services Secretary's Advisory Committee on Human Research Protections (SACHRP) questioned whether the proposal could even satisfy its goals of protecting participant autonomy. Members of the public who did not specifically identify as patients were more divided in their comments, but given that the government's rationale for many of the proposed changes regarding biospecimen research was ostensibly that public trust demanded them, it is important to note that 55 percent of the general public comments opposed one or more of the major proposed changes (a fact recognized in the preamble to the final rule). Sixty-one percent of the general public comments specifically opposed the expanded definition of "human subject" to include non-identified biospecimens, and opposition to that particular proposal was even higher among those who self-identified as patients and researchers; the same trend was noted in comments regarding broad consent for storage and secondary research with biospecimens. It is also essential to note that calculations offered by the American Society for Investigative Pathology indicate that the NPRM underestimated the true cost of its proposals to require broad consent for biospecimen research—which it estimated to be $1.2B annually—by a factor of at least ten (COGR and APLU 2016).

In light of these public comments on the NPRM, a final rule was published in January 2017 that did not significantly alter the regulatory approach to biospecimen research—at least for now. As explained more fully in chapter 1, the revised Common Rule will continue to treat research with non-identified specimens as outside the bounds of "human subjects" research, such that neither IRB review nor consent is required. However, the regulatory agencies, alongside appropriate experts, will periodically revisit the definition of identifiable every few years, such that more types of research with specimens could fall within the regulatory parameters in the future. Moreover, the final rule includes a provision that would allow research with *identifiable* specimens and data to proceed on the basis of broad, rather than specific, consent. Thus, the themes addressed by our authors, and the debate exemplified in the NPRM between greater control over biospecimens research and facilitating scientific advances for the public good, are enduring.

In view of the tension between the two main goals—patient autonomy and social benefit—any regulatory change will be unsatisfying to some. Increased protection of patient autonomy will necessarily increase burdens on conducting research with specimens and data, and minimizing such barriers will necessarily diminish protection of patient autonomy. That there is no consensus about the best path forward is about the only thing upon which all the authors in this volume can vigorously agree. In the pages that follow, experts from a range of disciplines and backgrounds explore the myriad ethical, social, and legal implications of biospecimen research with a focus on regulation and its consequences. Readers will learn about the prevailing practices for using biospecimens in research and will hear from a variety of viewpoints about the potential advantages and pitfalls of proposed changes. All of the chapters that follow will make reference in some way to the regulatory landscape, but they will also do much more. They will explore an imagined future for biospecimen research in which not only the rules are different, but also the expectations, opinions, and priorities of physicians, scientists, and patients.

The book is divided into five parts. Part I, introduced by Aaron Kesselheim, sets the stage for the rest of the volume by summarizing the legal, ethical, and historical foundations underlying the collection, ownership, control, and future use of biospecimens, and the implications for individual privacy and autonomy. While acknowledging that this volume relies upon and refers to US law, regulations, and customs, the exchange and sharing of biospecimen materials, research, and information are international undertakings. This first part of the book lays the foundation upon which further ethical, social, political, and scientific deliberations considered in the remainder of the volume rest.

In chapter 1, David Peloquin, Mark Barnes, and Barbara E. Bierer survey and analyze laws and regulations governing the research use of biospecimens around the world, with a focus on privacy, control, and return of research results. They begin with US law, exploring how the common law of property (exemplified in several foundational court cases), regulations specifically governing human subjects research (including 2017 revisions), and privacy laws are implicated in biospecimen research and biospecimen repositories. Next, they turn to European law, addressing both individual countries and Council of Europe standards for biospecimen research, and the particular challenge of biospecimens transferred outside of the European Union. They then address nuances of several other representative countries, including China, Brazil, and India, with an analysis of several significant challenges in international biobanking. After addressing a variety of recent regulatory changes, they conclude that the patchwork of national and international law governing biospecimens can hinder research.

In chapter 2, Russell Korobkin reviews many of the same cases referenced in chapter 1, but argues that these cases lay a foundation that counters the conclusion that there are no property rights that remain with the specimen donor. Indeed he argues that the donor may retain the right to their property, and he calls upon the informed consent document (and common law) to define the retained versus transferred property rights of the donor. He also introduces the concept of compensation, not only by consideration of value of the "gift" of the specimen but potentially by reference to patent law.

In chapter 3, Elisa Hurley, Kimberly Hensle Lowrance, and Avery Avrakotos step back from the legal and technical issues to discuss the role of public trust. They recount the history of several cases that have informed the current debate: those of Henrietta Lacks, the Havasupai, and the Newborn Screening Saves Lives Reauthorization Act of 2014. Each of these cases illustrates in its own way the tensions between the issues of privacy, informed consent, and the advancement of science. Hurley, Lowrance, and Avrakotos call for greater, expanded education and pubic engagement around biospecimen research, a theme that is further discussed later in the volume.

In part II, we are introduced to some of the thorniest and most important ethical challenges facing researchers, regulators, and bioethicists today. Introduced by Steven Joffe, the four chapters in this part explore the tension between three foundational principles of research ethics (respect for persons, beneficence, and justice) as they relate specifically to the use of human biospecimens for research. Chapters 4–6 address donors' rights and obligations, raising questions about the extent to which an individual's autonomy may or may not trump other important considerations, such as scientific advancement and public health. Chapter 7 focuses on the risk of privacy and questions assumptions about the true likelihood of harm to individuals whose biospecimens may be used in science.

In chapter 4, David Korn and Rachel Sachs argue that each person who benefits from advances in science and medicine has a duty to participate in the scientific enterprise by allowing his or her biospecimens to be used in research. Focusing on the often neglected principle of justice, Korn and Sachs assert that it is unfair for people who don't allow their biospecimens to be used to receive treatments developed by using materials derived from others. They conclude that the principle of respect for persons—as expressed by honoring autonomy through the mechanism of informed consent—should be balanced appropriately against the equally important and valid principle of justice.

In chapter 5, Barbara J. Evans and Eric M. Meslin also explore important questions about individuals' rights and obligations. They note that individuals seem more comfortable donating their biospecimens for science when the work is performed at academic and non-profit institutions than at for-profit entities, but they challenge the wisdom of that preference, noting that in many cases pharmaceutical companies may create significant public good through product development. Evans and Meslin approach these questions from a legal perspective and they promote the idea of a public benefit standard for evaluating research risks. Ultimately, they urge a balancing of individual and public interests for the purpose of mutually advantageous exchange.

In chapter 6, Suzanne M. Rivera and Heide Aungst examine what is known about why people are willing or unwilling to donate biospecimens for research, assess the various challenges (legal, ethical, and practical) associated with honoring donors' wishes, and evaluate to what extent we ought to be beholden to such wishes in the face of compelling scientific advancements that may be brought about by biospecimen research for the betterment of society. They draw analogies to the various ways in which individuals relinquish information about themselves on a daily basis in return for valuable services, via smartphones, social media, and Web searches, to demonstrate that perhaps biospecimens need not be treated so differently. Ultimately, Rivera and Aungst urge recognition of the social value of biospecimen research, rather than solely claims to individual autonomy, and call for a "different ethos, in which all people who could benefit from medical advances willingly contribute biospecimens for research" in the context of trust and collaboration with scientists.

In chapter 7, Ellen Wright Clayton and Bradley A. Malin take a different approach. They focus on the question of privacy by examining the risk of material harm to individuals from biospecimen use in research. They conclude, despite growing concerns about privacy risks from research, that the likelihood of potential harm to individuals is quite small and does not justify the proposed increases in oversight and control.

Part III focuses more specifically on the idea of informed consent—what it is meant to do and how it might be altered. Introduced by Pearl O'Rourke, the chapters in this part offer a further exploration of the different types of consent that may be utilized for biospecimen research (if consent is sought at all).

In chapter 8, Christine Grady and her co-authors recognize that various processes exist for obtaining consent for the future research use of biospecimens, and that there is confusion and uncertainty about appropriate and ethically permissible types of consent in this context. They focus on broad consent (as distinguished from both blanket and specific consent), which would permit an unspecified range of future research conditional on a few content or process restrictions and coupled with governance. Noting that various bodies have recently endorsed the broad consent approach, including the revised Common Rule, they endeavor to assess its ethical justifications and potential concerns. Reasons in its favor include respect for biospecimen donors, autonomy, and transparency, as well as some support for the approach in empirical studies to date and recognition that it is less burdensome than other potential approaches. The authors address various objections before concluding that broad consent is ethically appropriate and preferable to the alternatives in either direction. They recommend that broad consent include initial consent, oversight of future research projects, and mechanisms for communicating with donors.

In chapter 9, Nanibaa' A. Garrison discusses insights from researchers and participants in the age of broad consent and data sharing. In particular, Garrison focuses on the perspectives of several distinct groups, including certain indigenous tribes involved in recent litigation concerning genomics research and others with reservations about the ethical validity of broad consent. She describes the Havasupai controversy in detail and the desire of many indigenous peoples to have greater tribal control over research with their specimens than broad consent would allow. Garrison argues that it is essential to acknowledge that one implication of the trend toward broad consent may be that some communities that have been historically underrepresented in research may also be less willing to participate in research where their data are shared widely, leading to an inability to benefit from such research. She concludes that, for at least some groups, more engagement than is required by the regulations is needed to build respect and trust regarding biospecimen research.

In chapter 10, Ivor Pritchard and Julie Kaneshiro assess whether it is ethical to require subjects to agree to donate their specimens for future studies as a condition for being eligible to participate in a present study that offers the prospect of direct benefit. More specifically, they assess whether such "package deals" violate the regulatory prohibition on coercion and undue influence. Pritchard and Kaneshiro outline the relevant regulatory enforcement actions in analogous circumstances before concluding that, while some such package deals are ethical and some are not, in general package deals ought to be avoided because they might diminish the voluntariness of informed consent needed for ethical study participation.

Part IV, introduced by Pamela Gavin of the National Organization for Rare Disorders, examines how the general issues raised by specimen science have particular salience for special communities of patients (e.g., newborns, the underserved, those with rare diseases) and special communities of researchers (e.g., stem cell researchers).

In chapter 11, Aaron J. Goldenberg and Suzanne M. Rivera examine the implication of possible changes in the regulation of specimen science for precision medicine—the now-well-funded push to tailor treatment and prevention to subsets of patients on the basis of biological, behavioral, and social determinants of disease. They consider how such research efforts will be stymied if attention is not paid to the role of underserved and vulnerable populations. In particular, they focus on biorepository design and participant recruitment, specimen and data storage and management, and specimen and data research use and translation as obstacles to achieving truly representative national cohorts for biorepositories. They also discuss how to better enable members of these vulnerable and underserved groups to help in development and governance of biorepositories.

In chapter 12, Jeffrey Botkin, Erin Rothwell, Rebecca Anderson, and Aaron Goldenberg focus on a highly valued but controversial set of biospecimens: residual dried bloodspots (DBS) from newborns. They review the history of such programs and the controversies surrounding them, including lawsuits that have been brought in Indiana, in Minnesota, and in Texas. They also report findings from their large national survey (conducted in 2012) of public attitudes regarding the retention and use of DBS. Ultimately, they conclude that in this context a perfunctory consent process such as that used for compliance with the Health Insurance Portability and Accountability Act of 1996 will fail to rebuild the public's trust in the use of these biospecimens. They suggest that "the larger challenge is to develop more sophisticated approaches to information and education that foster a higher level of true understanding about how research is conducted and, potentially, the need to collaborate with communities to support research."

In chapter 13, Sara Chandros Hull takes a data-driven approach to controversies over the appropriate governance frameworks and consent models for biospecimen research. She systematically reviews the empirical literature on public attitudes toward broad consent for biospecimen banking, including controversies over the level of consensus in favor of broad consent. In keeping with the focus of this part of the book, she then reviews the more limited empirical work on attitudes of those with rare diseases and their family members. She finds that the "limited data available about the attitudes and preferences of potential participants in research on rare genetic diseases are generally consistent with robust frameworks of broad consent that have been described in recent policy proposals and are being used with increasing frequency." Using the example of researchers organizing a leukodystrophy research database, Hull shows how biospecimen repositories can use surveys and other empirical methods to tailor their governance frameworks and consent models to the needs and preferences of particular patient populations.

In chapter 14, Geoffrey Lomax and Heide Aungst use the experience of the California Institute for Regenerative Medicine (CIRM) as a case study in the challenges faced by some biorepositories. Among the issues CIRM faced were whether to require consent for commercial use of cell lines, the return of results to donors, the parameters and consent for re-contact regarding clinically significant events, and the management of donor withdrawal. Lomax and Aungst also discuss CIRM's decision making regarding "sensitive uses of derived lines," including reprogramming cells to produce gametes, cloning, and somatic cell nuclear transfer. They conclude the chapter by discussing the difficulties of setting policies for deriving pluripotent induced stem cells using previously banked research specimens.

Part V concludes the book with six chapters that explore practical, logistical, and operational aspects of changing norms with regard to biospecimen collection and use. Introduced by Barbara E. Bierer, the chapters in this part outline the current approach to governance and accountability of biospecimen repositories and detail specific operational obligations of creating and running a repository. In addition to institutional and organizational management, the responsibilities of the investigator are specifically elaborated. But the historical approach to governance, one that is reminiscent of institutional paternalism "protecting" the patients, is evolving: true partnerships with patients and their families, patient-led communities, and patient advocates are emerging and are leading to meaningful engagement.

In chapter 15, Karen Maschke focuses on biospecimen repository governance, summarizes the current approach to biospecimen repository management in the United States, and examines the concept of participatory governance as articulated by the working group for President Barack Obama's Precision Medicine Initiative (PMI). She also puts these concepts in broader context, explaining how previous attempts as participatory governance of biobanks, such as the attempt made in the United Kingdom, prepared the way for a new model of governance for biospecimen repositories in the United States. She concludes with an assertion that the PMI may increase the public's expectations about the role that specimen contributors should play in the research enterprise.

In chapter 16, Susan M. Wolf and Isaac S. Kohane argue that genomic science has caused a fundamental shift in the roles of patients and families in the conduct of research. With personal interests in driving research to generate new treatments for their diseases, patients and their families are setting agendas for the collection of data and specimens and are challenging traditional scientific norms that situated scientists at the center of decision making. Wolf and Kohane call this shift a "patient and family revolution" and explain how new technologies, including social media, have empowered individuals and groups to take a leadership role propelling research agendas. In historical context, Wolf and Kohane characterize today's level of engagement of patients and families in research as a movement from paternalism to partnership.

In chapter 17, Jane Perlmutter and Heide Aungst approach the aforementioned issues from the perspective of patients and patient advocates. For too long, they argue, patients and their advocates have not been involved in planning, communicating, or providing oversight for clinical research, nor have they been at the table in important policy deliberations. The patient's perspective is important to consider at all stages of research, beginning when the initial research question is being framed and continuing through dynamic consent and decisional input over the use of specimens. While the availability and the nature of information are changing rapidly, Perlmutter and Aungst argue that timely and comprehensive education is nevertheless important and should occur in advance of approaching any individual to donate rather than at a time of crisis, and that researchers should be sensitive to the health literacy and numeracy needs of the populations they study. Finally, greater understanding of and demonstrable respect for the contributions of the donors should be evident.

In chapter 18, Erin Rothwell and Erin Johnson reimagine the process of informed consent. Considering the literature showing that comprehension of informed consent has been disappointingly low, they explore alternatives to traditional informed consent and report the results of their empirical study to determine whether potential subjects would support new approaches, such as "broad" consent for future unspecified uses of specimens and the use of an "investigator oath" to foster trust in the researchers who will safeguard and study donated specimens. Their study, they report, showed that the notion of an investigator oath was well received, and that if paired with broad consent it might serve to increase transparency and foster trust.

In chapter 19, Quinn Ostrom and Jill Barnholtz-Sloan relate the operational details of establishing and running a repository, acknowledging the interdependencies of the institution, the clinician, the clinical investigator, and the basic researcher. Central to the multi-stakeholder analysis is the specimen donor upon whose contribution the biorepository depends. Those charged with the responsibility of running the biospecimen repository operate under strict standard operating procedures, including those for obtaining accurate and relevant clinical annotation, and an IRB-approved protocol that defines the process and the document for informed consent and the procedures for the collection, annotation, acquisition, and distribution of biospecimens both within and outside the institution. Proper implementation and governance help to ensure that biospecimens and data can be utilized in research so as to have the largest potential benefit.

The discussion of operational matters, including legal and institutional considerations, continues in the final chapter. Starting from a detailed review of the institutional research biospecimen repository protocol and appreciating the regulatory requirements in the United States, Kate Heffernan and her co-authors describe points to consider in designing how specimens and associated clinical data will be obtained and accepted, including how participants will be recruited; how informed consent will be obtained; how specimens and data will be stored, maintained, protected, and updated; how specimens and data will be released to data users; and quality control, auditing, and monitoring safeguards. They take a practical approach to considering the rights of participants, clinicians, researchers, institutions, and commercial and nonprofit sponsors. They call for education to increase participant engagement and to clarify, manage, and anticipate the needs and expectations of researchers and clinicians. They also offer suggestions for ensuring the financial sustainability of the institutional biospecimen repository.

This volume represents a variety of perspectives on the collection and use of biospecimens and their associated data for purposes of conducting research on important scientific questions with great relevance for human health. We have included chapters written by patient advocates, scientists, philosophers, legal scholars, regulators, and managers of biospecimen repositories. Their voices do not represent every viewpoint, but it is our hope that, together, they provide a rich and complex description of the current landscape. This is particularly important because the regulations governing research with specimens have just been revised for the first time in 26 years. The chapters in this volume suggest future directions for both practical and ethical considerations, which will require great care and attention as science and medicine continue to progress for the benefit of society.

## References

Council on Government Relations (COGR) and Association of Public and Land-grant Universities (APLU). 2016. Analysis of Public Comments on the Common Rule NPRM, http://www.cogr.edu/Human-Subjects-and-Animal-Research.

Presidential Commission for the Study of Bioethical Issues. 2011. "Ethically Impossible": STD Research in Guatemala from 1946–1948, http://bioethics.gov/cms/node/654.

# I Background and Foundations

# Introduction

Aaron S. Kesselheim

When a new area of inquiry opens up in the biomedical sciences, it is always accompanied by some trepidation. Widespread use of insufficiently tested novel approaches to studying or delivering health care can lead to unintended negative consequences, undermining the development of even the most promising innovations. After one infamous early misstep in the application of gene replacement—the Jesse Gelsinger tragedy at the University of Pennsylvania—fewer clinical trials opened and patient demand waned. The Gelsinger episode was widely recognized as having set the field back ten years or more as the complex array of contributors to the scientific research enterprise worked out safer ways of proceeding.

We are at a similar inflection point in the practice of using biospecimens to advance therapeutic science. The promise of this approach is undeniable, as technology related to gene identification combined with growing computing and data analytic power hold the prospect of unlocking new biomarkers, pathophysiologic bases for disease, and targets for therapeutic development. Achieving this potential, however, also requires healthy respect by all involved parties for the risks that investigation of patients' genetic material and tissues can engender. For example, unwanted manipulations of patients' material can lead to controversies such as the one related to the use of blood samples from members of the Havasupai tribe of Native Americans to investigate the prevalence of certain markers for psychiatric disease and other characteristics. Too many such controversies could snuff out the development of an approach that might transform our understanding and treatment of various diseases.

The initial line of protection against such problematic outcomes is the legal context in which specimen science operates. In the first chapter in this part, Peloquin, Barnes, and Bierer describe laws and regulations that relate to research involving biospecimens in the United States. They focus on the role played by privacy law and human subjects research protections enumerated by the Common Rule, both before and after a recent revision anticipated to take effect in January 2018. The authors evaluate whether these protections can defend patients against dangerous or unwanted disclosures of individual health information that could jeopardize the research enterprise. While they find some important safeguards currently in place, they also conclude that these protections are both not comprehensive and highly provincial, with far fewer protections in comparable settings outside the United States. Given the increasing interconnectedness of international research enterprises, there is real potential for activities—such as data mishandling—around the world to affect patients and the research enterprise in the United States.

Although Peloquin et al. assert that property law offers little real protection to patients in the context of biospecimen research, Korobkin takes a more optimistic view in his chapter, describing the legal basis on which patients may indeed rely on principles of property law to exert control over their biospecimens and hence protect themselves from misappropriation. Re-animating the concept of biospecimens as property provides additional authority to protect against risky outcomes relating to biospecimen research that privacy law and the Common Rule may not cover. It also has important implications that would strengthen a donor's right to compensation and the need for informed consent related to particular uses of biospecimens.

The importance of the analyses provided by Peloquin et al. and Korobkin is driven home in the final chapter in this part, in which Hurley, Lowrance, and Avrakotos wrestle with the legacy of Henrietta Lacks—the impoverished young black woman whose cervical cancer cells went on to be used in numerous scientific discoveries without sufficient acknowledgment of her role or her informed consent—and the possibility that similar cases could undercut the public trust on which specimen science depends. They conclude that privacy and informed consent can improve consumers' participation in the biospecimen research system and avoid emergence of more cases like that of Henrietta Lacks that could drag down a nascent research enterprise.

The authors of the chapters in this part also highlight a few ways in which responsible conduct of specimen science could be achieved. One such principle is consistency. For example, Peloquin et al. argue for more reliable rules across countries, while Korobkin points out the role of property in the legal treatment of biospecimens to provide a consistent logical thread among seemingly disparate results of cases in this area. The likelihood of unintended negative outcomes from specimen science is reduced when the rules are reliable and predictable.

Another principle is patient engagement. Hurley et al. explicitly call for a public discourse about how to handle controversial areas of biospecimen research, while Korobkin concludes that enhanced up-front communication is needed to clarify parties' rights in their property. These two approaches are pillars of a broader strategy to promote positive outcomes by aligning the expectations of the patient and research communities. Many of the conflicts in this area—including the Lacks case, the case of the Havasupai tribe, and the cases of newborn bloodspots evoked by Hurley et al.—could have been mitigated by assessing the perspectives of the community or the individuals at issue ahead of time.

These common principles suggest that the future of specimen science is put at risk in the current environment by the piecemeal approach to its oversight. including differing perspectives of local institutional review boards, a patchwork of court cases, and occasional legislation that either covers very narrow issues (such as the Newborn Screening Saving Lives Reauthorization Act) or were not written with biospecimens in mind (e.g., the Health Insurance Portability and Accountability Act's privacy rules). As a result, rules will fail to cover certain important issues, and application of those rules can vary widely.

Would the field be better served in the long run by a central oversight organization that might grow into a center of scientific expertise while also efficiently gathering and incorporating patients' perspectives and communities' needs? The US Food and Drug Administration was thrust into a similar role in the pharmaceutical market more than fifty years ago when concerns about the safety of prescription drugs led Congress to consolidate power over the market in a single regulatory agency and to give it gatekeeping authority before a new product could reach the market. A specimen science regulator may not need such a degree of authority, but could still be charged with developing and promulgating best practices related to managing biospecimens, addressing privacy, informed consent, and property issues, and assessing and diffusing potential conflicts among the parties involved.

If not a central regulator, the chapters in this part point to the need for specimen science to mature from a loosely interconnected group of investigators and institutions to an organized field with a rational governance structure that can help most effectively tap the enormous scientific potential that the field offers.

# 1 Legal and Regulatory Issues in Biospecimen Research: National and International Perspectives

David Peloquin, Mark Barnes, and Barbara E. Bierer

With recent advances in technology related to genetic testing, whole genome sequencing, and biomarker determinations, biospecimens collected during clinical research or remaining after routine medical care hold increasing scientific promise. While the utility of a biospecimen for research purposes is directly related to the clinical information with which it is associated, the utility is significantly increased by aggregating information derived from the comparative analysis of multiple specimens from multiple sites, domestic and international, representing diverse populations. Along with increased research potential, however, new concerns arise regarding the privacy of the individuals from whom specimens are derived, the extent to which they should be able to control the use of their biospecimens, and whether they should receive the results of future research on their biospecimens or share in any profits resulting from their use. Differences among international regulations and restrictions on import and export of biospecimens increase the complexity of research access and use. This chapter surveys and analyzes some of the key laws and regulations in this area. We focus first on the laws of the United States and then discuss the regulatory framework of Europe and representative states of the Global South, regions important to biomedical research where regulators have been grappling with biospecimen research issues.

### United States

In the United States, the use of biospecimens for research purposes implicates many areas of law, including the common law of property, regulations governing the protection of human subjects involved in research, and privacy laws regulating the use of demographic, phenotypic, and genotypic information either attached to biospecimens or generated during research on biospecimens. We explore the broad contours of each of these sets of laws below.

**Property Law**

Unless modified explicitly by contract, the common law of property in the United States has traditionally held that individuals have no property interest in the specimens collected from their bodies. This principle was most prominently set forth in the case of *Moore v. Regents of the University of California*, in which the California Supreme Court held that a patient who required a splenectomy had no property interest in his excised tissue (nor the cells that were generated from the tissue), and thus no claim under property law to a patented cell line derived from the spleen cells that had been created by the treating physician and subsequently commercialized.[1,2] The *Moore* court observed that "the laws governing such things as human tissues ... deal with human biological materials as objects *sui generis*, regulating their disposition to achieve policy goals rather than abandoning them to the general law of personal property." The *Moore* court recognized that if excised biological materials were considered personal property, the donor could bring an action under the common law tort of conversion against researchers using the specimen, leading to the undesirable result that "with every cell sample a researcher [would] purchase a ticket in a litigation lottery." The *Moore* court did, however, recognize a duty on the part of treating physicians to disclose, at the time of obtaining a patient's clinical procedural consent, any plans to perform research on such specimens, recognizing that such a duty arose because the potential plans "may affect [the physician's] medical judgment." Thus, despite the lack of property rights held by the patient, the court nonetheless sought to protect patients against potential conflicts of interest when the treating physician might benefit from a planned research use of the patient's specimen.

While the *Moore* case addressed the use of tissues collected during the course of clinical care, more recent case law has addressed the ownership of tissues donated for research purposes, reaching a similar result that patients do not retain property rights in such materials. In the case of *Washington University v. Catalona*, the US Court of Appeals for the Eighth Circuit addressed a declaratory judgment action brought by Washington University asserting that it had ownership of tumor specimens housed in an on-campus biorepository.[3] Washington University had filed the action to prevent a former Washington University researcher, Dr. Catalona, from moving the biospecimens to another university. In an attempt to legitimize the transfer, Dr. Catalona asked patients who had earlier consented for research use of their specimens to sign a release granting Dr. Catalona custody of the specimens.

In support of his argument that patients had a residual ownership interest in the specimens and thus the right to control the transfer thereof, Dr. Catalona pointed to the original informed consent language that permitted patients to withdraw consent and request that the specimens be destroyed. Washington University countered that the biospecimens were its property and that the patients had no residual right to release such specimens to Dr. Catalona. The Eighth Circuit ruled in favor of Washington University, holding that patients who had consented had completed an *inter vivos*[4] gift whereby they relinquished ownership of the biospecimens. The court acknowledged that while patients retained the ability to withdraw consent for such research use, this limited right to request destruction of the biospecimens did not suggest a broader right of residual ownership in the biospecimens. Indeed, the court noted, Dr. Catalona routinely discarded biospecimens that he no longer needed for research, and he did so without obtaining further consent from the patients, thereby undermining his argument that patients retained an ownership interest in the specimens.[5]

## Regulations on Research Involving Human Subjects

While the cases discussed above indicate that patients and research subjects generally do not retain a property interest in their biospecimens at common law, they also suggest the importance of the concept of informed consent in governing the use of biological materials for research purposes. At the center of the US regulatory regime lies the Common Rule, which governs human subjects research funded by a variety of federal agencies and departments that have adopted its protections in "common" and also provides the framework within which many US institutions conduct all of their research, regardless of funding source.[6] As the culmination of a process of proposed revisions that began in July 2011 with the issuance of an Advance Notice of Proposed Rulemaking ("ANPRM"), a substantial revision to these regulations was issued on January 19, 2017 (the "CR Final Rule"). Most of the revisions announced in the CR Final Rule will take effect on January 19, 2018.[7] We thus discuss here the Common Rule's approach to biospecimens before the revised draft was issued, which we refer to as the pre-2018 Common Rule, and we then discuss the key changes governing research involving biospecimens, which we refer to as the CR Final Rule.

The Common Rule applies to "research," defined in the pre-2018 Common Rule as "a systematic investigation, including research development, testing and evaluation, designed to develop or contribute to generalizable knowledge" that involves "human subjects," i.e., living individuals about whom the investigator conducting the research obtains "[d]ata through intervention or interaction with the individual" or "[i]dentifiable private information." The Common Rule considers private information to be identifiable if the identity of the subject may "readily be ascertained" by the investigator. Research activities involving human subjects under the Common Rule generally are required to undergo review and approval by an institutional review board (IRB) and, with some exceptions, can take place only after the written informed consent of the research subject has been obtained.

The Common Rule's pre-2018 definition of human subjects research has several implications for research involving biospecimens. First, because the Common Rule applies only to research involving living individuals, research on decedents' specimens is not subject to the rule's protections. Second, because the rule's protections attach only when the investigator obtains information from interacting directly with the subject or uses information from which the identity of the subject is *readily ascertainable by the investigator*, existing specimens that have been collected for another purpose can be removed from the ambit of the Common Rule by shedding them of identifiers.

To facilitate research with biospecimens, the US Department of Health and Human Services' Office for Human Research Protections (OHRP), which interprets and enforces the Common Rule, has issued guidance providing that specimens collected for purposes other than the present research may be labeled with a new unique identifier, or code, and still considered not to be identifiable in the hands of the researcher so long as (i) the researcher does not have access to the key that links the newly assigned code to the identity of the subject and (ii) there is a policy or agreement in place prohibiting such release (OHRP 2008). Importantly, specimens possessing such a re-identification code allow the investigator later to request additional phenotypic or clinical information pertaining to a particular specimen by supplying the codes for the specimen in question to the person holding the re-identification key.

The pre-2018 Common Rule further facilitates research involving biospecimens by providing that even if the investigator uses specimens that he or she can readily identify, and that are therefore defined as human subjects, the research may be "exempt" from the requirements of IRB review and informed consent if (i) the specimens already exist at the time of the research, and (ii) the investigator records information regarding the specimens in a manner that results in the data not being readily identifiable.[8] In addition, the pre-2018 Common Rule allows an IRB to waive the informed consent requirement if (i) the research involves no more than minimal risk to the subjects, (ii) the waiver will not adversely affect the rights and welfare of the subjects, (iii) the research could not practicably be carried out without the waiver or alteration, and (iv) whenever appropriate, the subjects will be provided with additional pertinent information after participation.[9] Accordingly, even when an investigator obtains and records identifiable information regarding specimens, the requirement to obtain informed consent of subjects can be avoided if the investigator demonstrates to the satisfaction of the cognizant IRB that the research satisfies the criteria for a waiver. This will often be the case when the specimens were collected in the past and it would be extremely difficult, if not impossible, to obtain informed consent from their source. Notably, research for which an IRB has waived the requirement of informed consent is not "exempt" research and therefore remains subject to the requirement of both initial and continuing review by the IRB.

Also of note to biospecimens research, the US Food and Drug Administration (FDA) has developed its own regulations governing human subjects research that apply to research involving products regulated by the FDA.[10] Like the Common Rule, these regulations generally require prospective IRB review and informed consent. Importantly, in the case of research involving medical devices, the FDA regulations define "subject" as a "human who participates in an investigation, either as an individual on whom *or on whose specimen* an investigational device is used or as a control." Thus research on a medical device that includes the use of human specimens would generally require IRB approval and informed consent, because the FDA's regulations do not generally permit an IRB to waive the requirement of informed consent.[11] The FDA has permitted one exception to the informed consent requirement for research involving existing specimens, issuing guidance stating that it exercises enforcement discretion to permit residual specimens collected in routine clinical care or earlier research studies to be used for research involving in vitro diagnostic devices (IVDs), e.g., clinical laboratory test kits, without informed consent if the research is approved by an IRB and the specimens are not "individually identifiable" (FDA 2006).

Taken together, the US research regulations currently provide for use of biospecimens with IRB oversight and informed consent (or, sometimes, the potential for waiver of informed consent) when identifiers are retained. However, when biospecimens initially collected for a purpose other than the current research project are de-identified, the research is not subject to regulation by the Common Rule. The CR Final Rule largely keeps this structure, in which the identifiability of the biospecimen is determinative of the applicable regulatory requirements, as discussed further below.

**Privacy Law**

The Health Insurance Portability and Accountability Act of 1996 (HIPAA) carries with it further implications for research involving biospecimens. HIPAA governs the use and disclosure of "protected health information" (PHI)[12] by "covered entities."[13,14] Because most biospecimen research in the US involves covered entities, HIPAA's requirements and protections must be considered with respect to research involving biospecimens.

PHI may be "de-identified" and thereby removed from the requirements of HIPAA through either (i) shedding eighteen "identifiers" from the information, which notably do not include DNA sequences or other genetic information, or (ii) obtaining the documented opinion of a statistical expert that the risk is "very small" that the data can be used by an anticipated recipient to identify an individual who is a subject of the information. Accordingly, a specimen that contains no label or only certain demographic or phenotypic information that is not among the list of HIPAA identifiers (e.g., weight, age in years, gender, and race) would, under normal circumstances, not be subject to HIPAA.

When a specimen is labeled with PHI, however, researchers and IRBs have a few options to permit the use of PHI in research. The first involves obtaining the authorization of the subject for the research use of his or her PHI. That authorization must contain several elements, including a description of the PHI to be used or disclosed, a description of each purpose of the requested use or disclosure of PHI, and the name of the person or class of persons authorized to use the information. Notably, the authorization may be combined with an informed consent form for research, and the form may permit subjects to authorize the use of their PHI for future research studies, so long as the authorization "adequately describes" the future research.[15] The authorization therefore permits the individual to authorize the use of PHI for a present research study, e.g., a clinical trial, and simultaneously authorize the storage of PHI generated in the study for future research, such as the storage in a biospecimen repository of specimens labeled with PHI.

HIPAA provides three possible exceptions to its authorization requirement. First, if the only two required HIPAA identifiers are certain dates related to the subject and limited geographic information, the PHI constitutes a "limited data set," and the researcher may use it if he enters into a "data use agreement" (DUA) with the party supplying the PHI. Second, if the research involves solely PHI of decedents, no authorization is needed so long as the researcher provides certain representations regarding use of the PHI to the covered entity, such as that the PHI will be used solely for research purposes and is necessary for the research. Third, similar to the Common Rule's provision for waiver, HIPAA permits an IRB or privacy board to waive the requirement that a HIPAA authorization be obtained from the research subject before the use of the subject's PHI for research.

**Recent Changes to the Common Rule**

One theme that applies to both the research and privacy regulations described above is that they apply only when biospecimens are considered "identifiable," meaning that specimens can be removed from their purview by dissociating the specimens from identifying information. This aspect of the regulations fails to consider, however, that genetic information contained in specimens may itself inherently be identifiable, a concern with which regulators have recently started to grapple. Two major questions in this regard include whether the identity of the person from whom a biospecimen is obtained can ever be considered *not* to be "readily identifiable" by an investigator; and the proper extent of consent requirements for research on biospecimens, whether identifiable or not (HHS 2011, 2015).

One response, which was proposed by the US Department of Health and Human Services (HHS) in its 2015 Notice of Proposed Rule Making (NPRM) to amend the Common Rule, the precursor of the CR Final Rule, would be to expand the definition of the term "human subject" to include a living individual about whom the investigator "obtains, uses, studies, or analyzes biospecimens," thereby requiring researchers to submit for IRB review and obtain informed consent from subjects for all research involving biospecimens, with very limited exceptions. For the most part, the research community in the United States opposed this proposal. A number of comments submitted in response to the NPRM cited concerns with the cost of obtaining and tracking consent forms, insufficient evidence of risk or public concern about the issue, a fear that the rule would result in fewer specimens collected from fewer sources, and a concern that the result would be retention of more identified biospecimens, thus resulting in greater privacy risks to subjects (HHS, 2017).

The CR Final Rule thus abandoned the proposal to revise the definition of "human subject" to include living persons whose biospecimens the investigator obtains, uses, studies, or analyzes. Instead, the CR Final Rule retains much the same framework as the pre-2018 Common Rule with respect to the rule's applicability to biospecimens, with certain modifications, the most important of which are discussed here.

The CR Final Rule adds a new defined term referred to as "identifiable biospecimen," defined as a biospecimen for which the identity of the subject is or may readily be ascertained by the investigator or associated with the biospecimen, and modifies the definition of "human subject" to include a living individual about whom an investigator (i) obtains information *or biospecimens* through interaction with the individual, or (ii) obtains, uses, studies, analyzes, or generates identifiable private information or *identifiable biospecimens*. While the pre-2018 Common Rule's definition of "identifiable private information" had encompassed identifiable biospecimens, the CR Final Rule's modifications make this explicit, thus recognizing the increasing importance of research involving existing collections of biospecimens. The CR Final Rule's preamble explains that a biospecimen becomes identifiable, for example, when it is "tagged with the name or other information that indicates the person from whom the biospecimen was obtained" (HHS 2017).

Notably, while the CR Final Rule continues the pre-2018 Common Rule's policy of differentiating between the treatment of identifiable and non-identifiable biospecimens, it does introduce a process by which Common Rule departments and agencies can regularly assess, in consultation with data matching and re-identification experts, whether new technological developments merit reconsideration of how identifiability of information or biospecimens should be interpreted in the context of research. If this process results in a determination that a particular technology or technique, when applied to non-identifiable biospecimens, could produce identifiable information or render the biospecimens identifiable, the technology will be placed on a list of technologies or techniques meeting such determination, and a recommendation may be made as to consent, privacy, and data security protections for use of such technologies or techniques, including possibly requiring the consent of the human subject before the technology or technique is applied to the non-identifiable biospecimen. In addition, the CR Final Rule provides that the Common Rule departments and agencies will reexamine the meaning of the terms "identifiable private information" and "identifiable biospecimen" from time to time, and may alter the interpretation of these terms, including through the issuance of guidance.

While maintaining the same options as existing law to permit the use of identifiable biospecimens for research purposes, that is, obtaining the consent of the subject for such use, removing identifiers from the specimens to render them non-identifiable, or obtaining a waiver of informed consent from an IRB, the CR Final Rule introduces a fourth option: obtaining the "broad" consent of the subject for storage of the identifiable biospecimens for secondary research uses and the use of such biospecimens for secondary research purposes.[16] The broad consent must contain several elements, including a general description of the types of research that may be conducted with information and biospecimens, the types of information or biospecimens that might be used in the research, and the types of institutions that might conduct research with the biospecimens or information. When specimens are stored pursuant to a broad consent, full IRB review is not required, but rather, the IRB reviews solely whether the broad consent contains the required elements of broad consent, the process for obtaining the broad consent, and whether adequate provisions are in place to protect the privacy of subjects and maintain the confidentiality of data. When biospecimens stored pursuant to broad consent are released for specific research projects, the IRB also conducts only a limited review, including whether adequate provisions are in place to protect the privacy of subjects and confidentiality of data and whether the proposed secondary analysis is within the parameters of the broad consent that was obtained for secondary research purposes. The availability of this new broad consent option provides increased flexibility for researchers seeking to use identifiable biospecimens in research.

An important question raised in the context of secondary research performed pursuant to a broad consent is how to handle results gleaned from secondary research that might be clinically relevant to the individual from whom the biospecimen was derived. Should specific-consent and IRB review be required in order for results to be returned to such individuals? Or should broad consent for the unspecified secondary research suffice? The CR Final Rule, consistent with the earlier NPRM, requires that a full IRB review of the research generally be conducted if individual research results will be returned to subjects, including a review of whether results are returned in an appropriate manner. Thus if researchers wish to qualify for the limited IRB review described above for secondary uses of identifiable biospecimens pursuant to a broad consent, the consent form should inform subjects that results will not be returned as part of the research. This approach has certain logistical advantages due to the potential difficulty of locating donors down the road. If the researcher had not planned to return individual research results from specimens collected pursuant to a broad consent but later finds a result that he or she wishes to return because it may be important to the individual's health, the CR Final Rule revisions would require the researcher to obtain review and approval from an IRB of the plan for returning results to the subjects.

Another important issue raised by the ability to use a broad consent to store identifiable biospecimens for future research and use such biospecimens for future research is whether people should be permitted to *refuse* to allow their identifiable specimens to be used for future research. Moreover, such refusals would have to be tracked in perpetuity to ensure that specimens are used only for purposes that had been consented to. This sort of tracking could prove to be difficult and expensive for research institutions. Nonetheless, HHS included this concept in the CR Final Rule: if an individual were offered but refused to provide broad consent, an IRB would not be allowed to waive informed consent with regard to that individual's identifiable specimens down the road. Thus, offers, acceptances, and refusals of broad consent for research uses of identifiable biospecimens would have to be tracked throughout the life of that individual. Such tracking, it should be noted, requires the annotation of identifiers with the specimen in perpetuity (to ensure that it is never used for research) and thus increases the risk of inadvertent or advertent re-identification.

The provision of the CR Final Rule thus may reduce the frequency with which the research community avails itself of the broad consent option, since offering broad consent carries with it the risk that the subject will refuse to provide such consent and thus the subject's identifiable specimens will never be able to be used for research purposes (unless they are rendered non-identifiable). In contrast, if such a consent were not offered, the research community could still utilize the IRB waiver provisions to perform research on identifiable specimens in the absence of the subject's consent.

The NPRM would have also significantly curtailed the ability of an IRB to waive informed consent, requiring that (i) there be compelling scientific reasons for the research uses of the biospecimens, and (ii) the research could not be conducted with other biospecimens for which informed consent was or could be obtained. These changes were rejected in the CR Final Rule; however, the rule does add an additional element to the informed consent waiver criteria, i.e., that the research could not practicably be carried out without accessing or using identifiers. This criterion is similar to that found in the HIPAA Privacy Rule's requirements for a waiver of HIPAA authorization (discussed above) except that because the definition of nonidentifiable information under the Common Rule is less strict than the definition of "de-identified" information under HIPAA, this criterion may be satisfied more often in the Common Rule context than it is in the HIPAA context. Nevertheless, the purpose of this requirement appears to be to encourage researchers to use non-identifiable information whenever possible, and it may decrease the amount of research conducted on large collections of identifiable biospecimens for which consent (be it broad or specific) cannot be obtained.

A final question as we consider different regulatory models to potentially govern biospecimen research is whether there should be any distinction between research with biospecimens and research with identifiable private health information (i.e., data). The possibility of compromised autonomy and potential dignitary harm appears to be equivalent for both types of research, suggesting that a uniform regulatory approach would make the most sense. While in the NPRM HHS proposed to regulate research with biospecimens more stringently than research with data (Lynch et al. 2016), the CR Final Rule eliminated this approach and decided to regulate the two type of research in an equivalent fashion.

### Legal Framework of Biospecimen Repositories

Whereas part V of this volume focuses extensively on biospecimen repository structure and governance, we touch here briefly on how repositories fit within the legal and regulatory structure discussed above. A major legal question is the extent to which the act of creating a repository is itself an activity that constitutes "research" within the meaning of applicable regulations. Some institutions take the position that the creation of the biospecimen repository is not itself "research" requiring IRB oversight because the mere act of creating the repository is not intended to contribute to "generalizable knowledge"; rather, these institutions maintain that it is only once biospecimens are proposed to be or have been released to investigators for use in an actual research protocol that any "research" takes place. In the context of federally funded research, however, OHRP has long advised that if a research repository contains identifiable biospecimens, the collection of tissue specimens for the biospecimen repository as well as the actual storage of the biospecimens is "research" that should be governed by a research protocol and subject to IRB review (OPRR 1997). This view appears to be confirmed by the CR Final Rule, with its introduction of an avenue to obtain broad consent for storage of biospecimens for secondary research purposes.

Many institutions thus have established protocols for biospecimen repositories subject to initial and continuing IRB review and oversight that govern the contribution of biospecimens to the repository and the process by which researchers can request such biospecimens for particular research uses.

In recent years, many industry sponsors of research have amassed large collections of biospecimens from their own sponsored research and from others willing to provide biospecimens for industry research. Such biospecimens are often placed into internal company repositories for use in intracompany research. Because such research does not take place at a HIPAA covered entity and is not federally funded, this research is not subject to the requirements of HIPAA or the Common Rule, and because the results of such research often are not submitted to the FDA in support of a marketing application, this research also is not subject to the requirements of FDA's human subject protection regulations. Thus, there is a subset of biospecimen research that is not presently subject to any federal regulatory requirements.[17]

**Europe**

European countries are also grappling with the privacy and ethical implications of the use of biospecimens in research. While individual countries continue to regulate biospecimen research pursuant to their own national laws, guidelines for the development of such laws have been issued by the Council of Europe (COE), a 47-member state cooperative organization that establishes human rights standards. In 2006, the Council of Europe issued recommendations on the regulation of research uses of biospecimens (COE 2006). An updated version of this guidance was issued for public comment in 2014 (COE 2014) and finalized with revisions in May 2016 (COE 2016). These documents are intended to provide a baseline of protections to which individual COE member states can add additional requirements in their national laws.

As in the United States, the COE's guidance differentiates between "identifiable biological materials" and "non-identifiable biological materials." Although the preamble to the 2016 guidance document states that it was developed "considering that new developments in the field of biomedical research, in particular in the field of genetics, increase issues regarding protection of privacy" it nevertheless continues to take the position that biospecimens can be rendered non-identifiable. In a break with the US approach, in which coded specimens are not considered identifiable in the hands of the investigator if he or she lacks access to the key to the code,

# Legal and Regulatory Issues 37

the EU guidance indicates that *all* coded specimens should be considered identifiable, regardless of who holds the key to the code.

When biospecimens are collected expressly for storage for future research, the COE's 2016 guidance states that the consent of the subject should be obtained, and such consent should be "as precise as possible" with regard to any "envisaged" research uses and possible choices that the subject could exercise. The COE has thus not endorsed the concept of a "broad" consent for future use of biospecimens, such as that adopted in the CR Final Rule, which would permit a "general description" of the types of research uses that will be made of a specimen and does not contemplate participants limiting the types of research that may be conducted using their specimens.

The COE's 2006 guidance states that if identifiable biospecimens that have been collected for other purposes are used for future research purposes not falling within the scope of the consent pursuant to which they were collected, "reasonable efforts" must be made to contact the subject in order to obtain consent for the proposed use. If such reasonable efforts fail, the biospecimens can be used for research only if (i) the research addresses an important scientific interest, (ii) the aims of the research could not reasonably be achieved using materials for which consent can be obtained, and (iii) there is no evidence that the subject has expressly opposed the research uses in question. If biospecimens collected for other purposes have been anonymized, they may be used for secondary research provided that the research use does not violate any restrictions placed by the subject before anonymization.

The 2016 guidance heightens the consent requirements regarding residual use of biospecimens originally collected for another purpose by providing that consent *must* be obtained before identifiable biological materials can be stored for future research purposes. The 2016 guidance continues to permit anonymized biospecimens to be stored for research purposes without the consent of the subject provided that such use is authorized by local law. The 2016 guidance also introduces a new requirement that "clear policies" be developed regarding the return to individuals of findings that are significant to their health, and that subjects be provided an opportunity to opt-out of receiving any such results.

As in the United States, where use of biospecimens is subject to both human subjects research regulations and the HIPAA privacy regime, the privacy laws of the European Union carry implications for those conducting research on biospecimens. The EU's privacy framework, Directive 95/46/EC (the "Data Protection Directive") protects the processing of "personal data," which includes "any information relating to an identified or identifiable natural person" taking into account factors such as physical, physiological, mental, economic, cultural and social identity.[18] Given this broad facts-and-circumstances definition of personal data, it is difficult to be certain whether data are "anonymized" and thus no longer subject to the requirements of the Data Protection Directive. This carries with it implications for research involving biospecimens because it may be difficult to know with certainty when sufficient identifiers have been removed from specimens such that they may be considered "anonymized" and thus outside the ambit of the Data Protection Directive. The Data Protection Directive is in the process of being replaced by a General Data Protection Regulation, approved in December 2015 and issued in final form in May 2016, which will continue with the facts and circumstances definition of "anonymization," but will probably be interpreted by regulators so as to permit data subjects to provide a broad consent for future use of their "personal data" for research (General Data Protection Regulation 2016).

EU privacy law introduces further complications for biospecimen research because it limits the extent to which personal data may be transferred outside of the EU, providing that such data may be transferred only to countries that ensure an "adequate" level of data protection. The EU has recognized only a handful of countries as providing an adequate level of protection, and thus transfers of personal data to other countries, including the United States, require a legal basis under EU law. Permissible bases for transfer include the use of certain model contractual clauses that bind the data recipient through contract to use sufficient data protections, or obtaining the "unambiguous consent" of the research subject to the transfer. Certain bilateral agreements permitting transfer of personal data outside of the EU have also been negotiated, such as the EU-US Privacy Shield program, though there is some uncertainty regarding the future of such programs after the EU Court of Justice invalidated the EU-US Privacy Shield's predecessor program, the EU-US Safe Harbor, in October 2015 (*Schrems v. Commissioner*). It appears that there remains opposition to such programs in the EU and future litigation challenging their legitimacy likely lies ahead. These data transfer requirements introduce additional steps that must be taken by researchers wishing to draw biospecimens labeled with or accompanied by "personal data" out of the EU.

### China

Chinese law imposes unique challenges for those doing research on biospecimens. The governing law, the "Interim Measures for the Administration of Human Genetic Resources," (the "Interim Measures") enacted in 1998, applies to the use of "human genetic resources," which includes "genetic materials," e.g., human tissues, as well as to "information related to such genetic materials."[19] Accordingly, the law applies to the export of *genetic data* from China even if the specimens from which such data are gathered remain in China. Under the Interim Measures, if "human genetic resources" are to be exported from China, an application for an export permit must be submitted to the Human Genetic Resources Administration of China (HGRAC). Where research is conducted as a collaborative project between a Chinese institution and a foreign institution, the Chinese institution must submit the application to the HGRAC. The HGRAC evaluates several factors when deciding whether or not to approve the transfer, including (i) whether the informed consent of the subject has been obtained, (ii) whether the research project for which the specimen will be used has a precise objective or purpose, (iii) whether the apportionment of ownership and intellectual property rights in the specimen and research results is fair, and (iv) whether the foreign institution possesses adequate research capability to carry out the proposed research.

In October 2012 the Chinese State Council published a draft *Regulation on Administration of Human Genetic Resources* (the "2012 Regulation") that is intended to replace the Interim Measures. The 2012 Regulation attempts to account for technological advances since the issuance of the Interim Measures, making clear, for example, that the cross-border transmission via electronic means of information derived from "human genetic resources" requires approval of HGRAC in the same manner as shipment of hard copy materials across borders. The 2012 Regulation further specifies that anyone collecting "human genetic resources" for research purposes must be a legal entity established in China with a clear reason to collect the information and must have obtained research ethics committee approval. Before any biospecimen is collected, the informed consent and ethic committee approval must be reviewed and approved by the provincial science and technology administrative departments. Notably, in addition to the export requirements specified in the Interim Measures, the 2012 Regulation would require that the application for export of specimens define the period during which the biological specimens will be used as well as the method by which the samples will be disposed of when the research is complete.

## Brazil

Brazil has one of the most developed sets of laws surrounding biospecimen research around the world. In 2005, Brazil's National Health Council adopted Resolution 347/05, which regulates the use of human biological materials in research projects, and in 2011, the Ministry of Health published National Guidelines for Biorepositories and Biobanks (the "National Guidelines"). The National Guidelines draw a distinction between a "biobank," defined as an institutional facility dedicated to the systematic collection of human biological material to support multiple, future studies, and a "biorepository," defined as a collection of samples collected for a single, specific research project. Under the National Guidelines, before biospecimens can be contributed to a biorepository, a specific consent for the contemplated use of the biospecimen must be obtained. By contrast, the informed consent for a biobank may consist of a general consent for future research purposes. A biobank informed consent must be a "tiered" consent that provides research subjects the option of (i) authorizing their specimens to be used in *any* future research project or (ii) requiring that the research subject be contacted for reauthorization before his or her specimen is used in a specific project.

The informed consent for biobanking must also contain several other elements, including informing the subject that ethics committee review will be obtained before use of the biospecimens, offering subjects a choice of whether or not to receive genetic data generated from the use of their biospecimens, offering subjects a choice of to whom genetic data derived from the biospecimens may be released in the event of the subject's death or disabling condition, and the steps that will be taken to ensure the anonymity of information derived from the use of the biospecimens. From an organizational standpoint, the National Guidelines require that biobanks be governed by a research protocol approved by a research ethics committee that sets standard operating procedures of the biobank. In addition, the National Guidelines require that ethics committee approval be obtained before specimens are anonymized, since anonymization deprives a subject of the ability to receive genetic information generated through use of his or her samples. The National Guidelines permit biospecimens to be transferred abroad for research purposes so long as adequate steps are taken to protect subject privacy and the specimens are used in a protocol approved by a research ethics committee.

### India

In addition to the Drugs and Cosmetics Act of 1940 and its amendments giving regulatory authority to the Central Drugs Standard Control Organization (CDSCO), the Indian Council of Medical Research (ICMR) promotes policies and guidance over the ethical conduct of biomedical research in India (the Drugs and Cosmetics Act and Rules). In general, before enrolling a participant in a clinical research study, the investigator is required to provide and obtain informed consent from the participant, including making available an informational sheet about the study.[20] In addition to providing the participant with information about the risks of participation in the research, the investigator must delineate anticipated current and future uses of any collected specimen, any risks of participation and of biologically sensitive information that might be discovered, and establish safeguards for confidentiality. The participant must be informed about the potential benefit, if any, of the use of the specimen; has the right to withhold and/or withdraw consent for research use of his or her biospecimen; must be informed as to the storage period of the specimen and be given the right to deny permission to store; must be given a choice as to future use; must be allowed to claim or disclaim commercial benefit resulting from use of the specimen; and must be protected from coercion of participation. Further if genetic or HIV testing is conducted, counseling must be provided. (CDSCO, Good Clinical Practices for Clinical Research in India 2001; ICMR, Ethical Guidelines for Biomedical Research on Human Participants 2006).[21]

ICMR has further issued guidance on the transfer of human biological material for research and/or commercial purposes that is applicable to both materials collected in clinical trials and other research studies, and materials collected in the course of clinical care. This guidance requires review of an application for such transfer, including a copy of the ethics committee review and approval of the research, the informed consent document and patient information sheet, the signed material transfer agreement, the import certificate issued by the foreign regulatory authority to the importing institution, and the signed memorandum of understanding defining Indian and foreign commercial benefits to each party, among other items (Government of India Ministry of Health & Family Welfare 1997). A committee constituted by ICMR for foreign and commercial collaborations, or constituted by the appropriate authority, reviews each application; the basis for the case-by-case approval has not been described.

**Challenges in International Biobanking**

While this survey of laws governing specimen use in foreign countries is brief, it demonstrates that there is considerable diversity among such laws—and they are in substantial flux. Definitions of biospecimens and of human biological material, standards of and for identifiability, informed consent requirements, and import/export rules differ. In the age of multi-site, multinational research studies, stakeholders such as research sponsors and collaborating institutions must understand the laws of all countries in which biospecimens are collected for the research in order to ensure that the specimens were collected pursuant to the informed consent, ethics committee review, and other requirements of local law in order to ensure that unimpeachable rights exist to use the specimens in further research. Once biospecimens are received in a centralized biobank or for a specific research project, the institution overseeing the biobank or research must continue to track the terms pursuant to which the biospecimens were obtained to ensure that they are used only in conformance with the terms and conditions pursuant to which they were collected. For example, if the biospecimens were collected pursuant to a broad consent in the United States CR Final Rule, they could be used for a wide variety of research purposes without obtaining the re-consent of the research subject, whereas if the biospecimens were collected in a member state of the COE that has implemented the 2016 recommendation that consent be as "precise" as possible regarding envisaged future uses of the specimens, care should be taken that any research involving the biospecimen is consistent with the type of research authorized in the informed consent form.

The discussion above illustrates the importance for those conducting biospecimen research of understanding the provenance of any specimens used in the research, particularly if such specimens have been collected in foreign jurisdictions. Research institutions should require those providing specimens to explain how specimens were collected, represent and warrant in a material transfer agreement that provision of such specimens for research purposes is consistent with the laws, regulations, and informed consent terms under which such specimens were collected, and list any restrictions on the types of research that may be conducted using the specimens. Tracking these requirements requires a sophisticated administrative infrastructure managed by persons with knowledge of the pertinent issues discussed above.

## Conclusion

As more frequent discoveries are made through the use of biospecimens, biospecimens have become an increasingly important and valuable tool for researchers worldwide. This increased interest in collecting and storing biospecimens for research purposes has come to the attention of policy makers and regulators, who are struggling to map existing concepts of informed consent, research ethics committee review, and privacy law onto research involving the storage and use of biospecimens. These efforts result inevitably in a patchwork of laws both national and international with which researchers and institutions must familiarize themselves.

## Notes

1. *Moore v. Regents of the Univ. of Cal.*, 51 Cal.3d 120 (Cal. 1990). The *Moore* case is discussed further in chapter 2.

2. While *Moore* was a California state court case, it has been cited by many other state and federal courts for its expression of the common law rule pertaining to ownership of biological specimens. See, for example, *Doe v. Healthpartners, Inc.*, No.A06–1169, 2007 WL 1412936, at * 2 (Minn. Ct. App. May 15, 2007); *Janicki v. Hosp. of St. Raphael*, 46 Conn. Supp. 204, 216–17 (2007).

3. *Washington Univ. v. Catalona*, 490 F.3d 667 (8th Cir. 2007).

4. *Inter vivos* refers to a gift or transfer made during one's lifetime (as opposed to one that takes effect at death).

5. Similarly, in the case of *Greenberg v. Miami Children's Hospital Research Institute, Inc.*, 264 F. Supp. 2d 1064 (S.D. Fla. 2003), a federal district court held that patients did not retain a property interest in body tissue and genetic information that they had voluntarily donated for research purposes. The *Greenberg* court also held that the researchers did not need to disclose any economic benefit they might derive from use of the tissue to the donors, as unlike in *Moore*, where the researcher was also the patient's treating physician, there was no treatment relationship in the *Greenberg* case and thus no fiduciary duty owed by the researcher to the patient. This case is discussed extensively in chapter 2 of this volume.

6. 45 C.F.R. pt. 46 (2016).

7. We note that the revisions to the regulation were released one day before a new presidential administration began in the United States. At the time we went to press, it was not clear whether the new administration would attempt to block some or all of the revisions. However, the fact that the CR Final Rule rejected some of the more controversial proposals made earlier during the revisions process makes it less likely that the rule will be revoked by the new administration.

8. 45 C.F.R. § 46.101(b)(4).

9. 45 C.F.R. § 46.116(c).

10. 21 C.F.R. §§ 50, 56.

11. On December 13, 2016, President Barack Obama signed into law the 21st Century Cures Act, which amends the Federal Food, Drug & Cosmetic Act to permit IRBs to waive or alter informed consent requirements for FDA-regulated clinical investigations involving no more than "minimal risk," so long as appropriate safeguards are in effect to protect the rights, safety, and welfare of the research subject. While FDA has yet to promulgate regulations to implement these statutory revisions, it seems likely that it will adopt a waiver provision similar to that which appears in the Common Rule.

12. PHI is defined broadly to include any information, including genetic information, that is created or received by a covered entity and that (i) relates to the past, present, or future physical or mental health or condition of an individual and (ii) identifies the individual, or with respect to which there is a reasonable basis to believe that the information can be used to identify the individual.

13. This term is defined to include, among others, health-care providers that engage in certain billing transactions electronically, and thus includes most health-care practitioners and academic medical centers located in the United States.

14. 45 C.F.R. §§ 160, 164.

15. Whether future research—which by its very nature continuously evolves both rapidly and unpredictably—can ever be "adequately described" has been challenged in the application of both the HIPAA Privacy Rule and the concept of broad consent for future research. The 21st Century Cures Act, recognizing this difficulty with respect to HIPAA, requires HHS to issue guidance setting forth criteria to help the research community determine the circumstances under which a HIPAA authorization for the use and disclosure of PHI for future research purposes contains a sufficient description of the future use or disclosure.

16. The term "secondary research" in this context means "re-using identifiable information and identifiable biospecimens that are collected for some other 'primary' or 'initial' activity" (HHS 2017).

17. Note that the NPRM, if finalized in its proposed form, would have altered this understanding for certain biospecimens collected in research subject to the Common Rule by requiring that certain protections be flowed down to recipients of biospecimens collected in research subject to the Common Rule (Barnes et al. 2015).

18. The Data Protection Directive applies in the 28 member states of the EU plus Iceland, Norway, and Liechtenstein (an area collectively known as the European Economic Area), and it provides broad principles of data protection that member states are required to "transpose" into their national laws.

19. While the "Interim Measures" were intended to be temporary when first passed in 1998, they continue in effect today due to the failure thus far to adopt a replacement.

20. The Indian regulatory framework does not differentiate between the expectations of the investigators enrolling potential participants in a clinical trial and that of potential donors of biospecimens.

21. The ICMR Ethical Guidelines for Biomedical Research on Human Participants provide that the informed consent of a research subject may be waived for "[r]esearch on anonymised biological samples from deceased individuals, left over samples after clinical investigation, cell lines or cell free derivatives like viral isolates, DNA or RNA from recognized institutions or qualified investigators, samples or data from repositories or registries." This suggests that the informed consent requirements outlined in the text above would apply to uses of an individual's biospecimens for research, unless (i) the research meets these waiver requirements, and (ii) a waiver is in fact granted by the cognizant research ethics committee.

## References

Barnes, Mark et al. 2015. Impact of Proposed Federal Research Regulation Amendments (the Common Rule NPRM) on Life Sciences Companies. Bloomberg BNA Life Sciences Law & Industry Report.

COE (Council of Europe). 2006. Recommendation of the Committee of Ministers to Member States on Research on Biological Materials of Human Origin.

COE. 2014. Working Document on Research on Biological Materials of Human Origin.

COE. 2016. Recommendation of the Committee of Ministers to Member States on Research on Biological Materials of Human Origin.

FDA (US Food and Drug Administration). 2006. Guidance for Sponsors, Institutional Review Boards, Clinical Investigators, and FDA Staff: Guidance on Informed Consent for In Vitro Diagnostic Device Studies Using Leftover Human Specimens.

General Data Protection Regulation. 2016. Regulation (EU) 2016/679 of the European Parliament and of the Council of 27 April 2016 on the Protection of Natural Persons with Regard to the Processing of Personal Data and on the Free Movement of Such Data, and Repealing Directive 95/46/EC (General Data Protection Regulation). Official Journal of the European Union.

Government of India Ministry of Health & Family Welfare. 1997. Office Memorandum, Guidelines for Exchange of Human Biological Material for Biomedical Research Purposes. http://www.icmr.nic.in/min.htm

Government of India Ministry of Health & Family Welfare. 2001. Central Drugs Standard Control Organization. Good Clinical Practices for Clinical Research in India.

Government of India Ministry of Health & Family Welfare. 2008. The Drugs and Cosmetics Act and Rules.

*Greenberg v. Miami Children's Hospital Research Institute, Inc.*, 264 F. Supp. 2d 1064 (S.D. Fla. 2003).

HHS (US Department of Health and Human Services). 2011. Human Subjects Research Protections, Advanced Notice of Proposed Rulemaking, 76 Fed. Reg. 44,512, 44,525.

HHS. 2015. Federal Policy for the Protection of Human Subjects, Notice of Proposed Rulemaking, 80 Fed. Reg. 53,933.

HHS. 2017. Federal Policy for the Protection of Human Subjects, Final Rule, 82 Fed. Reg. 7,149.

Indian Council of Medical Research, International Health Division. 1997. Guidance on the Transfer of Human Biological Materials for Commercial Purposes and/or Research for Development of Commercial Products. www.icmr.nic.in/ihd/ihd.htm.

Indian Council of Medical Research. 2006. Ethical Guidelines for Biomedical Research on Human Participants.

Lynch, H. F., B. E. Bierer, and I. G. Cohen. 2016. Confronting biospecimen exceptionalism in proposed revisions to the common rule. *Hastings Center Report* 46 (1): 4–5. doi:10.1002/hast.528.

*Moore v. Regents of the University of California*, 793 P.2d 479 (Cal. 1990).

OHRP (US Department of Health and Human Services Office for Human Research Protections). 2008. *Guidance on Research Involving Coded Private Information or Biological Specimens*.

OPRR. 1997. Issues to Consider in the Research Use of Stored Data or Tissues. Last modified November 7, 1997. http://www.hhs.gov/ohrp/policy/reposit.html.

*Schrems v. Data Protection Commissioner*, Case C-362/14 (EU Court of Justice 2015).

*Washington University v. Catalona*, 490 F.3d 667 (8th Cir. 2007).

# 2 Property Rights and the Control of Human Biospecimens

Russell Korobkin[1]

As genomic science rapidly advances, human biospecimens currently maintained in biospecimen repositories will undoubtedly become even more important to the study of disease and creation of treatments. Uncertainty and contention around a series of related legal questions—who if anyone owns the biosamples, what limitations are there on their use and who gets to decide, and who can profit from their use in medical research—can result in disputes between specimen donors, researchers, and repositories and impede the effective and efficient use of tissue samples. This chapter addresses these complex legal issues, with particular attention to three leading judicial opinions concerning the use of tissue samples: *Moore v. Regents of the University of California*, *Greenberg v. Miami Children's Hospital Research Institute, Inc.*, and *Washington University v. Catalona*.

## Tissues as Chattels

In the past, the courts—especially in England, mostly in dicta, and mostly in cases concerning disputes over corpses—often stated that there are no property rights in the human body (Heng 2003). Other courts fashioned an ill-defined category of "quasi-property," without clearly defining the concept, which further confused the issue.[2] And unfortunately, the most famous modern American legal opinion dealing with parts of the human body, *Moore v. Regents*, contains some language that can be read as suggesting body parts or tissues might not be considered property even when removed from the body. In that case, the California Supreme Court held that John Moore, who suffered from hairy cell leukemia, lacked property rights in the cells derived from his enlarged spleen, which was surgically removed from his body in order to save his life.[3] As a consequence, a litany of scholars has concluded that the law does not recognize property rights in the human body and/or parts of the human body (Swain and Marusyk 1990; Rao 2000, 373; Seeney 1998, 1165).

The place to begin any discussion of ownership and control of biospecimens is by recognizing that this common conclusion about the actual state of the law is incorrect. There is, in fact, no real dispute that human tissues or organs *can* be property, at least as the word "property" is understood today. Courts have held, for example, that a researcher was liable for the tort of conversion (essentially theft of property) when he took and destroyed human cells being cultured by a colleague;[4] that a jury could find conversion when a doctor intentionally destroyed the plaintiff couple's gametes held in frozen storage for in vitro fertilization;[5] and that defendants were guilty of criminal theft for taking their own urine (*R v. Welsh* [1974] RTR 478) and blood (*R v. Rothery* [1976] 550) samples that they had earlier given to the police. As Justice Broussard pointed out in dissent in *Moore*, if a thief had purloined John Moore's spleen from his physician's laboratory, he most certainly would have been liable under the law of conversion.[6] Although confusing, the majority opinion in *Moore* does not stand for the contrary position. It is better read for the proposition that John Moore's disembodied cells did not remain *Moore's* property after ownership was transferred to a physician/researcher than for the proposition that the cells cannot be *anyone's* property.

Modern legal theory conceptualizes property as a variable collection of "sticks" in a bundle. A full bundle of rights would include unfettered rights to use, exclude, and transfer the item in question, but property rights are rarely so extensive. Zoning laws limit the uses to which I can put my house, but no one would say that my house is not property, as I can generally exclude others from using it and I can put it to many uses for myself. Nor does the rule against perpetuities, which limits restrictions I can put on the transfer of my house after my death, render the house incapable of being property. It is similarly true that law provides some limitations on the dominion that individuals—be they donors, researchers, or institutions—can exercise over bodily tissues, but these limitations are hardly serious enough to conclude that those tissues are not, or cannot be, property.

There are relatively few restrictions on the extent of dominion with regard to most tissues. Possessors of bodily materials can exclude others from using them. This is obviously the case when they remain encased in a live body, as verified in *McFall v. Shimp*, where a court refused to grant the plaintiff's request to his cousin's bone marrow, even though the value to plaintiff (who faced death without a transplant) far outweighed the cost of sharing to the defendant.[7] And the right to exclude clearly extends to circumstances in which tissues are disembodied, as the cases described above indicate. I have as much right to exclude others from using a vial of blood stored in my refrigerator as I do to exclude them from using the same amount of blood stored in my person. Despite occasional modern cases that state the contrary, the law is nearly as clear when the tissues in question reside in a deceased body. American judicial decisions, most notably *Brotherton v. Cleveland*,[8] have found that the coroner cannot remove tissues from a dead body (corneas in that case) against the wishes of the next of kin.

For most bodily materials, possessors have considerable latitude in transferring them to others. The Uniform Anatomical Gift Act (UAGA), a model law enacted in all 50 states, recognizes a right of an individual to bequeath the entire body or "body parts" upon death for research, transplant, or education.[9] There are active commercial markets for some types of biospecimens. Men sell sperm and women sell ova for reproductive uses without legal interference, for example. Ted Slavin, a hemophiliac whose blood contained unusually high levels of hepatitis B antibodies, famously earned a living selling his blood to pharmaceutical companies for up to $10 a milliliter (Skloot 2006).

In some situations, discussed in greater detail below, the law prohibits a compensated transfer of human body parts, which does constitute a notable limitation on the extent of property rights. The UAGA prohibits some sales of body parts, but notably, those prohibitions are circumscribed.[10] The National Organ Transplantation Act (NOTA) forbids the sales of solid organs for human transplant purposes, but not for research purposes.[11] The careful delineation of limitations on alienability support, rather than undermine, the observation that tissues are property. Limits written into the law are the exceptions that prove the general rule.

The most significant restriction on control over bodies and their parts is the generally accepted principle that one cannot sell a live person, or complete and permanent dominion over a live person (even oneself). But even in this extreme case, the broad rights an individual otherwise has to use one's own body and to exclude others form using it suggest that, conceptually, even entire live bodies comfortably with within a broad, modern definition of property.

## The Information Within

Human biospecimens, as chattels, are subject to private ownership as property, but the information that exists within those tissues lives in the public domain, beyond the monopolization of both donors and researchers. The United States Constitution recognizes the need for copyright and patent protection of intellectual property in order to incentivize the production of written works and useful inventions,[12] but facts about the world, importantly, cannot be owned.

*Greenberg v. Miami Children's Hospital Research Institute, Inc.* concerned tissues of children suffering from Canavan disease. The parents of the children provided tissue samples, as well as health records, to Dr. Reuben Matalon and his hospital, for the purpose of attempting to identify the genetic mutation that caused the disease. A dispute arose between the researchers and the donors when Matalon and Miami Children's not only identified the Canavan gene but then patented it and subsequently demanded that companies wishing to use genetic tests to diagnose the disease pay a licensing fee.[13] Unlike in *Moore*, the Greenberg plaintiffs did not object to the researchers' use of the children's tissue samples, but to the use to which the researchers put the knowledge they gained about the Canavan gene and how to determine who would suffer from the disease in the future.

At the time of the *Greenberg* dispute, the United States Patent and Trademark Office (USPTO), with the approval of the US Court of Appeals for the Federal Circuit, routinely issue patents on isolated genes.[14] Although patent law doctrine has long been clear that "products of nature" are ineligible for patent protection,[15] courts long avoided this apparent obstacle to gene patents by asserting a chemical distinction between a gene as it exists in human cells and a gene as it exists in the laboratory, where its physical makeup is laid bare.[16] In *Association for Molecular Pathology v. Myriad*, the United States Supreme Court reversed a line of lower court precedent and held that, because what makes a gene patent valuable is the information contained within the gene, and that the information is the same whether or not a gene is encased in longer DNA strands, genes are "discovered" rather than "invented" by scientists and thus are not a proper subject of a patent.[17]

After *Myriad*, neither biospecimen donors nor researchers can claim a property right in the information about human functioning and wellness encased in the specimens. If Dr. Matalon successfully isolated the Canavan gene today, he could patent a particular diagnostic test, assuming it met the standard requirements of patentability such as non-obviousness. He could not, however, claim ownership of the gene itself, which would enable him to charge a royalty for the use of any type of diagnostic test for the gene.

## The Transfer of Property Rights

Human specimens preserved in biospecimen repositories no doubt begin their journeys as the property of the person in whom they originally resided. A research facility could no more demand extraction of a tissue sample from Shimp than could his cousin, McFall, no matter how valuable that tissue might be to the research community. But what is the property status of specimens that are housed in repositories after being excised from the body with permission?

Subject to certain specific limitations, tissue donors may transfer their property rights to an individual or institution. Just as I might make a gift of money to an academic research institution, a hospital, or even an individual researcher, I may convey ownership of my tissue by making a gift—that is, a transfer of property for no consideration in return.[18] I may choose to donate tissues while I am still alive, in the form of an *inter vivos* gift. Making such a gift legally effective traditionally requires an intent to convey ownership, delivery of the property, and acceptance of the gift (Penalver 2010, 198). Alternatively, I may donate tissues on the occasion of my death, in the form of a bequest.[19]

Most *inter vivos* gifts and bequests are unconditional; that is to say that, once the gift is made, the donor forfeits all rights of control of the corpus of the gift. Once I write a check to the American Red Cross, for example, I can't stop the charity from using my money for flood victims because I don't find them as sympathetic as earthquake victims, or insist that my contribution must be used for food and not for blankets. And if I decide I don't like the way the organization is using my donation after it is made, I have no legal standing to demand a refund, although, of course, I am free to choose not to donate again. After an unconditional gift is made, the donor lacks control over the future use of the corpus that was gifted, and the donee enjoys the full bundle of property rights that the donor used to have (Baron 1989).

That most gifts of human tissues are unconditional is best understood as a default rule, subject to change by the donor at the time of donation and not, to use a legal term, a "mandatory" or "immutable" rule (Ayres and Gertner 1989). Donors are perfectly free to make their gifts limited in scope, restricted to particular uses, or conditional on future events. For example, gifts often specify that the corpus of the gift may be used only for certain purposes or may not used for other purposes. If the proposed conditions are too limiting or too onerous for the recipient to follow, or if the recipient would prefer not to abide by them, the recipient may reject the gift. But when a conditional gift is accepted, the law requires the recipient to abide by the conditions. The failure of the recipient to abide by a condition that limits the use of chattel can create liability for conversion.[20]

Whether a gift is conditional or unconditional requires a particularized inquiry into any language used in describing the nature of the gift and also the surrounding context. The gift of an engagement ring is most often interpreted by courts as being conditional on the marriage taking place, meaning that the recipient must return the jewelry if the nuptials is canceled, even though the conditional nature is rarely expressed orally, much less in writing (Tushnet 1998). The gift of a ring on the occasion of the recipient's birthday, in contrast, is usually interpreted as unconditional.[21] But either of these presumptions can be overcome if language or circumstances manifests a contrary intent on the part of the donor.

Interpreting the breadth of a donation's condition can be challenging, and this was the problem that made *Washington University v. Catalona* a difficult case. *Catalona* concerned two types of tissue samples: tissues that were excised from the body of donors by Washington University physicians for therapeutic purposes (i.e., prostate tissue samples excised during surgery) and subsequently used for research, and tissues that were removed from the body exclusively for research purposes (blood and serum samples provided by non-patients of Washington University physicians).[22] All donors had signed two forms, provided to them by Washington University primarily for the purpose of documenting the informed consent of patients/donors. The forms indicated that the signatories were donating the tissues for research, were forfeiting any claims to those tissues, and were doing so with no expectation of monetary payments or rights in any medical products that might result from the research.[23]

Although Dr. Catalona claimed that he, not Washington University, was the recipient of the donors' *inter vivos* gifts of tissues, this argument was barely colorable. Both the informed consent document language and tissue donation norms make clear that the more reasonable interpretation was that the tissues were donated to the University's biospecimen repository rather than to a single, particular researcher. The harder question was whether, when making that gift, the donors retained the right to redirect their tissue samples at a later time.

By way of the informed consent documents, the *Catalona* donors expressly retained their right to "withdraw [their] consent at any time," and in the case of some (but not all) of the donations documents, the donors "could request destruction of their biological materials if they changed their minds about participating in the study" but "recognized that it would not be possible to destroy or recall any research results already obtained."[24] All donors also received a brochure upon making their donation that specified they would retain the right to have their tissues destroyed at any time in the future should they change their minds about study participation.[25]

When Dr. Catalona moved from Washington University to Northwestern University, he asked the patients to sign a request to Washington University that their tissues be transferred to him. Thousands did so, but Washington University refused to part with the specimens. Correctly noting that *inter vivos* gifts may be made with conditions attached, the Eighth Circuit ruled that, by the language of the written documents, the donors retained the right to stop further use of their tissues for research but not the right to have the tissues returned or transferred to another biospecimen repository.[26]

Given that it would have been highly unusual for Washington University to ship a prostate tissue sample back to its donor, the court's narrow interpretation of the extent of the limitations on the donors' *inter vivos* gifts seems defensible. A broader interpretation, however, would have also been plausible: that a donor's clearly specified right to stop the further use of his tissues in studies conducted under the aegis of the Washington University repository *implied* an accompanying right to direct a transfer of the tissues to another institution.

One way of understanding the *Catalona* decision is that the court not only determined that an unconditional transfer of rights is the default interpretation of *inter vivos* gifts of human tissues, but also that it established a fairly strict "altering rule"—that is, the steps that private parties must take to avoid the default rule (Ayres 2012). Specifically, to recognize a condition of a gift, the court required an express statement of that condition; it was unwilling to search for contextual clues that might imply a condition or limitation. The *Greenberg* court's decision can be understood similarly. When the *Greenberg* defendants patented the Canavan gene and charged licensing fees for Canavan diagnostic tests, contrary to the plaintiffs alleged "understanding" that diagnostic tests would be provided on an affordable basis, the court rejected the plaintiffs' conversion (i.e., theft of property) claim due to the lack of any express condition accompanying the tissue transfer. According to the court, "[t]he property right in blood and tissue samples ... evaporates once the sample is voluntarily given to a third party" in the absence of allegations that the defendants used plaintiff's materials for an "expressly unauthorized act."[27] A court more willing to explore the context surrounding the gift might well have determined that the defendants' actions were *implicitly* unauthorized.

Property considered valuable by its owner is usually transferred, if at all, by gift or sale. But property considered valueless—or, more precisely, property that comes with liabilities that exceed its benefits—is often abandoned by one party and claimed by another. On the final days of the academic year, the sidewalks in front of apartment buildings close to the UCLA campus, not far from my house, are often strewn with used furniture in various states of repair, put there by departing students who judged the items to be less valuable than the cost of moving or storing them. Once abandoned, which under the common law requires the owner to dispose of chattels with the intent to disclaim his rights, the property legally belongs to the first person who claims it (Penalver 2010). If property is abandoned and not claimed (think "trash"), modern governments often hold the abandoner responsible for disposal (think "fines for littering"), but when what is one person's trash is another's treasure, the common law system still operates fairly well.

The somewhat archaic law of abandonment provides a lens with which we might understand California Supreme Court's decision in *Moore*. John Moore's treating physician, Dr. David Golde, removed his patient's diseased spleen for therapeutic reasons—Moore would have certainly died otherwise—and then used the disembodied organ to develop the "Mo" cell line, which was particularly valuable given the unusual blood cells residing in Moore's diseased spleen.[28] The court denied Moore's conversion claim on the ground that he lacked property rights in the spleen, but it failed to clearly explain the basis for the conclusion. Since it is clear that it is possible to have property rights in disembodied tissue, the unexplained mystery of the *Moore* opinion is how, exactly, those property rights were transferred from the patient to his physician.

John Moore claimed, apparently without contradiction, that he had no idea that the spleen would be used for medical research,[29] so it seems difficult to argue that he made an *inter vivos* gift of the tissue, which would require an intent to transfer ownership to Golde. But Moore did know that the spleen would be removed from his body, and we can infer from his lack of concern over its fate at the time that he intended to sever his relationship with it. No one suggested that Moore asked the UCLA hospital to return the spleen to him when he left the hospital or expressed any interest at all in his doctor's plans for the spleen's disposal. In fact, the consent form he signed before the surgery, which he might or might not have read, purported to give permission to the hospital to dispose of organs removed during surgical procedures (Andrews 1986, 28).

The facts of *Moore* bear a logical similarity to those of *Venner v. State of Maryland*. In that case, state police recovered narcotics filled balloons from the bedpan of the hospital patient, Venner, who then sought exclusion of the incriminating evidence on the grounds of illegal search. Holding that the police lawfully obtained the evidence, the court found that, while Venner enjoyed property rights in the waste that was once contained in his body, "it is all but universal human custom and human experience" that the waste is, "in a legal sense, abandoned by the person from whom [it] emanate[s]," and was, in fact, abandoned by Venner.[30]

To Moore, his diseased spleen was medical waste with a negative value, which he expected to never see again. Golde no more "stole" Moore's property than the police stole the contents of Venner's bedpan or fortune hunters who troll the streets adjacent to college dorms on moving day steal the beaten up old sofas left on the sidewalk. The facts that Golde placed a positive value on Moore's spleen, that the police placed a positive value on the contents of Venner's bed pan, or that the fortune hunter places a positive value on an old sofa, do not change these conclusions.

## Compensation

In a free-market society, many property rights are transferred not by gift or as a result of abandonment, but by way of a sale—that is, traded for money or other valuable consideration. In this case, contract law generally enforces the sale on the terms agreed to by the parties absent fraud or duress. In a limited category of circumstances, however, the law prohibits an owner from selling certain property rights, even when those rights may be transferred altruistically as a gift. Numerous scholars have advanced the argument that human tissues should fall into this category, thus implicating the question of what values are implicated in the sale of human tissues that are different from those implicated in other transactions in chattels. Supporters of prohibitions on compensation ("prohibitionists") often contend that markets in human tissue are different than other types of markets because they indicate a lack of respect for human life. Alternatively, prohibitionists sometimes contend that permitting payment for tissues might unduly coerce the economically disadvantage to part with necessary tissues or submit to dangerous medical procedures in order to provide the tissues (Korobkin 2007a, 51).

The prohibitionist argument has had some effect on the law in the United States regarding human tissues directly, and perhaps a greater indirect effect. As noted above, the NOTA prohibits the buying and selling of human organs (not including renewal tissues such as blood) for use in human transplantation,[31] and the UAGA prohibits the sale of body parts for transplantation or therapy if the parts are to be removed "after the death of the decedent."[32] Notably, neither of these broadly applicable statutes would apply to the sales of tissues for research purposes. This distinction perhaps is due to the fact that sales of tissues for transplant implicate issues of equity among potential tissue recipients, and thus reflects popular distaste with the idea that transplantable organs could be purchased only by the very rich—an issue of far less relevance to biomedical research—rather than broader concerns with commodification of the human body or coercion of donors.

A small number of states have enacted prohibitionist statutes concerning human biospecimens, however, that are broad enough to encompass research uses, although some of these exempt renewable tissues (sometimes defined to include ova along with blood and sperm even though human ova are not, strictly speaking, renewable) (Korobkin 2007, 48–49). And there are some examples of state laws that prohibit payment to specific types of tissue donors beyond compensation for expenses incurred, such as California and Massachusetts prohibitions against compensating women who donate ova for purposes of stem cell research.[33]

Where human tissue samples are concerned, the greatest legal impediment to payments might be indirect, rather than direct. Various ethical guidelines promulgated by professional organizations call for placing limits on payments to human research subjects in order to avoid what they consider potential coercion of tissue donors (and other research participants). The Council for International Organizations of Medical Sciences recently proposed, for example, an ethical guideline stating that compensation for biomedical research participants "must not be so large as to induce potential participants to consent to participate in the research against their better judgment ("undue inducement") (CIOMS). Such guidelines lack the direct force of law, but federal law does require research institutions that obtain federal funding to comply with the regulation adopted by 18 agencies known as the "Common Rule."[34] The Common Rule, in turn, requires researchers at covered institutions to obtain local institutional review board (IRB) approval before conducting research. When local IRBs decide to approve only research proposals consistent with ethical guidelines promulgated by professional societies that propose limits on research subject compensation, the Common Rule can indirectly limit the amount of compensation that can be provided in consideration for tissue donations (Emanuel 2004).

Interestingly, other scholars and some research organizations have advanced an argument that is precisely opposite to the claim that tissue sales should be prohibited entirely or payment amounts limited: specifically, that compensation for tissues should be required and completely altruistic transfer prohibited. Supporters of mandatory compensation contend that when donor specimens are used for socially valuable purposes, fairness demands that donor share in the benefits. They often argue for some type of profit sharing or in-kind compensation arrangements with either individual donors or communities of donors (HUGO Ethics Committee 2000). This position is sometimes framed as a claim that donors should be found to have property rights in their disembodied tissues, but as it is usually clear that property rights have been transferred, the claim is better understood as one that compensation should be recognized as an implicit or constructive term of the transaction. Although this "pro-compensation" principle has been adopted in non-binding organizational guidelines and recommendations (Austin et al. 2005), I know of no American court or legislature that has established this principle as a matter of law.

In most circumstances, then, compensation of human tissue donors is neither legally required nor proscribed. In these circumstances, the most important legal questions become "what is the default rule concerning compensation that will be applied if the donor and recipient do not reach an alternative agreement?" and, relatedly, "what is the altering rule?"

John Moore and his physicians had no discussion about whether compensation would be provided for the use of Moore's spleen, either as a fixed payment or conditional on the physicians successfully using the tissue to produce a profitable cell line. When Moore sought compensation, the California Supreme Court ruled that he was not entitled. One interpretation of this outcome is that the legal default rule is one of "no compensation" for biospecimens that are either abandoned or, alternatively, donated (Korobkin 2007).

Unfortunately, the *Moore* court did not discuss the relevant altering rule; that is, what conversation or agreement would have been necessary before John Moore's splenectomy to entitle him to payment. Presumably, if Moore had told Dr. Golde that the doctor could use his spleen to create a cell line conditional on Golde paying Moore 10 percent of any profits, this express language would have been sufficient to avoid the default rule, but the legal significance of half-measures were left undefined. What if, for example, Moore had merely said, before being wheeled into surgery, that it would be nice if his unusual spleen turned out to have medicinal properties that he could profit from, and Golde responded with a vague shrug?

In *Greenberg*, the plaintiffs alleged that when they had provided their children's tissues to Dr. Matalon for his research they did so with "the understanding" that any discoveries would benefit "the population at large" and that diagnostic tests would "be provided on an affordable and accessible basis."[35] Although the court ruled that this inchoate "understanding" did not constitute a condition on the use of the tissues, it also found that this allegation, if proved to a jury, would constitute a valid claim for "unjust enrichment," defined as a situation in which "it would be inequitable for the defendant to retain the benefits [of the tissue donation] *without paying for it.*"[36]

*Greenberg* was decided by a federal trial court in Florida and was not appealed (the parties settled out of court following the ruling), and thus the opinion has no formal precedential value. This noted, if the court's decision to reject the plaintiffs' conversion claim while simultaneously upholding their unjust enrichment claim is to be interpreted in a way that is internally consistent, it should be understood to stand for the principle that even hazy discussions about a researcher providing some modicum of consideration for a tissue donation—even if the consideration is conditional and even if it would flow, if at all, to third parties—could overcome the default presumption that donations are altruistic.

In light of the uncertainty surrounding how courts will interpret vague discussions of compensation or benefit sharing, careful researchers and biospecimen repositories are wise to ask donors to explicitly disclaim any right to future compensation, as Washington University did in connection with Dr. Catalona's prostate tissue samples and is standard procedure at most repositories (Marchant 2005). Donors who wish to condition tissue donations on individual or class-wide compensation should likewise make this understanding express. As an example, and in clear contrast to the *Greenberg* plaintiffs, the families of children suffering from the rare genetic condition of pseudoxanthoma elasticum formed a nonprofit organization and contracted with researchers to provide blood and tissue samples in return for a share of profits resulting from research involving the tissues (Terry and Boyd 2001).

## The Complicating Interaction of Informed Consent Law and Privacy Concerns

The extent to which donors of biospecimens have the legal right to control or veto research uses is complicated by the overlap between questions concerning property rights and the distinct legal concept of informed consent. Confusion often results when commentators ask whether, in certain situations, tissue donors have "property rights" in tissues, although it is clear that they have transferred their property rights but questions might be raised as to whether they have provided appropriate consent for particular uses. The issue is further complicated by a lack of clarity over whether "informed consent" concerns are rooted in the value of autonomy or privacy.

The legal concept of informed consent is relevant to biospecimen research in two ways. First, the common law of informed consent requires that physicians convey the risks associated with a medical procedure before obtaining a patient's consent to the treatment.[37] This means that if a therapeutic medical treatment involves excising tissues from the body that will subsequently be used for research purposes, such as in the case of John Moore's splenectomy or the patients of Dr. Catalona's prostate surgeries, the treating physician must explain to the patient health risks that will arise from the surgical intervention in the process of obtaining consent for the surgery.

Courts have extended the common law of informed consent to require treating physicians to inform patients of any profit motive they might have in performing a therapeutic procedure (besides being paid for the procedure itself, which is presumably obvious).[38] The reasoning is that the existence of a conflict of interest might compromise the doctor's judgment and cause a reasonable patient to be more skeptical of his advice to undergo a treatment or procedure, which might then affect whether the patient would consent (Andrews 1986, 31). On the basis of this logic, the California Supreme Court held that Dr. Golde violated his obligation to obtain John Moore's informed consent to the surgery by failing to disclose his desire to use Moore's spleen for research.[39]

This reasoning, based as it is on the concern that a patient be able to appreciate a medical procedure's risks before agreeing to undergo the procedure, has several implications. First, it suggests that Golde would not have had an obligation to obtain Moore's consent—informed or otherwise—to use the spleen for research purposes if he had no plan to do so before performing the splenectomy. A research plan hatched post-surgery, after all, could not create any risk to Moore's physical health or well-being. Second, it implies that Golde had no obligation to explain to Moore precisely what type of research he wished to perform on the excised spleen—for example, that he hoped to create a cell line for the treatment of hepatitis B—because it was the existence of Golde's profit motive rather than the research plan *per se* that was potentially relevant to Moore's determination as to whether the risks of the surgery would exceed its potential benefits. It follows from this analysis of the nature of informed consent that this principle does not provide a basis for any donor control over a particular use of a biospecimen obtained with consent that is informed by knowledge of the risks inherent in the exaction of tissue.

Independent of the common law, however, the Common Rule,[40] along with a similar US Food and Drug Administration rule,[41] requires researchers to obtain informed consent when conducting "human subject" research. Unlike the common law of informed consent, these regulations require researchers to disclose the "description" and "purpose" of a research study to participants before obtaining their consent.[42] Most IRBs consequently will require disclosure of a study's specific details to donors as a condition of the study's approval. But conflicts still may arise when a researchers wish to conduct secondary research on a biospecimen—that is, when the specimen was collected as part of a therapeutic procedure at an earlier time at which there was no clear research plan, or when researchers wish to use a portion of a sample originally obtained for a specific research project at a later time for different and unrelated research. If there was no plan to conduct research at the time the tissue was extracted, or if a researcher obtained "broad consent" or "blanket consent"[43] to unknown future research uses and the donor later disapproves of one particular use, should the donor have grounds to object based on the claim that he did not provide *informed* consent the research in question? The answer is important because biospecimen researchers could find themselves having to re-contact past donors whenever they conceived of new research usages, which often would prove costly and complicated, and in some cases make research a practical impossibility.

Under the common law, the transfer of chattels via gift, abandonment or sale does not require that the recipient inform the provider of the former's intended use. If I give my old car to a friend unconditionally and he later drives it to Las Vegas, I cannot object to this use on the ground that I disapprove of gambling. What makes the transfer of biospecimens unique is that they continue to carry with them, in the DNA of each cell, an enormous quantity of personal information, over which the donor will often wish to maintain privacy protection. Any effort to provide some manner of input or control to tissue donors concerning research unanticipated at the time of donation should thus be focused on their interests in privacy, not in property or in decisional autonomy.

That privacy rather than autonomy concerns should be central to the question of whether donors should be permitted to retain any control over uses after unconditional transfers of property rights is, in fact, implicit in the Common Rule's structure. Under the current rule, researchers need not obtain "informed consent" at all before conducting research on disembodied tissues that have been stripped of identifying information. The textual explanation is that the study of previously disembodied tissues constitutes "human subject" research under the rule if the donor is identifiable but not if the donor cannot be identified.[44] The only logical basis for this distinction is that privacy concerns disappear if the source of the tissue cannot be determined. The Notice of Proposed Rulemaking, published in September 2015, proposed expanding the reach of the Common Rule, and thus of the informed consent requirement for secondary biospecimen research, to even de-identified specimens. But this proposal, eventually tabled, was not inconsistent with the understanding that privacy rather than decisional autonomy is the rule's central concern, because the proposal's primary justification was that the presence of unique genetic markers combined with technological advances call into question whether tissues stripped of external identifiers can ever really be de-identified (Largent 2016).

## Conclusion

In spite of confusion sown by the trio of notable judicial opinions, biospecimens are chattels subject to property rights that can be transferred through the processes of gift, abandonment, and—subject to some limitations—sale. The default rule is that such transfers are unconditional, meaning that a successful transfer reallocates all of the original holder's property rights, but they can be made conditional or limited in scope. The proper interpretation depends on both the language of the transfer and the content of the altering rule. Whether a donor retains any property rights in tissues that have been transferred is often conflated with the question of whether a donor has independent grounds to veto certain research uses of which he or she disapproves even after unconditionally transferring all property rights. Federal research regulations cast this issue as one of informed consent, but an understanding of the scope of the common law principle of informed consent for therapeutic medical procedures and of the fundamental difference between human biospecimens and other chattels suggest that the case for donor control following the unconditional transfer of property rights is, and should be, rooted in privacy rather than autonomy concerns.

**Notes**

1. Erin Malone provided indispensable research assistance.

2. *Pierce v. Props. of Swan Point Cemetery*, 10 R.I. 227 (R.I. 1872); *Georgia Lions Eye Bank, Inc. v. Lavant*, 255 Ga. 60 (Ga. 1985).

3. *Moore v. Regents of Univ. of Cal.*, 51 Cal. 3d 120, 125–127 (Cal. 1990).

4. *United States v. Prince Kumar Arora*, 860 F. Supp. 1091 (D. Md. 1994).

5. *Del Zio v. The Presbyterian Hosp. in the City of N.Y.*, No. 74 Civ. 3588 U.S. Dist. Lexis 14450 (S.D.N.Y. Nov. 14, 1978).

6. *Moore*, 51 Cal. 3d at 153.

7. *McFall v. Shimp*, 10 Pa. D. & C.3d 90 (Ct. Com. Pl. 1978).

8. *Brotherton v. Cleveland*, 923 F.2d 477 (6th Cir.1991).

9. Unif. Anatomical Gift Act §6(a) (1987).

10. Revised Unif. Anatomical Gift Act §§16, 17 (2006).

11. Nat'l Organ Transplantation Act, 42 U.S.C. §274(e) (2013).

12. U.S. Const. art. I, § 8, cl. 8.

13. Greenberg v. Miami Children's Hosp. Research Inst., Inc., 264 F. Supp. 2d 1064 (S.D. Fla. 2003).

14. Id. at 1067.

15. *Am. Fruit Growers v. Brogdex Co.*, 51 S.Ct. 328 (1931).

16. *Diamond v. Chakrabarty*, 100 S.Ct. 2204 (1980).

17. *Ass'n for Molecular Pathology v. Myriad Genetics, Inc.*, 133 S.Ct. 2107 (2013).

18. *Washington Univ. v. Catalona*, 490 F.3d 667, 674 (8th Cir. 2007).

19. Unif. Anatomical Gift Act §§1–5 (2006).

20. Restatement (Second) Torts §§227–228 (1965).

21. Restatement (Second) of Property §31.1 (1991).

22. *Catalona*, 490 F.3d at 672.

23. Id. at 671.

24. Id.

25. Id.

26. Id. at 675.

27. *Greenberg*, 264 F. Supp. 2d at 1075–76.

28. *Moore*, 51 Cal.3d at 148.

29. Id. at 152.

30. *Venner v. State*, 354 A.2d 483 (Md. Ct. Spec. App. 1976).

31. 42 U.S.C. §274(e)(c)(1).

32. UNIF. ANATOMICAL GIFT ACT §6(a) (1987).

33. CAL. HEALTH AND SAFETY CODE §125355 (2007); MASS. CODE REGS. TIT. 105 §960.006.

34. 45 C.F.R. § 46.

35. *Greenberg*, 264 F. Supp. 2d at 1067.

36. Id. at 1072 (emphasis added).

37. *Salgo v. Leland Stanford Jr. Univ. Bd. Of Trustees*, 154 Cal. App. 2d 560 (Cal. Ct. App. 1957).

38. *Darke v. Estate of Isner*, 17 Mass. L. Reptr. 698 (2004).

39. *Moore*, 51 Cal.3d at 150–153.

40. 45 C.F.R. § 46.

41. 21 C.F.R. §§ 50, 56.

42. 45 C.F.R. § 46.116.

43. "Blanket consent" constitutes unconditional permission to use. "Broad consent" is usually understood as permission to use for a variety of research purposes, but with some oversight, such as IRB review (Grady et al. 2015).

44. 45 C.F.R. § 46.102(f).

## References

Allen, Monica, Michelle Powers, K. Scott Gronowski, and Ann M. Gronowski. 2010. Human tissue ownership and use in research: What laboratorians and researchers should know. *Clinical Chemistry* 56 (11): 1675–1682.

*American Fruit Growers v. Brogdex Co.* 51 S.Ct. 328 (1931).

Andrews, Lori B. 1986. My body, my property. *Hastings Center Report* 16 (5): 28–38.

*Ass'n for Molecular Pathology v. Myriad Genetics, Inc.* 133 S.Ct. 2107 (2013).

Austin, Melissa A., Julia Crouch, and Alyssa DiGiacomo. 2005. Applying international guidelines on ethical, legal, and social issues to new international genebanks. *Jurimetrics* 45 (2): 115–134.

Ayres, Ian. 2012. Regulating opt-out: An economic theory of altering rules. *Yale Law Journal* 121 (8): 2032–16.

Ayres, Ian, and Robert Gertner. 1989. Filling gaps in incomplete contracts: An economic theory of default rules. *Yale Law Journal* 99: 87–130.

Baron, Jane. 1989. Gifts, bargains, and form. *Indiana Law Journal* 64: 155.

*Brotherton v. Cleveland*, 923 F.2d 477 (6th Cir. 1991).

CIOMS (Council for International Organizations of Medical Sciences). Guideline 13: Reimbursement and Compensation for Research Purposes. CIOMS Working Group.

*Darke v. Estate of Isner*, 17 Mass. L. Reptr. 698 (2004).

*Del Zio v. The Presbyterian Hospital in the City of New York*, No. 74 Civ. 3588 U.S. Dist. Lexis 14450 (S.D.N.Y. Nov. 14, 1978).

Department of Health and Human Services. 2015. Federal policy for the protection of human subjects. Notice of proposed rulemaking, 80 *Federal Register* 53: 933.

*Diamond v. Chakrabarty*, 100 S.Ct. 2204 (1980).

Emanuel, Ezekiel J. 2004. Ending concerns about undue inducement. *Journal of Law, Medicine & Ethics* 32 (1): 100–105.

*Georgia Lions Eye Bank, Inc. v. Lavant*, 255 Ga. 60 (Ga. 1985).

Grady, Christine, Lisa Eckstein, Ben Berkman, et al. 2015. Broad consent for research with biological samples: Workshop conclusions. *American Journal of Bioethics* 15: 34.

*Greenberg v. Miami Children's Hospital Research Institute, Inc.*, 264 F.Supp.2d 1064 (S.D. Fla. 2003).

Heng, Tan Boon. 2003. Property rights in human tissues—Call "a spade, a spade." *Singapore Academy Law Journal* 15: 61–125.

HUGO Ethics Committee. 2000. Statement on Benefit-Sharing.

Korobkin, Russell. 2007a. Buying and selling human tissues for stem cell research. *Arizona Law Review* 49: 45–67.

Korobkin, Russell. 2007b. "No compensation" or "pro compensation": *Moore v. Regents* and default rules for human tissue donations. *Journal of Health Law* 40 (1): 1–27.

Korobkin, Russell. 2007c. Autonomy and informed consent in biomedical research. *UCLA Law Review* 54 (3): 605–630.

Largent, Emily A. 2016. Recently proposed changes to legal and ethical guidelines governing human subjects research. *Journal of Law and the Biosciences* 3 (1): 206–216.

Marchant, Gary. 2005. Property rights and benefit-sharing for DNA donors? *Jurimetrics* 45 (2): 153–178.

*McFall v. Shimp*, 10 Pa. D. & C.3d 90 (Ct. Com. Pl. 1978).

*Moore v. Regents of University of California*, 51 Cal.3d 120, 125–127 (Cal. 1990).

National Bioethics Advisory Commission. 1999. Research Involving Human Biological Materials: Ethical Issues and Policy Guidance. Report and Recommendations of the National Bioethics Advisory Commission I.

National Organ Transplant Act, 42 U.S.C. §274(e) (2013).

Penalver, Eduardo M. 2010. The illusory right to abandon. *Michigan Law Review* 109 (2): 191–291.

*Pierce v. Properties of Swan Point Cemetery*, 10 R.I. 227 (R.I. 1872).

Protection of Human Subjects, 45 C.F.R. §46 (2009).

*R v. Rothery* [1976] 550.

*R v. Welsh* [1974] RTR 478.

Rao, Radhika. 2000. Property, privacy, and the human body. *Boston University Law Review* 80 (2): 359–460.

Restatement (Second) of Property §31.1 (1991).

Restatement (Second) Torts §§227–228 (1965).

Revised Uniform Anatomical Gift Act §§1–27 (2006).

*Salgo v. Leland Stanford Jr. Univ. Bd. Of Trustees*, 154 Cal. App.2d 560 (Cal. Ct. App. 1957).

Seeney, Eric B. 1998. Moore 10 years later—Still trying to fill the gap: Creating a personal property right in genetic material. *New England Law Review* 32 (4): 1131–1191.

Skloot, Rebecca. 2006. Taking the least of you. *New York Times Magazine*, April 16.

Stewart, Douglas, and Joseph DeMarco. 2005. An economic theory of patient decision-making. *Journal of Bioethical Inquiry* 2 (3): 153–164.

Swain, Margaret S., and Randy W. Marusyk. 1990. An alternative to property rights in human tissue. *Hastings Center Report* 20 (5): 12–15. doi:10.2307/3562526.

Terry, Sharon F., and Charles D. Boyd. 2001. Researching the biology of PXE: Partnering in the process. *American Journal of Medical Genetics* 106 (3): 177–184.

Tushnet, Rebecca. 1998. Rules of engagement. *Yale Law Journal* 107 (8): 2583–2618.

Uniform Anatomical Gift Act §§1–17 (1987).

*United States v. Prince Kumar Arora*, 860 F. Supp. 1091 (D. Md. 1994).

U.S. Const. art. I, § 8, cl. 8.

*Venner v. State*, 354 A.2d 483 (Md. Ct. Spec. App. 1976).

*Washington University v. Catalona*, 490 F.3d 667, 674 (8th Cir. 2007).

# 3 Research with Biospecimens: Tensions, Tradeoffs, and Trust

Elisa A. Hurley, Kimberly Hensle Lowrance, and Avery Avrakotos

In 2010, a science writer named Rebecca Skloot published *The Immortal Life of Henrietta Lacks*, the story of a poor black woman who sought treatment for cervical cancer at Johns Hopkins Hospital in 1951. After a clinically indicated operation, it was discovered that the cells from Lacks' collected tumor tissue could grow indefinitely in the laboratory, making them the first human cells successfully grown outside the body. These "HeLa" cells later became the most widely used cell line in the world. The book prompted widespread public condemnation of how the research establishment treated Henrietta Lacks and her descendants, who learned about the HeLa cell line 20 years after Lacks' death. Skloot's telling of the story implied that it was unethical for the doctor who took Lacks' cancer cells not to have asked her permission or at least informed her family about the existence of her cells and their subsequent uses (Nisbet and Fahy 2013). The tenor of the discourse in the aftermath of the book's publication suggested that the public largely agreed.

However, when Lacks sought care for cervical cancer, no law or regulation required doctors to ask permission before taking tissue from a patient for research or any other purpose (Skloot 2010, 89). Nor was it common practice for doctors to tell family members of patients about taking their tissues, or for what those tissues would be used (Javitt 2010, 718). In fact, had the current federal regulations governing human subjects research been in place in 1951, they would have permitted the collection and use of Lacks' cells in the same manner without necessarily seeking her informed consent.[1]

Nevertheless, Lacks' story became a flashpoint in the contemporary conversation about biospecimen research. For one thing, it brought biospecimen research to widespread public attention. For another, it prompted some in the research establishment to call for a rethinking of the regulations governing such research. In 2013, leadership at the National Institutes of Health (NIH) suggested that "the furore around HeLa cells brought the absence of consent requirements for some biospecimen research to public attention," and that Lacks' story was "catalyzing enduring changes in policy" (Hudson and Collins 2013, 142), namely, the revisions to the regulations governing biospecimen research proposed in the 2015 Notice of Proposed Rulemaking (NPRM) that sought to revise the Federal Policy for the Protection of Human Subjects, or Common Rule. As explained by Peloquin et al. in chapter 1 of this volume, the NPRM proposed to redefine research with human subjects to include research with non-identified biospecimens and, therefore, to require consent for such research for the first time. The final revised Common Rule, released in January of 2017, ultimately did not include such a sweeping change.

Perhaps the most significant impact of Lacks' story, though, is that it revealed a level of public mistrust around biospecimens research. A theme of much of the public conversation about Lacks—influenced, it seems, by Skloot's telling of the story—is that there was something nefarious or deceitful about the taking of Lacks' cancer cells for research purposes without her or her family's permission or knowledge. Much of the ensuing public conversation seems to suggest that this is just one more example of researchers callously doing bad things to good people in the name of science, and hence one more reason for people—especially poor people and people of color—not to trust in the research enterprise (Nisbet and Fahy 2013).

Though not necessarily agreeing that the treatment of Lacks was unethical, even by today's standards, ethicists, policy makers, and scientists have taken note of this mistrust; many now agree that realizing the tremendous promise of biospecimen research will require proactive measures to secure the public's trust (Beskow and Dean 2008; Weil et al. 2013; Rodriguez et al. 2013). This is because, while most of the potential benefits of biospecimen research—including advances in diagnostics and improved medical treatments—are likely to be realized in the future and are likely to redound to society as a whole, most of the risks—including, predominantly, risks to privacy—fall on individuals who provide their biological materials (Bioethics Commission 2012, 16).

This tension between advancing science for the collective good and accommodating individual interests arises for all research involving human subjects. It is, however, particularly salient in the context of biospecimen research for two reasons. First, biospecimen research is especially unlikely to yield benefits for individuals but raises the specter of risks that are not well articulated or understood due to the complexity and rapid progress of genomic science. Second, large-scale genomic research will be most successful, with respect to its collective benefits, when more people willingly participate; the more biospecimens, with their rich genomic information, that can be collected, stored, and made available for future research, the greater the potential to determine the genomic contributors to disease and its treatment.

Thus, in order to realize the promise of biospecimen research, individuals must be willing to assume the risks of participation; that willingness, in turn, depends on the scientific community's ability to secure and maintain the public's trust (Rodriguez et al. 2013, 275). How, then, to build trust in biospecimen research? In what follows, we explore two mechanisms for securing public trust that have received significant attention in this domain: informed consent and privacy protections.[2] Although informed consent and privacy protections are both familiar concepts applicable to all research with human subjects, their application to biospecimen research is particularly complicated and comes with tradeoffs that may threaten the advancement of biospecimen science. We examine some of those complications in order to shed light on both the role of public trust in fulfilling the promise of biospecimen research and the challenges the biospecimen research community faces in securing that trust.

## Informed Consent

Voluntary informed consent has long been a core tenant of ethical human subjects research. In the United States, the requirement to seek and obtain valid informed consent from potential research subjects has its foundations in the Belmont Report, which outlines three ethical principles—respect for persons, beneficence, and justice—that should guide all research with human subjects. In explaining the principle of respect for persons, the Belmont Report makes clear that individuals must be treated as autonomous agents, a requirement that is operationalized through the process of seeking and obtaining their informed consent (National Commission 1979). Informed consent is intended to honor individuals' "interest in making significant decisions about [their] lives for [themselves] and according to [their] own conceptions of a good life, and then to be free...to act on those decisions without interference from others" (Brock 2008, 606).

Informed consent has also historically served as a mechanism through which to foster public trust in research (Brock 2008). The seeds of the ethical requirement for informed consent can be traced to past research abuses (Blacksher and Moreno 2008). The requirement for "voluntary consent" in the research context was first codified in the 1949 Nuremberg

Code, which was promulgated in response to atrocities committed by Nazi medical doctors during World War II. In the United States, the Tuskegee Study of Untreated Syphilis in the Negro Male, a US Public Health Service (PHS) study that involved withholding effective treatment for syphilis from 400 poor black men for 20 years in order to study the natural history of the disease, was revealed in 1972 to public uproar. Along with other research abuses that came to light in the 1960s and the 1970s, the PHS Syphilis Study contributed to a general sense of mistrust in the research enterprise, especially with respect to the treatment of vulnerable and marginalized populations such as the sick, poor, and imprisoned. The ethical requirement to first inform people about and ask their *permission* for their involvement in research, rather than involving them without their knowledge and consent, became an important mechanism for ameliorating that mistrust.

The revelations about the PHS Syphilis Study ultimately led to the creation of the National Research Act, that, in turn, resulted in the drafting of the Belmont Report and the promulgation of federal regulations governing human subjects research based on its principles, now known as the Common Rule. Under the Common Rule, investigators must obtain an individual's "legally effective informed consent" before involving them in research. Among other things, this involves providing potential subjects with information about the nature of the research, including its risks and benefits, making clear that participation is voluntary, and outlining the extent to which confidentiality will be maintained.

Interestingly, however, according to the regulations, not all research with biospecimens requires prospective voluntary informed consent. As explained by Peloquin et al., if a biospecimen was collected for purposes other than the current research, it is permissible to conduct research on that biospecimen (known as "secondary research") without consent, as long the specimen is not individually identifiable, because such an activity does not meet the regulatory definition of human subjects research (45 C.F.R. § 46.102; OHRP 2008). Conversely, however, under the Common Rule, *prospective* collection of biospecimens for research typically constitutes human subjects research and is thus subject to the requirements for informed consent.

Lacks' story has prompted some to question the justification for this regulatory distinction between prospective research and secondary research with biospecimens, and the corresponding differences in consent requirements, on the grounds that the principle of respect for persons demands that individuals be allowed to make informed decisions regarding potential uses of their biospecimens, whether identified or not (Skloot 2015; Hudson and Collins 2013; Grady et al. 2015). Others argue that a requirement to obtain informed consent for all uses of a given biospecimen, as was proposed in the NPRM, is unnecessary and would stymie medical research (Rivera 2014). These disparate views illustrate one of the central tensions that arise in the context of biospecimen research: how to honor the principle of respect for persons while allowing for the advancement of science for the public good.

This tension is also reflected in recent empirical research. Research suggests that the public is generally supportive of biospecimen research and willing to contribute in the interest of advancing science *when they trust that their privacy is being adequately protected* (Trinidad et al. 2011, 2012). However, the literature also suggests that people would like to be asked to contribute their biospecimens to future research, even if the request is very broad, thereby exerting some measure of control over use of their biospecimens and allowing their contributions to science to be acknowledged and respected (Grady et al. 2015; Javitt 2010). It is worth noting that this same literature shows that, when people are provided with some education and explanation about what is involved in biospecimen research, and then asked whether they would want the research to proceed even if informed consent was not possible or entailed not conducting the research, they opt for the research to proceed.

In an attempt to strike the right balance, ethicists and others in the research community have proposed a number of mechanisms for obtaining consent for future research use of biospecimens. Many of these are discussed in more detail by Grady et al., in chapter 8 of this volume. Specific consent, the most restrictive option, involves re-contacting and seeking consent from donors for each "new use of their specimen or for information that is outside the scope of their original consent" (Mello and Wolf 2010). While serving the principle of respect for persons by providing subjects with considerable information about and control over use of their specimens, specific consent has largely been dismissed as making research infeasible, due to the burdens and costs associated with having to re-contact donors for every proposed new use (ibid).

Other suggested mechanisms include tiered consent, which allows individuals to select from different options about how their specimens might be used (e.g., general permission for future use, consent only for future uses similar to the original study topic, etc.), and presumed consent, which assumes consent unless permission is expressly denied at the time of collection (Mello and Wolf 2010). These, too, come with tradeoffs. Tiered consent honors individuals' autonomy interests by allowing them to restrict the use of their specimens to certain classes of research, but it is logistically burdensome, requiring tracking mechanisms and possibly cutting off access to some specimens (ibid.). Conversely, the model of presumed consent may better advance the public good, but, by making the default access to specimens for research and requiring individuals to expressly deny permission to use their specimens, it may undermine public trust in research.

A fourth and much-discussed mechanism is the use of broad consent, that is, "consent for an unspecified range of future research subject to a few content and/or process restrictions" (Grady et al. 2015, 35). On the model of broad consent, at the time specimens are collected, individuals are asked to agree to use of their samples for a wide range of future research purposes. This is intended to honor an individual's autonomy interests while minimizing burdens on subjects and researchers and facilitating the extensive use of those biospecimens for future research (Grady et al., chapter 8 in the present volume; Mello and Wolf 2010). However, a common criticism of broad consent is that it does not provide potential subjects with sufficient information to make informed decisions about research participation. Specifically, because broad consent does not provide information about how or by whom an individual's specimens will be used in the future, it does not constitute meaningful consent (Mello and Wolf 2010; Rivera 2014). One case that illuminates this concern is that of the Havasupai Indian Tribe.

In the 1980s, high rates of Type 2 diabetes among the Havasupai prompted tribal leaders to contact researchers from Arizona State University (ASU) for assistance (Harmon 2010). Between 1990 and 1994, the ASU researchers collected more than 200 blood samples from the Havasupai. Although there are inconsistent accounts of the consent process used, by at least some accounts, the researchers sought broad consent via forms that said blood samples would be used to study "causes of behavioral/medical disorders." When the Havasupai later learned their samples were being used not just for diabetes research but for research on schizophrenia, inbreeding, and population migration—research they later said that they would never have agreed to had they known its aims—they objected and sued ASU (Drabiak-Syed 2010).

Despite the salient problems illuminated by the Havasupai case, discussed in more detail by Garrison in chapter 9, the 2015 NPRM included provisions that would have mandated broad consent for essentially all biospecimen research. The prospect of a sweeping regulatory requirement raised additional concerns about broad consent, specifically, concerns about the tradeoffs it might entail with respect to the public benefits of biospecimen research. Had such a mandate been adopted, all facilities collecting biospecimens for research—from large academic medical centers to small community-based clinics—would have needed to have the infrastructure in place for obtaining and tracking consent related to each potential biospecimen collection. This would have been burdensome and costly for most institutions, and for those that are less well-resourced would likely have been impracticable, threatening to make many clinically collected biospecimens, and in particular those from underserved communities, inaccessible.[3]

Where, then, does this leave the prospects for relying on informed consent as a mechanism for building trust and fostering participation in biospecimen research? An ethical or regulatory requirement to seek individual informed consent—whether specific, tiered, or broad—for all future biospecimen research would, we believe, result in insurmountable barriers to the conduct of that research. However, fostering trust in research requires that the public, as potential research subjects, believe that they and their interests will be respected. To address this tension, other avenues for respecting persons and fostering trust should be explored. In the context of biospecimens research, it may be that deliberately engaging in a public conversation about the role in and importance to research of biospecimens better strikes the balance between individual interest and the collective good. We return to this idea in our conclusion.

### Privacy and Identifiability

The foundation for concern about privacy in the context of human subjects research, like that for informed consent, can be found in the Belmont principle of respect for persons. As the Presidential Commission for the Study of Bioethical Issues has articulated, respect for persons includes respect for the dignity and *privacy* of individuals (Bioethics Commission 2012, 36). Privacy refers to a person's interest in determining what information he or she shares about him or herself, and with whom. Thus privacy also invokes an individual's interest in controlling who has access to personal information, as well as controlling how that information is used (ibid., 43).

The imperative to protect and respect individual research subjects' privacy is codified in several places in the Common Rule. The criteria for institutional review board (IRB) review at 45 C.F.R. 46.111 makes clear that research protocols must make "adequate provisions" to protect the privacy of research subjects. Additionally, informed consent forms must describe the extent to which the confidentiality of records identifying the subject—and thus the subject's privacy with respect to the information such records might contain—will be maintained.

Privacy is closely tied to, though not synonymous with, the concept of identifiability. Broadly, "identifiability" refers to the potential for a biological specimen (or information) to be associated with a specific individual (Lowrance and Collins 2007, 600). In the regulatory context, "identifiability" of a specimen refers specifically to whether an investigator may "readily ascertain" the identity of the subject to whom the specimen pertains. If an investigator cannot readily link the specimen to an individual, either directly or through a coding system, then that specimen is not considered readily identifiable, and research conducted with it is not considered human subjects research (45 C.F.R. § 46.102; HHS 2008). In other words, under the Common Rule, research protections (including around privacy) are triggered by identifiability (Gutmann and Wagner 2013, 16).

Concerns about privacy and identifiability are particularly salient in the context of biospecimens research. First, biospecimens contain genomic data that are "private, intimate, and sensitive," and unique to an individual (Lin et al. 2004, 183). Second, biospecimens are most scientifically useful when they are linked to demographic, clinical, and other phenotypic information about the specimen source (McGuire and Beskow 2010, 1). That is to say, what makes biospecimens especially useful for the purposes of research is linking them to clinical information—including potentially sensitive and intimate information—related to disease and mental health status, disability, origins, etc., that a person might prefer to keep private (Bioethics Commission 2012, 2). Access by others to such information raises concerns about legal or financial ramifications, stigmatization, and insurance or employment discrimination. Furthermore, the more information that is linked, the greater the potential that the individual source of the specimen and information will be identified (Gutmann and Wagner 2013, 15). Third, the study and sharing of genomic data and linked information raise all of these concerns not just for individual specimen donors, but for blood relatives and members of relatively homogenous genetic populations (Weil et al. 2013, 1001).

Such privacy concerns are compounded by the rapidly evolving nature of genomic science. As technology progresses, unanticipated genomic information might be extracted from existing biospecimens (Bioethics Commission 2012; Therrell et al. 2011). Furthermore, as genomic technologies have advanced, data sharing mechanisms (including publically available databases) proliferated, and the amount of genomic and associated information readily available increased, questions have been raised about whether complete de-identification of biospecimens is still realistic or even possible. Re-identification of the individual from de-identified data typically requires triangulating that data with other data sources. Notably, however, individuals are posting and releasing significant amounts of information about themselves voluntarily (e.g., Facebook and other social media sites), making the data available through third party sources ever richer. In recent years, a small number of studies have demonstrated the possibility of re-identifying data that were previously thought to be de-identified through combining data from multiple sources (Gymrek et al. 2013).

In light of these developments, some are calling for a reconsideration of "whether a simplistic distinction between identifiability and non-identifiability remains adequate as a metric for describing expectations about participant protections," and suggesting that the risk of identifiability is better framed as lying along a continuum rather than as an absolute (Rodriguez et al. 2013, 276; Lowrance and Collins 2007; Gutmann and Wagner 2013). The implication is that, in the context of biospecimen research, privacy risks may be unavoidable, and individuals take on more such risks as the science evolves and as more information becomes publically accessible. But realizing the promise of large-scale biospecimen research requires broad public participation. The willingness of individuals and communities to assume risks in order to participate in research depends on the scientific community's ability to maintain the public's trust that they will protect privacy to the extent possible (Rodriguez et al. 2013).

To that end, some suggest that, when thinking about protecting people from risks, we shift our focus from what makes something identifiable, to what keeps something private (Gutmann and Wagner 2013, 16). Two recent cases demonstrate how addressing concerns about access to private information is central to maintaining trust in research.

In 2013, decades after Henrietta Lacks' cells were taken, German researchers posted the genome of a HeLa cell line on an open-access database without consulting the Lacks family. Although permissible according to the regulations, the move was criticized by the Lacks family and advocates (Hudson and Collins 2013). Amid public pressure, the researchers removed the sequence from the database in order to allow the NIH time to consider how to resolve the situation in a way that would provide access to the HeLa genome data and protect the Lacks family's privacy (Collins 2013).

In the end, NIH leadership forged an unprecedented agreement with the Lacks family: DNA sequences derived from HeLa cells were put into NIH's database of phenotypes and genotypes (dbGaP), and researchers' requests for access to the data would be reviewed by a working group that included members of the Lacks family (Hudson and Collins 2013).

Most agree that while the arrangement with the Lacks family represents a reasonable solution to a difficult and very public situation, it should not be seen as precedent-setting. Giving every specimen donor control on a case-by case basis of who has access to their genomic information is not practicable as a general approach for protecting the privacy of all biospecimen donors and would bring research to a halt (Hudson and Collins 2013). However, it seems clear that, in the interest of fostering trust and participation in biospecimen research, information about data sharing options should be communicated during the informed consent process (Weil et al. 2013).

Also relevant here are recent developments around residual bloodspots from newborn screening programs, whereby small samples of blood are taken from infants at birth to test for a number of serious but treatable genetic conditions. In most US states, these newborn screening programs are mandatory and do not require parental consent. The resulting bloodspots are often retained for a number of purposes, including research.

When the Newborn Screening Saves Lives Reauthorization Act of 2014 was signed into law, it radically changed the regulatory framework for research involving residual newborn bloodspots. A provision requiring that informed consent be sought for any federally funded research using residual newborn bloodspots, regardless of the identifiability of the specimens, was added to the legislation at the last minute (OHRP 2015). The legislation also eliminated the ability of the IRB to alter or waive informed consent requirements. The inclusion of these requirements, which contradict those in the Common Rule, were driven by special interests groups playing on public fears about government access to private information about children and the "possible discrimination, psychological harm, identification of paternity, and social injustice" that could result (Couzin-Frankel 2015; see also Therrell et al. 2011; Bachmann 2014).[4]

Following the passage of the legislation, research with the bloodspots was halted, as the research community sought to evaluate the feasibility of continuing to use residual newborn bloodspots in light of the new requirements (Lewis et al. 2012; Drabiak-Syed 2010). Most agree this has been a major setback to the research enterprise. This outcome highlights what happens when too much weight is placed on individual interests—in this case, overblown concerns about access to and misuse of private information—and the balance between those interests and the public good is thrown off. Unfortunately, this was not an isolated incident. In chapter 12, Botkin et al. highlight how large collections of residual newborn bloodspots were destroyed in various states following a series of lawsuits in the 2010s.

Several less drastic approaches to privacy protections have recently been proposed. The 2015 NRPM's redefinition of human subjects research to include all biospecimens research rested on the assumption that it was impossible to guarantee that an individual could not be identified from a supposedly non-identified biospecimen. Thus, in addition to the informed consent requirements discussed above, the NPRM proposed that all such research be subject to privacy "safeguards" that would "reasonably protect against anticipated threats or hazards to the security or integrity of the information or biospecimens, as well as reasonably protect the information and biospecimens from any intentional or unintentional use, release, or disclosure" (HHS 2015, 53941). However, the 2017 published rule removes references to a standardized set of privacy safeguards for research involving biospecimens and information, leaving it to IRBs, as in the 1991 rule, to asses "what provisions are adequate to protect the privacy of subjects and confidentiality of data," but also promising future guidance to assist IRBs with that assessment (HHS 2017, 7264).

In addition to regulatory mechanisms for protecting privacy, some, including the Secretary's Advisory Committee on Human Research Protections, have suggested legal approaches to ensuring privacy protections, specifically, the promulgation of laws that would penalize, and thus deter, unauthorized and inappropriate access, re-identification, and misuse of biospecimens and genetic data (Lowrance and Collins 2007; Gutmann and Wagner 2013; OHRP 2011).

However, solutions for engendering trust with respect to privacy cannot focus solely on individuals. When biospecimen research is designed to study a small population group or a defined community, it has the potential to result in unwanted findings regarding that group, even if individual samples are de-identified (Drabiak-Syed 2010, 209). Those who are members of the group might thereby be harmed, for instance, by stigma and discrimination or by psychological stress that the findings may cause. For example, the Havasupai were confronted with research that suggested their "ancestors had crossed the frozen Bering Sea to arrive in North America," a claim that directly contradicted the group's origin story and sense of identity (Harmon 2010; Tilousi 2010). This suggests that privacy protections might need to encompass mechanisms for protecting identifiable populations from unwanted access to or disclosure of information about them as a group. This is an underdeveloped area of research and policy, made more complicated by questions about what constitutes a relevant group and the fact that the regulations are silent on the issue of such "group harms." One mechanism that has been suggested for addressing privacy concerns in this domain is IRB analysis of the potential for group harms during review of biospecimens research (OHRP 2011).

Also of relevance to the discussion of the role of privacy protections in fostering public trust in biospecimen research is the fact that attitudes and expectations around privacy generally appear to be changing (Rodriguez et al. 2013; Rivera 2014). The proliferation of mobile technologies and social media platforms that allow for information to be shared quickly and continuously may "underscore a cultural shift toward greater openness about personal information" (Weil et al. 2013, 1001). The willingness to "move formerly private information into a public sphere" (Rivera 2014, 256) extends to medical records and genomic data through platforms like the 1000 Genomes Project (www.1000genomes.org) and PatientsLikeMe (www.patientslikeme.com). What might such shifts mean with respect to trust in the biospecimen research enterprise? Perhaps, in the future, engendering trust in research will be less a matter of assuring subjects that their information is safe in the hands of researchers, and more a matter of demonstrating that subjects will be treated as respected partners in the research process.

**Conclusion**

The promise of biospecimen research will only be realized when mechanisms to foster trust in such research—for example, informed consent and privacy protections—rest on a commitment on the part of researchers, ethicists, sponsors, patient advocates, and government agencies to meaningfully engage with the public.

Empirical evidence suggests that public understanding of biospecimen research is varied (Hull et al. 2008; Lemke et al. 2010). Engagement thus begins with concerted efforts to educate the public about the nature, individual risks, and potential collective benefits of biospecimen research. Additionally, such efforts would promote an understanding of biospecimen research as most successful when there is widespread participation from all different kinds of people. Engagement also involves making clear to what extent, how, and to whom findings from biospecimens research will (or will not) be communicated, and why.

In a research context characterized by transparency and shared understanding of the promise and perils of such research, as well as by appropriate privacy protections and ethical oversight, the public is, by design, both seen and treated as valued contributors to and stakeholders in biospecimen research—they are respected as persons. These are the conditions that will engender public trust in biospecimen research and ultimately allow it to flourish.

## Notes

1. The naming convention used to arrive at "HeLa" would not be allowed by the regulations today, since it does not adequately de-identify the source of the cells. Furthermore, today, IRB oversight would be required, unless the cell line was de-identified or anonymized. In addition, the leaking of Lacks' name as the source of the cells in the 1970s would have triggered regulatory or legal protections.

2. To be sure, considerations of justice—whose biospecimens are collected for future study, who shares in the benefits of biospecimen research, and whether particularly vulnerable groups require special protections in the context of biospecimen research—are also important to building trust. Many of these considerations are addressed in chapter 4 and part IV of this volume. Also key to fostering trust are sound governance and oversight of biospecimen repositories; these issues are discussed in part V.

3. It should be noted that the 2017 rule does add broad consent as an option when conducting research on *identifiable* biospecimens. According to the 2017 rule, when broad consent is used and other limited requirements are met, such research may be exempt from the regulations.

4. Some empirical research suggests that parents' concerns about research with newborn bloodspots are not *primarily* about these privacy risks, though it was those concerns that largely drove the public discourse surrounding the legislation. Rather, parents object in principle to the idea that their children's samples would be used for research without their knowledge or permission (Lewis et al. 2012).

## References

Bachmann, Michele. 2014. Statement to the House Committee on Education and Labor, the Committee on Energy and Commerce, and the Committee on Health, Education, Labor, and Pensions. *Newborn Screening Saves Lives Reauthorization Act of*

*2014*. June 24. 113th Cong. 2nd Session. US Congress, *Congressional Record*, Volume 160, issue 99: H5696–H5699. https://www.gpo.gov/fdsys/pkg/CREC-2014-06-24/html/CREC-2014-06-24-pt1-PgH5696.htm.

Beskow, Laura M., and Elizabeth Dean. 2008. Informed consent for biorepositories: Assessing prospective participants' understanding and opinions. *Cancer Epidemiology, Biomarkers & Prevention* 17 (6): 1440–1451.

Bioethics Commission (Presidential Commission for the Study of Bioethical Issues). 2012. Privacy and Progress in Whole Genomic Sequencing.

Blacksher, Erica, and Jonathan D. Moreno. 2008. A history of informed consent in clinical research. In *The Oxford Textbook of Clinical Research Ethics*, ed. Ezekiel J. Emanuel et al. Oxford University Press.

Botkin, Jeffrey R., Erin Rothwell, Rebecca Anderson, Louisa Stark, Aaron Goldenberg, Michelle Lewis, Matthew Burbank, and Bob Wong. 2012. Public attitudes regarding the use of residual newborn screening specimens for research. *Pediatrics* 129 (2): 231–238.

Brock, Dan W. 2008. Philosophical justifications of informed consent in research. In *The Oxford Textbook of Clinical Research Ethics*, ed. Ezekiel J. Emanuel et al. Oxford University Press.

Collins, Francis. 2013. HeLa cells: A new chapter in an enduring story. NIH Directors Blog, August 7. https://directorsblog.nih.gov/2013/08/07/hela-cells-a-new-chapter-in-an-enduring-story/.

Couzin-Frankel, Jennifer. 2015. Newborn screening collides with privacy fears. *Science* 348 (6236): 740–741.

Department of Homeland Security et al. 2015. Notice of Proposed Rulemaking: Federal policy for the protection of human subjects. *Federal Register* 80: 53931–54061.

Drabiak-Syed, Katherine. 2010. Lessons from *Havasupai Tribe v. Arizona State University Board of Regents*: Recognizing group, cultural, and dignitary harms as legitimate risks warranting integration into research practice. *Journal of Health & Biomedical Law* 6 (2): 175–225.

Grady, Christine, Lisa Eckstein, Dan Brock Ben Berkman, Robert Cook-Deegan, Stephanie M. Fullerton, Hank Greely, Mats G. Hansson, et al. 2015. Broad consent for research with biological samples: Workshop conclusions. *American Journal of Bioethics* 15 (9): 34–42.

Gutmann, Amy, and James W. Wagner. 2013. Found your DNA on the Web: Reconciling privacy and progress. *Hastings Center Report* 43: 15–16.

Gymrek, Melissa, Amy L, McGuire, David Golan, Eran Halperin, and Yaniv Erlich. 2013. Identifying personal genomes by surname inference. *Science* 339: 321–324.

Harmon, Amy. 2010. Indian tribe wins fight to limit research of its DNA. *New York Times*, April 21. http://www.nytimes.com/2010/04/22/us/22dna.html.

HHS (US Department of Health and Human Services). 2009. Code of Federal Regulations. Title 45. Basic HHS policy for the protection of human research subjects. *Federal Register* 56: 28012–28018.

HHS (US Department of Health and Human Services). 2017. Federal policy for the protection of human subjects. *Federal Register* 82: 7149–7274.

Hudson, Kathy L., and Francis S. Collins. 2013. Family matters. *Nature* 500: 141–142.

Hull, Sara Chandros, Richard R. Sharp, Jeffrey R. Botkin, Mark Brown, Mark Hughes, Jeremy Sugarman, Debra Schwinn, et al. 2008. Patients' views on identifiability of samples and informed consent for genetic research. *American Journal of Bioethics* 8 (10): 62–70.

Javitt, Gail. 2010. Why not take all of me? Reflection on *The Immortal Life of Henrietta Lacks* and the status of participants in research using human specimens. *Minnesota Journal of Law, Science & Technology* 11 (2): 713–755.

Lemke, Amy A., Wendy A. Wolf, Jennifer Hebert-Beirne, and Maureen E. Smith. 2010. Public and biobank participant attitudes toward genetic research participation and data sharing. *Public Health Genomics* 13 (6): 368–377.

Lewis, Michelle Huckaby, Michael E. Scheurer, Robert C. Green, and Amy L. McGuire. 2012. Research results: Preserving newborn blood samples. *Science Translational Medicine* 4 (159): 1–3.

Lin, Zhen, Art B. Owen, and Russ B. Altman. 2004. Genomic research and human subject privacy. *Science* 305: 183.

Lowrance, William. W. and Francis S. Collins. 2007. Identifiability in genomic research. *Science* 317: 600–602.

McGuire, Amy L., and Laura M. Beskow. 2010. Informed consent in genomics and genetics research. *Annual Review of Genomics and Human Genetics* 11: 361–381.

Mello, Michelle M., and Leslie E. Wolf. 2010. The Havasupai Indian Tribe case—Lessons for research involving stored biologic samples. *New England Journal of Medicine* 363: 204–207.

National Commission (National Commission for the Protection of Human Subjects of Biomedical and Behavioral Research). 1979. The Belmont Report: Ethical Principles and Guidelines for the Protection of Human Subjects of Research.

Nisbet, Matthew C., and Declan Fahy. 2013. Bioethics in popular science: Evaluating the media impact of *The Immortal Life of Henrietta Lacks* on the biobank debate. http://bmcmedethics.biomedcentral.com/articles/10.1186/1472-6939-14-10.

OHRP (Office for Human Research Protections). 2008. Guidance on research involving coded private information on biological specimens. US Department of Health and Human Services. October 16. http://www.hhs.gov/ohrp/policy/cdebiol.html.

OHRP (Office for Human Research Protections Secretary's Advisory Committee on Human Research Protections). 2011. SACHRP Letter to the HHS Secretary: SACHRP ANPRM Comments. October 13. http://www.hhs.gov/ohrp/sachrp/commsec/sachrpanprmcommentsfinal.pdf

OHRP (Office for Human Research Protections Secretary's Advisory Committee on Human Research Protections). 2015. Recommendations Regarding Research Uses of Newborn Dried Bloodspots and the Newborn Screening Saves Lives Reauthorization Act of 2014. US Department of Health and Human Services. March 17. http://www.hhs.gov/ohrp/sachrp/commsec/researchusesofnewborndriedbloodspots&thenewbornscreening.html.

Rivera, Suzanne M. 2014. Reconsidering Privacy Protections for Human Research. In *Human Subjects Research Regulation: Perspectives on the Future*, ed. I. Glenn Cohen and Holly Fernandez Lynch. MIT Press.

Rodriguez, Laura L., Lisa D. Brooks, Judith H. Greenberg, and Eric D. Green. 2013. The complexities of genomic identifiability. *Science* 339: 275–276.

Rothwell, Erin, Rebecca Anderson, Aaron Goldenberg, Michelle H. Lewis, Louisa Stark, Matthew Burbank, Bob Wonga, and Jeffrey R. Botkin. 2012. Assessing public attitudes on the retention and use of residual newborn screening blood samples: A focus group study. *Social Science & Medicine* 74 (8): 1305–1309.

Skloot, Rebecca. 2010. *The Immortal Life of Henrietta Lacks*. Crown.

Skloot, Rebecca. 2015. Your Cells. Their Research. Your Permission? *The New York Times*. December 30. http://www.nytimes.com/2015/12/30/opinion/your-cells-their-research-your-permission.html.

Therrell, Bradford L., W. Harry Hannon, Donald B. Bailey, Edward B. Goldman, Jana Monaco, Bent Norgaard-Pedersen, Sharon F. Terry, Alissa Johnson, and R. Rodney Howell. 2011. Committee report: Considerations and recommendations for national guidance regarding the retention and use of residual dried blood spot specimens after newborn screening. *Genetics in Medicine* 13 (7): 621–624.

Tilousi, Carletta. 2010. Making sense of community: Responses to tissue research. Panel Presentation at 2010 Advancing Ethical Research Conference, Public Responsibility in Medicine and Research, San Diego.

Trinidad, Susan Brown, Stephanie M. Fullerton, E.J. Ludman, Gail P. Jarvick, Eric B. Larson, and Wylie Burke. 2011. Research practice and participant preferences: The growing gulf. *Science* 331 (6015): 287–288.

Trinidad, Susan Brown, Stephanie M. Fullerton, Julie M. Bares, Gail P. Jarvick, Eric B. Larson, and Wylie Burke. 2012. Informed consent in genome-scale research: What do prospective participants think? *AJOB Primary Research* 3 (3): 3–11.

Weil, Carol J., Leah E. Mechanic, Tiffany Green, Christopher Kinsinger, Nicole C, Lockhart, Stephanie A. Nelson, Laura L. Rodriguez, and Laura D. Buccini. 2013. NCI think tank concerning the identifiability of biospecimens and "omic" data. *Genetics in Medicine* 15 (12): 997–1003.

## II  Roots of the Debate: Autonomy, Justice, and Privacy

# Introduction

Steven Joffe

Canonical codes of research ethics, such as the Nuremberg Code, the Declaration of Helsinki, and the Belmont Report, address investigators and research sponsors as their primary audience. These codes articulate the principles that investigators and sponsors must respect, and the specific duties they must fulfill, to satisfy their ethical obligations to research participants. Key principles, such as respect for participants' autonomy through the mechanism of informed consent, beneficence through maximizing benefits for participants and society and minimizing risks to participants, and justice through fairly distributing the benefits and burdens of research, have served the research enterprise well for the past half-century. Despite broad areas of agreement about the ethics of research, however, significant controversies about specific classes of research remain. How to apply these principles to research with human biospecimens is a particular point of contention.

The chapters in part II advance the debate by focusing on the perspectives of stakeholders beyond investigators and sponsors. Korn and Sachs, Evans and Meslin, and Rivera and Aungst address the obligations of patients from whom excess biospecimens are derived. Although they take different argumentative lines, they all ask how patients' interests in controlling the uses of their specimens fit with their duties to contribute—as potential or actual beneficiaries—to the advancement of biomedical knowledge. Clayton and Malin tackle the likelihood that individuals will be re-identified from genomic data and suffer harm. They consider the incentives of data sharers to protect data, the incentives and costs faced by those who seek to re-identify, and the laws that seek to protect individuals. They conclude that the potential harms to individuals are less than many fear.

Korn and Sachs argue that each of us "will one day depend upon and benefit from ... medical knowledge," and therefore that we all have a duty to participate when the opportunity arises and risks are sufficiently low. Because research involving excess clinically collected biospecimens that would otherwise be discarded satisfies these criteria, they contend that each of us is obligated to allow our specimens to be used in research. They locate this duty in the principle of justice, reasoning that investigators' "ethical obligation not to target particular disadvantaged populations to bear the burdens of research" implies a corollary duty on individuals to allow their excess biospecimens to be used in research. Individuals who refuse use of their specimens for research, they argue, burden others while reaping the benefits themselves. Others have arrived at the same conclusion. John Harris, for example, appeals to basic fairness to ground the obligation to participate in biomedical research. Similar to Korn and Sachs, Harris (2005) argues that "since we accept these benefits [of medical research], we have an obligation in justice to contribute to the social practice which produces them." G. Owen Schaefer, Ezekiel Emanuel, and Alan Wertheimer (2009) distinguish between obligations to avoid free-riding, which increases burdens on others, and obligations to contribute to public goods, which emphasize reciprocity between the individual and the community. Subtle differences in philosophical grounding notwithstanding, these authors agree that a *prima facie* duty to contribute our excess biospecimens to research provides one argument against an absolute right of consent.

Evans and Meslin start from the observation that, in surveys, people are generally more willing to allow their specimens to be used in academic than in commercial research. They infer from this that people want research involving their specimens to confer public benefits, and that this preference for academic over commercial research reflects a view—mistaken, Evans and Meslin believe—that academic research is more likely than commercial research to achieve this goal. They show that various federal commissions whose work underpins the regulatory regime governing research have taken the public benefit requirement seriously, and argue that considerations of public benefit limit the constraining power of individual autonomy. They then look to laws governing the taking of property for public purposes—laws of eminent domain—for insight and guidance regarding how to balance individual interests in privacy and control through consent with our collective interest in mutually advantageous arrangements. This line of argument leads inexorably to the difficult question, which Evans and Meslin ask but do not answer, of who can legitimately decide how to balance these interests when questions of requiring or waiving consent arise.

Rivera and Aungst ask what motivates patients to donate, or to decline to donate, their biospecimens for research. They marshal data to support altruism, trust, and self-interest as critical determinants of willingness to donate. Their review of the empirical literature suggests that many, perhaps most, people want some degree of control over the use of their specimens through some form of consent. They do not, however, take these empirical data to be determinative in the debate over the role of consent in biospecimen research, but rather as one consideration among many. Logistical difficulties, the possibility of error in honoring wishes, and the likelihood of forgone scientific benefit provide arguments against maximizing individual control through consent. Ultimately, they advocate a "communitarian view of specimen donation" that echoes the public benefit-driven frameworks advanced by Korn and Sachs and by Evans and Meslin.

The first three chapters in this part address the ethical values and principles—especially those that weigh against reliance on individual consent—that policy makers defining governance frameworks for biospecimen research must take into account. Clayton and Malin take a different approach. Rather than focusing on counterarguments to consent, they question the ethical grounding of the consent requirement itself. One reason to emphasize consent is that research involving biospecimens involves the possibility of privacy breaches and informational harm, particularly if specimens are associated with the identity of the person from whom they are obtained. People should not have to bear the risk of informational harm, the argument goes, unless they make an informed decision to accept it. Removal of traditional identifiers would seem to mitigate such risks, but—genetically identical siblings aside—the uniqueness of each individual's DNA raises the possibility of re-identification (Erlich and Narayanan 2014; Gymrek et al. 2013; Shringarpure and Bustamante 2015). Though technically possible, how great is the threat? Clayton and Malin draw on evidence from modeling and game-theory exercises to show that, under most realistic scenarios, the costs of re-identification attempts to "adversaries" outweigh any plausible benefits to them. They argue further that imposition of modest penalties for re-identification can make the benefit-cost calculus, from would-be adversaries' point of view, definitively unfavorable. Of course, other normative considerations in favor of consent—particularly our autonomy-based interest in choosing the objectives and activities to which we contribute—remain.

In considering the ethics of human biospecimen research, the chapters in this part highlight the missing element in the ethical principles of respect for autonomy, beneficence and justice. In this traditional framework, one facet of beneficence (the production of socially valuable knowledge) defines the goal of research, whereas respect for autonomy, justice, and another facet of beneficence (the promotion of individual benefit and prevention of harm), define the constraints on the pursuit of that goal (Joffe and Miller 2008). This conception is insufficient because it speaks only to those who conduct or govern research; it has nothing to say to those who would participate in or benefit from research. Yet research is by nature a collective enterprise, one that benefits all of us and that cannot succeed without our broad participation. Coherent support for the goals of research therefore entails accepting the principle of solidarity, which Bruce Jennings and Angus Dawson (2015) define in part as "mutuality of interdependence, care, and concern for others and for their relational human flourishing." By adding the principle of solidarity to those of respect for autonomy, beneficence and justice, we take an important step toward a comprehensive account of the ethics of human research.

## References

Erlich, Y., and A. Narayanan. 2014. Routes for breaching and protecting genetic privacy. *Nature Reviews. Genetics* 15 (6): 409–421.

Gymrek, M., A. L. McGuire, D. Golan, E. Halperin, and Y. Erlich. 2013. Identifying personal genomes by surname inference. *Science* 339 (6117): 321–324.

Harris, J. 2005. Scientific research is a moral duty. *Journal of Medical Ethics* 31 (4): 242–248.

Jennings, B., and A. Dawson. 2015. Solidarity in the moral imagination of bioethics. *Hastings Center Report* 45 (5): 31–38.

Joffe, S., and F. G. Miller. 2008. Bench to bedside: Mapping the moral terrain of clinical research. *Hastings Center Report* 32 (2): 30–42.

Schaefer, G. O., E. J. Emanuel, and A. Wertheimer. 2009. The obligation to participate in biomedical research. *Journal of the American Medical Association* 302 (1): 67–72.

Shringarpure, S. S., and C. D. Bustamante. 2015. Privacy risks from genomic data-sharing beacons. *American Journal of Human Genetics* 97 (5): 631–646.

# 4 Research on Human Tissue Samples: Balancing Autonomy vs. Justice

David Korn and Rachel E. Sachs

Human-tissue-based research has for nearly 150 years made major contributions to medical knowledge, often asking questions unimaginable, using technologies unknown, at the time of tissue extraction. Recently, vocal ethicists and lawyers have argued for finely specified *ex ante* consent procedures when obtaining specimens, resting their position on their balancing of the three foundational Belmont Principles: Autonomy, Beneficence, and Justice. We strongly disagree. In our view, these advocates' arguments place far too much emphasis on Autonomy to the neglect of Justice, enshrining the self at the expense of society. Since all of us and all our loved ones will eventually depend upon and benefit from the robustness of medical knowledge, we argue that, when opportunities arise and risk can be minimized, all of us should recognize our ethical obligation to contribute to the continued expansion of medical knowledge by permitting our biospecimens to be used for IRB-approved research. Concerns expressed by the aforementioned advocates about public trust and engagement are important but are more fruitfully dealt with through the research review process, and by implementing "best practices" for the conduct of specimen research.

## Background

The faint origins of what may be considered modern medicine have been dated by some to the thirteenth century, when Italian medical and law professors began to examine the organs of their dead patients and clients in an effort to better understand their disease processes and causes of death (Malkin 1993). But it was not until the middle of the eighteenth century, when Giovanni Morgagni published his classic tome *De Sedibus et Causis Morborum per Anatomen Indagatis* (*On the Seats and Causes of Diseases as Investigated by Autopsy*), and the middle of the nineteenth century, when

Rudolf Virchow applied light microscopy to a systematic examination of ultrathin specimens of diseased human tissues and published his transformative work, *The Cellular Basis of Disease*, that a robust scientific foundation of medicine was established, from which modern medical science would emerge. Ever since, histopathological and cytopathological examinations of excised or aspirated tissue specimens have been dispositive in establishing surgical diagnoses, and often medical diagnoses. Over the decades, major teaching hospitals across the United States have accumulated hundreds of millions of pathological specimens, most of them fixed in formalin (or other preservative) and embedded in paraffin blocks, but increasing numbers of them fresh-frozen. The extraordinary breadth of these collections has made them for generations a trove of incomparable value.

Remarkably, these aged, fixed, and even paraffin-embedded specimens are often capable of yielding valuable information when examined by modern-day technologies, for example, to assess genomic pathologies or abnormalities of gene expression, or to reveal markers of what prove to be clinically significant and heretofore unrecognized tumor subtypes. An especially dramatic example occurred in 1996 when Jeffrey Taubenberger and his team at the Armed Forces Institute of Pathology were able to partially reconstruct the genome of the infamous "Spanish Flu" virus from formalin-fixed lung specimens removed at autopsy in 1918 from a young male victim of that lethal global pandemic. Building on that finding, complete reconstruction of the "Spanish Flu" genome was accomplished by another research team shortly thereafter from pieces of lung removed from male Inuit victims buried in northern Canada in permafrost, a remarkably good preservative.

Of course, not all contributions from specimen research are so dramatic, but specimens have played and continue to play a uniquely important role in advancing medical knowledge. Specimen research has been valuable in establishing the natural history of diseases and their long-term responsiveness to different therapeutic strategies; in recognizing novel, clinically significant disease variants and sub-types; in discovering novel disease markers with prognostic and/or therapeutic significance; and in assessing the effectiveness of novel diagnostic technologies. Even when these findings are to be confirmed prospectively, new studies will be shaped by these findings and may never have been conceived without them.

It is critical to the central focus of this paper, which is whether and when to require informed consent from tissue sources, to recognize that the longevity of these specimens enables research studies directed at questions inconceivable, employing technologies unimaginable, at the time of specimen collection.

Thus, we posit (1) that research using human tissue specimens will continue to be a key contributor to the advancement of medical knowledge and practice, (2) that all members of society sooner or later will be beneficiaries of this steadily expanding store of medical experience and wisdom, and (3) that continuing expansion and refinement of medical knowledge in robust ways, minimizing sampling errors, require that scientists continue to have access to the widest possible range of human tissue specimens—not only those already banked, but also those that will accrue in coming years.

Historically, banked tissues have been obtained from surgical excisions and from autopsies, and of course permissions are required for both. In teaching hospitals, these permissions have routinely noted that portions of excised tissues not needed for diagnosis and care might be saved for teaching and research, but the focus of conversations with patients or next of kin was appropriately on the imminent procedures, not the fate of excised tissues.

Researchers' access to human tissue specimens, whether in teaching hospitals or from biospecimen repositories, requires approval from the designated custodian of the repository (in teaching hospitals, typically the pathology department, but individual academic clinicians often manage their own disease-specific tissue collections) and from the designated institutional review board (IRB). Historically, research access to the central tissue repository and related clinical records did not require explicit informed consent from their sources. However, subsequent to the 2001 Health Insurance Portability and Accountability Act regulation that requires explicit informed consent for access to patients' identified or identifiable clinical information, such information required for archival tissue studies or clinical research of any kind is commonly provided to investigators without individual identifiers. Of course, tissue specimens collected by individual academic clinicians from their own patients for specific research purposes must be obtained with informed consent—and autopsies must be done with permission of next of kin. Finally, in the United States, as we shall address later, tissue sources generally have no legally recognized property right in their excised specimens.

Recently, vocal ethicists, journalists, and lawyers have challenged this system, arguing that there is an ethically dictated need for *ex ante* consents from specimen sources, and not just a broad consent for research, but one that is explicit and detailed, specifying what kinds of research may be done with the specimen and by whom, whether the specimens can be shared with other researchers, etc. In the next section we will detail these arguments more precisely, consider the ethical basis for this view, and argue that it reflects a distortion and unbalancing of the values of American bioethics, privileging Autonomy at the expense of Justice. But it is important here to consider the serious practical consequences of this position for whether portions of tissues excised for clinical reasons, but not needed for diagnosis or care, may be saved and used for research.

As we have pointed out, at the time of specimen collection, questions that have been successfully addressed in tissue specimen research often have not yet arisen, or been unimaginable, and the technologies used not yet in existence. If researchers must find ever-aging sources each time their samples are to be used for further research, and also whenever new scientific advancements or methodologies make it possible to ask new questions and gain new information from long-ago excised, fixed or frozen tissue specimens, the effect would surely be to restrict severely the potential range of subject matter and the time span of permissible research. Moreover, a unique value of central tissue repositories has been their broad inclusiveness. Thus, an investigator might have access to a large deceased population with a particular disease, treated within the institution medically and/or surgically over a long span of time. Such broad access reduces (but does not eliminate) the chances of erroneous conclusions due to uncontrolled statistical biases in the study populations.

Moreover, most patients facing excision of their organs and tissues by surgery, or even biopsy, whether in their caregivers' offices or on entering hospitals, are distracted by their illnesses and worried about what is to happen to them. Most will know little or nothing *ex ante* about the technical details of their pending procedures; the methods involved in arriving at pathological diagnoses; or what "excess tissue specimens" or "tissue-based research" actually means. Nor are they likely to be receptive to tutorials in those anxiety-fraught settings.

Accordingly, we fear that typical pre-operative patients cannot truly provide "informed consent" on this matter, and if forced to decide, will too often opt to "just say no," or impose arbitrary, restrictive conditions on specimen use. Further, explaining in detail in the pre-operative or procedural setting the benefits and risks of tissue-based research to naive, anxious and distracted patients would be, we posit, too often an empty ritual. This concern is not merely "paternalistic": since the balancing of risks versus benefits, beyond the purely hypothetical, are only now being argued among professionals, we fear meaningful risk-benefit analysis may be beyond the capability of many clinicians, as well. Finally, there are numerous examples in society in which the benefits to all of rules imposed without consultation far outweigh the costs to individuals; one common example is speed limits.

In sum, we believe the position of the advocates of explicit consent could jeopardize scientists' ability to use existing central institutional repositories at all and will raise significant and unnecessary new barriers to performing research on banked specimens in future years.

**Balancing Autonomy with Justice**

Bioethics' focus on the centrality of Autonomy dates back to the founding documents of the field, all of which enshrine the importance of consent. Indeed, the very first sentence of the Nuremberg Code, a set of research ethics principles composed in 1947 in response to atrocities committed in the name of "experimentation" by Nazi physicians during World War II, avers that "the voluntary consent of the human subject is absolutely essential" (Nuremberg Code 1947). Autonomy similarly took precedence in the 1964 Declaration of Helsinki, adopted by the World Medical Association, that particularly emphasized and added content to the requirement that the patient not merely give consent, but give fully free, informed consent (World Medical Association 1964).

But the US biomedical research establishment would not adopt its own set of ethics principles until the Belmont Report, issued in 1978 by the National Commission for the Protection of Human Subjects of Biomedical and Behavioral Research. In the wake of Henry Beecher's famous 1966 expose in the *New England Journal of Medicine* (Beecher 1966) drawing attention to American research abuses of human subjects, and the 1972 revelations of the Tuskegee Syphilis Study (Jones 1993), the Belmont Report similarly focused on the importance of the ethical principle of Autonomy, phrased here as "respect for persons" (Belmont Report 1979).

When first promulgated, none of these three documents referred specifically to research involving human biospecimens, referring only more generally to research involving human subjects. That has changed only very recently, with the latest revision of the Declaration of Helsinki in 2013. The Declaration now devotes a single paragraph to research on human tissues, noting that "[f]or medical research using identifiable human material or data, such as research on material or data contained in biobanks or similar repositories, physicians must seek informed consent for its collection, storage and/or reuse. There may be exceptional situations where consent would be impossible or impracticable to obtain for such research. In such situations the research may be done only after consideration and approval of a research ethics committee" (World Medical Association 2013).

This passage encapsulates the general view of many bioethicists on the topic of research involving human biospecimens. The problem, of course, is that key terms in the paragraph are open to debate and dispute. In particular, scholars and policy makers have hotly debated what it means to give "informed consent" for future research on a given biospecimen. Views about the proper definition of this term span a continuum from highly specified, iterative requirements to little or no requirements at all.

Some scholars stand close to one end of the spectrum, advocating highly specified consent forms that would inform potential research subjects not only that their tissues might someday be used for research, but that would list specific ways in which those tissues might be used and require the subjects to select, in checklist fashion, only those uses to which they consent (Tomlinson 2013; Arnason 2004). Also specified may be the researchers or institutions that will have access to the specimens, and with whom they could or could not be shared. Many of these authors would also require researchers to gain consent anew each time they wished to use an identified specimen, a requirement that becomes more difficult (and then impossible or meaningless) the older the specimen.

This position is fundamentally rooted in the value of Autonomy. In the views of these scholars, taking the principle of Autonomy seriously requires that specimen sources complete this set of finely specified consent procedures. For them, the value of Autonomy is not simply a formality to be observed. Giving it meaning requires potential research subjects to be informed not merely that their tissues may be used, but when and how they will be used, by whom, and to what end. In practice, this would require repeated notices and consent forms to be signed, and it isn't clear whether the specimen would be accessible at all for research after the death of the source. As we discussed in the previous section, such detailed consent poses the threat of insurmountable barriers to the permissible scope and time span of potential research on human specimens.

Rejecting the implications of this extreme position, many ethicists and policy makers have adopted a less restrictive interpretation of what informed consent requires. These views range from the argument that a general, broad consent at the time of donation is sufficient to the view that even no consent may often be appropriate, possibly in opposition to the Declaration of Helsinki's recent pronouncement (Grady et al. 2015; Edwards et al. 2014; Wendler 2013).

We agree with those rejecting the position requiring specific consent for different research uses of banked tissue. But we write to provide another justification for our position: specifically, we argue that advocates for the most extreme informed consent requirements have placed far too much emphasis on the principle of Autonomy embodied in the founding documents of bioethics, to the exclusion of the principle of Justice. Like Autonomy, Justice is one of the four principles of American bioethics famously articulated by Beauchamp and Childress (Beauchamp and Childress 2012), along with Beneficence and Non-maleficence. Although in our view these advocates also ignore the Principle of Beneficence, the *obligation* to act in the best interests of others, this topic is taken up elsewhere in this volume.

With their focus on Autonomy above all else, these advocates have seriously distorted the integrity of the Belmont Principles and shattered the carefully reasoned balance intended by the Belmont Report's framers. They ignore that the principle of Justice also has pride of place in the Belmont Report, which frames the concept as requiring the equitable distribution of the burdens and benefits of scientific experimentation. One of the Report's paradigmatic examples of injustice is the Tuskegee syphilis study, which not only "used disadvantaged, rural black men to study the untreated course of a disease that is by no means confined to that population" (Belmont Report 1979) but continued even after the discovery of penicillin made effective treatment of syphilis possible. More generally, where the benefits of a particular scientific study accrue broadly, it is unjust to permit the burdens of research to fall on disadvantaged societal groups, which in the United States have often been prisoners, the mentally disabled, racial minorities, or institutionalized children.

The principle of Justice has a straightforward application in research based on human tissue specimens. Typically, this research is performed to increase the store of medical knowledge that will redound to the benefit of society as a whole. Sometimes a particular population must be targeted because they are directly or uniquely relevant to the problem being addressed, or because they would reap particular benefits from the research. But when that is not true, researchers have an ethical obligation not to target particular disadvantaged populations to bear the burdens of research.

In our view, this duty of researchers imposes a reciprocal ethical duty on potential subjects to participate in research when opportunities arise that do not thereby put them at undue risk. Put simply, because every individual will inevitably benefit from the accumulation of medical knowledge from research on human tissues, we are all obligated to avoid letting the burdens of that research fall solely on others. This duty is surely limited to cases in which opportunities arise—we do not contend that individuals are obligated to make regular donations to tissue banks (excluding blood banks) or undergo more invasive procedures such as bone marrow donation *sua sponte*. But where individuals are having or have had tissue removed from their bodies and banked, as many of us have, we argue that refusal to consent to that tissue's use in research is typically unjust.

Although our argument largely rejects the prioritization of Autonomy above all, it is useful to consider a real-world example in which our views would reach the same ethical conclusions, although for different reasons: the Havasupai case. As previous chapters have noted, the rate of diabetes among the Havasupai is extremely high, and researchers have been interested in studying the tribe to learn more about the various potential causes of this disease, with the ultimate aim of contributing to the development of preventive behavioral practices and possibly effective therapies. In the early 1990s, the Havasupai tribe consented to the collection and use of their blood samples for a set of research projects on diabetes. But in 2003, they were dismayed to find that researchers had used their samples in addition to study migration patterns, reaching conclusions that contradicted the Havasupai's traditional spiritual beliefs about the site of the tribe's origin.

In 2004, the tribe sued the researchers involved. Although the case was settled and did not result in a precedential opinion addressing the legal rights of each party, we agree with other scholars that the researchers who used the tribe's samples for migration research without consent acted unethically. In our view, though, the reason is not simply that members of the tribe were not asked for permission to conduct the study, but also that the principle of Justice imposes heightened consent requirements in cases involving populations who are marginalized or hold particular cultural and religious beliefs. Especially in the case of research specifically involving Native American tribes, respect requires researchers to obtain consent from the tribal leadership before approaching individual members of the tribe. In the case of the diabetes research, the tribe itself stood to gain from the knowledge gained about the genetic causes of the disease, even though that benefit would obviously be distributed more broadly. But in the case of the migration research, the tribe gained nothing—indeed, it was harmed—solely in pursuit of knowledge.

In this case, the researchers exploited the fact that the samples were accessible to them and displayed a lack of sensitivity to and respect for the

Havasupai and their beliefs, as well as respect for their tribal governance. We do not argue there is never a need for any type of consent in research with stored samples. But in our view, "informed consent" must be given content not merely by the principle of Autonomy but also by the principle of Justice. And the term can be interpreted permissively, rather than narrowly, to permit a scope of research on these specimens that is much broader than commonly believed. As we explain in the next section, any concerns about the legitimacy of research or the privacy policies implemented can be mitigated through a robust IRB review process that not only assures privacy, but can also serve as a surrogate for incomplete or missing consent.

## The Legal Landscape

The legal cases regarding ownership and control of human tissue specimens are more closely aligned with our recommendation than they are with the views of scholars urging a more extreme version of informed consent. The published decisions in the well-known cases involving John Moore, the Greenberg family, and Dr. William Catalona all agree with our view of the absence of residual rights possessed by individuals in tissues removed from their bodies. As each of these cases has already been highlighted in previous chapters, we provide here only the most basic background.

The facts of the John Moore case are well-known to lawyers and ethicists alike. Moore's physician, in the course of treating him for hairy-cell leukemia, used tissue removed from Moore in repeated visits to produce what became a multi-billion-dollar cell line—all without Moore's knowledge or permission (*Moore v. Regents of University of California*, 125). Moore sued on a range of legal theories, most notably arguing that his physician should have disclosed his financial interest in Moore's tissues, on theories of breach of fiduciary duty and informed consent, and that his physician should be liable for "conversion," as he had deprived Moore of a property interest in his tissues.

The California Supreme Court's 1990 opinion ruled in Moore's favor on the first count, and against him on the second. First, the court held that treating physicians have a fiduciary obligation to their patients to disclose any research or financial interests they may have in those patients (*Moore v. Regents of University of California*, 131). Importantly, though, this duty did not extend to other researchers working with Moore's physician who were not themselves involved in patient care. Second, the court went on to hold that Moore lacked a sufficient ownership interest in his excised cells to support a claim for conversion. Not only did the court conclude that existing law did not grant Moore such an interest; it went on to hold that creating such an interest would have deleterious policy consequences, impeding "activities that are important to society, such as research" (ibid., 146).

The next canonical case, *Greenberg v. Miami Children's Hospital*, involved more agency on the part of the research subjects. A group of families affected by Canavan disease had donated their tissues (as well as financial resources through related nonprofits) to researchers interested in studying the disease and specifically, in creating a prenatal test for the condition. When the test was developed, and was patented by the researchers' institution and made available only on a limited basis, the families and nonprofits sued (*Greenberg v. Miami Children's Hospital*, 1067).

In the *Greenberg* case, much as in the *Moore* case, the federal district court concluded that the researchers involved, who were not the plaintiffs' treating physicians, did not have a duty to disclose their economic interests to the plaintiffs. In the court's view, doing so would be "unworkable and would chill medical research as it would mandate that researchers constantly evaluate whether a discloseable event has occurred" (*Greenberg*, 1070). Similarly, the court ruled that the plaintiffs had "no cognizable property interest in body tissue and genetic matter donated for research" (*Greenberg*, 1074). The decision was not entirely in the researchers' favor—the court declined to dismiss the plaintiffs' allegations of unjust enrichment, which the plaintiffs then used to extract a favorable settlement (Colaianni et al. 2010).

More recently, the Court of Appeals for the Eighth Circuit decided *Washington University v. Catalona*. During his decades as a urologist and researcher at Washington University, Dr. William Catalona had initiated and overseen a prostate cancer tissue repository that had amassed many thousands of surgically excised specimens that he and other members of the faculty used in their many independent research projects on prostate cancer. When Dr. Catalona accepted a position at Northwestern University, he—without any IRB approval—sent forms to his former patients requesting their consent to transfer their stored biological materials to Northwestern. Washington University sued Catalona to enjoin him from taking the biorepository to Northwestern, because the collection had become a widely used institutional research asset at Washington University.

The Eighth Circuit put the legal question presented by the case as follows: "whether individuals who make an informed decision to contribute their biological materials voluntarily to a particular research institution for the purpose of medical research retain an ownership interest allowing the individuals to direct or authorize the transfer of such materials to a third party" (*Catalona*, 673). The answer, quite simply, was that they do not. The research subjects surely had the right to decline to donate more materials in the future or to decline to answer other research queries, but in the court's view, they retained no right to control their tissues after donation.

These several courts were tasked with resolving questions of law and policy, not of ethics. It may be that the legality of the situation does not always accord with the ethics involved. Accordingly, we do not argue that our ethical argument is necessarily correct because it leads to the same conclusions reached by these courts. However, to the extent that laws and legal decisions often track the ethical views of society, the consistency of courts on this question over the past several decades is noteworthy.

A recent legal development—the September 2015 release by the Office of Human Research Protection (OHRP) of a Notice of Proposed Rulemaking (NPRM) to amend the Common Rule—is more overtly linked to the ethical questions we address here. The term Common Rule refers not only to the legal rules protecting human subjects involved in research, but also is generally thought to instantiate the ethical principles promulgated in the Belmont Report. The NPRM aimed comprehensively to overhaul the rules governing human research with biospecimens, as well as human subjects research more generally.

In some ways, the NPRM's proposed rules regarding biospecimen research were more restrictive even than the Declaration of Helsinki, as they proposed to require informed consent for research "even if the investigator is not being given information that would enable him or her to identify whose biospecimen it is" (NPRM 2015). Because of the rapid pace of technological progress, especially in genomics, and the fact that even de-identified data can be used to identify individuals in ways which were "simply not possible, or even imaginable" when the Common Rule was first adopted, this provision sought to protect individuals out of an abundance of caution. Importantly, the final rule did not include this provision about non-identified biospecimens.

But, critically, even the NPRM did not adopt the extreme view encouraged by many scholars of what "informed consent" requires. Specifically, for de-identified biospecimen research, informed consent "could be obtained using a 'broad' consent form in which a person would give consent to future unspecified research uses." The NPRM explicitly stated that consent does not need to be obtained for each specific study seeking to use a given biospecimen, as long as general consent has been provided up front. In reaching this conclusion, the NPRM specifically considered the ways in which the principle of Autonomy relates to the principle of Justice.

## Sustaining Public Support through Rigorous Oversight and Enforcement of Standards

We have explained in some detail our opposition to requiring informed consent for de-identified biospecimen research in particular and to declaring use of such specimens to qualify as human subjects research more generally. But we recognize that maintaining public support for biomedical research is essential. If a danger to be avoided is re-identification of de-identified specimens by employing powerful new technologies, the most effective approach by far, in our view, and one that would *per se* keep pace with continuing rapid advancements in these technologies, would be to prohibit attempts at re-identification and to punish transgressing investigators severely enough to serve as effective deterrence for all. These prohibitions might best be promulgated in a Code of Best Practices for Human Biospecimen Research, to be developed by the OHRP in consultation with the Secretary's Advisory Committee on Human Research Protection (SACHRP). We believe such a Code would serve the research community well by assuring the public that the scientists recognize that performing research on human tissue specimens is a privilege that they, as well as federal funding agencies, respect and take seriously.

Further reassurance of the public and the funding agencies would be provided by requiring that for research awards involving the use of human biospecimens, the agencies' award letters must contain a pledge to be signed by the researchers, that they have read and will abide by the Code of Best Practices, and in particular will make no effort to re-identify the sources of de-identified specimens. The signed pledges would be countersigned by cognizant institutional officials and returned to the agency as a condition of award. Finally, at present, most human subjects research—including research on human biospecimens—is subject to oversight by an IRB, which in this instance serves to ensure that the research will be conducted ethically by qualified investigators seeking to answer significant scientific questions. Currently, research involving de-identified human biospecimens is generally (though not always) exempt from IRB oversight under the Common Rule. As noted, the NPRM would have required consent for performance of this research and given IRBs responsibility for reviewing the consent process, but not for overseeing the research itself. Because of public concerns about privacy and re-identification, we believe that limited IRB review of the research plan and full review of the proposed privacy protections would be beneficial to the researcher and the institution, as well as provide further reassurance to the public.

Relatedly, we call on the NIH and the DHHS to rethink their approach to responding to episodes of research misconduct, especially when widely publicized. All too frequently, in our view, such episodes are met not only with sanctions on the researchers involved, but also with a broad tightening of access, for example, to patient records, for all future researchers. The agencies' motivation seems to be that preservation of public trust requires not only punishing the wrongdoers but also ensuring that the relevant actions will not recur, which arguably can be most effectively accomplished by tightening the rules for all.

In our view, this latter approach was evident in the NPRM, and is a mistake. We certainly agree that the proper remedy for wrongdoing researchers is to punish them with swiftness and certainty, and make clear to the research community that such misconduct is not tolerated. But we cannot agree that one or even a few individuals' wrongdoing should trigger putatively preventive agency actions that we believe may or may not reduce the risk of misconduct but most assuredly add burden on all scientists, as well as their institutions, in this area of research. Research on biospecimens has contributed much to our understanding of human pathology. Now there is apparently fear that, because of astonishing progress in genomics and big data research, it may be possible for researchers to re-identify the sources of unidentified tissue specimens. The public trust may appear to be served in the short term by a system-wide crackdown, but in the longer term, that trust is best served by demonstrating to the public the benefits of the scientific progress that has been and will continue to be made through legitimate research inquiries.

Our argument here also has implications for other hotly debated topics within bioethics and the law. Just as researchers should never attempt to re-identify de-identified tissue specimens, they should never attempt to contact sources directly to return research findings, which may or may not be incidental, that appear to bear on the health of the sources. Examples would be increased risks for particular diseases unrelated to the research with which specific mutations have been associated even slightly. The concern here is not simply the re-identification that would need to happen for such results to be returned directly, or even indirectly through a trusted intermediary. It is also a very practical concern relating to the all too frequent unreliability of incidental findings from research laboratories, as compared to findings from accredited and licensed clinical laboratories, which are held to standards of precision, accuracy and record keeping that generally far surpass those of research laboratories.

Thus, purely from the perspective of quality of clinical care, it is a bad idea ever to return results from a research laboratory directly to specimen sources, and this should be another prohibited behavior in our proposed Code of Conduct. However, there may be instances in which test results obtained in research are provided by a licensed clinical laboratory and may be of such potential import to warrant communication. In such instances, there should be a way for the researcher to contact a trusted intermediary, for example, the hospital or clinic where a procedure was performed, and the intermediary could convey the information to the source's physician. We recognize that this approach is cumbersome, but protecting the source's identity from the researcher would appear to be the government's highest priority.

## Conclusion

Bioethicists have long debated the most appropriate ways to secure informed consent from participants in human subjects research. But only recently have they begun to separate out ethical concerns arising in the context of research involving biospecimens from research performed directly on human subjects themselves. In our view, the bioethics community, the Office of Human Research Protections, the National Institutes of Health, and the Public Health Service should give much more attention to the principle of Justice and retreat from what has seemed to be an increasingly and astonishingly single-minded focus on Autonomy. Restoring balance in the interpretation and application of the Belmont Principles, as intended by its framers, would enrich our understanding of what investigators and human research subjects properly owe each other. Particularly where reinforced through a Code of Best Practices, this restored balance would be welcomed by the research community and research institutions.

## References

Arnason, Vilhjalmur. 2004. Coding and consent: Moral challenges of the database project in Iceland. *Bioethics* 18 (1): 27–49.

Beauchamp, Tom L., and James F. Childress. 2012. *Principles of Biomedical Ethics*, seventh edition. Oxford University Press.

Beecher, Henry K. 1966. Ethics and clinical research. *New England Journal of Medicine* 274 (24): 1354–1360.

Colaianni, Alessandra, Subhashini Chandrasekharan, and Robert Cook-Deegan. 2010. Impact of gene patents and licensing practices on access to genetic testing and carrier screening for Tay-Sachs and Canavan Disease. *Genetics in Medicine* 12 (4): S5–S14.

Edwards, Teresa P., R. Jean Cadigan, James P. Evans, and Gail E. Henderson. 2014. Biobanks containing clinical specimens: Defining characteristics, policies, and practices. *Clinical Biochemistry* 47 (0): 245–251.

Grady, Christine, Lisa Eckstein, Dan Brock Ben Berkman, Robert Cook-Deegan, Stephanie M. Fullerton, Hank Greely, Mats G. Hansson, et al. 2015. Broad consent for research with biological samples: Workshop conclusions. *American Journal of Bioethics* 15 (9): 34–42.

*Greenberg v. Miami Children's Hospital Research Institute, Inc.*, 264 F. Supp.2d 1064 (S.D. Fla. 2003).

Jones, James H. 1993. *Bad Blood: The Tuskegee Syphilis Experiment.* Free Press.

Malkin, H. M. 1993. *The Foundation of Medicine and Modern Pathology During the Nineteenth Century.* Vesalius Books.

*Moore v. Regents of University of California*, 51 Cal.3d 120, 793 P.2d 479 (1990).

National Commission for the Protection of Human Subjects of Biomedical and Behavioral Research. 1979. *The Belmont Report: Ethical Principles and Guidelines for the Protection of Human Subjects of Research.*

Notice of Proposed Rulemaking: Federal Policy for the Protection of Human Subjects. 2015. *Federal Register* 80: 53936.

Nuremberg Code. 1949. *Trials of War Criminals before the Nuremberg Military Tribunals under Control Council Law No. 10*, volume 2. Government Printing Office.

Tomlinson, Tom. Respecting Donors to Biobank Research. 2013. *Hastings Center Report* 43 (1): 41–47.

*Washington University v. Catalona*, 490 F.3d 667 (8th Cir. 2007)

Wendler, David. 2013. Broad versus blanket consent for research with human biological samples. *Hastings Center Report* 43 (5): 3–4.

World Medical Association. 1964. Declaration of Helsinki. Helsinki, Finland.

World Medical Association. 2013. Declaration of Helsinki. Fortaleza, Brazil.

# 5 Biospecimens, Commercial Research, and the Elusive Public Benefit Standard

Barbara J. Evans and Eric M. Meslin[1]

The term "informational studies" refers to a variety of research and public health activities that use people's health records or data derived by testing their previously collected biospecimens. The data subjects—the persons who supplied the information—are not directly and personally involved in the study; only their data and specimens are used. Informational studies occupy an increasingly important place alongside traditional interventional (or clinical) studies that perturb participants' bodies—for example, by giving the person an experimental drug—in order to observe how they respond to the perturbation. Survey data portray members of the public as generally willing to let their data and specimens be used in biomedical research and public health activities (Institute of Medicine 2009; Topol 2015; Kish and Topol 2015; Health Data Exploration Project 2014). Many people, however, indicate that they are less comfortable with commercial uses of their biospecimens than with uses by academic and non-profit institutions and public health agencies (Institute of Medicine 2009).

This chapter explores whether the perceived distinction between commercial and non-commercial uses of biospecimens is a meaningful one. We argue that what most people want, consistent with the survey data, is for all uses of their specimens to produce benefits to future patients or the general public. In effect, people want the researchers who use their biospecimens to be held to a public benefit standard—a requirement that the research should advance socially beneficial aims and offer a prospect of useful discoveries that benefit others.

A public benefit standard for biospecimen research would be conceptually similar to the public use criterion seen in eminent domain jurisprudence, which requires takings of property to fulfill a socially beneficial, rather than a purely private, purpose (Evans 2011). Bioethical discourse dating back to the 1970s has identified public benefit as an important criterion when assessing whether a given use of identifiable health data and biospecimens should be allowed (see discussion *infra*). Unfortunately, a public benefit standard has never been successfully operationalized in federal regulations that govern biospecimen access.

In this chapter we argue that the perceived distinction between academic and commercial research may be serving as a proxy for a public benefit standard. In the public mind, commercial uses may be identified with impermissible private uses. If that is true, then it may be possible to enunciate a more nuanced set of criteria that would distinguish—and preserve biospecimen access for—the subset of commercial research that offers a potential for public benefit. Incorporating such criteria into privacy and human-subject protection regulations seemingly would support better decisions about access to biospecimens and could enhance public trust. Instead of debating whether commercially sponsored informational research should be allowed at all, this approach would focus on whether a proposed commercial use of data or biospecimens offers public benefits. Some commercial uses of biospecimens do deliver significant benefits to the public in the form of new diagnostics and therapies available for clinical use. A public benefit standard aims to identify and facilitate biospecimen access for such uses. We do not presume by this argument that the identification of *some* public benefit from commercial uses of data or biospecimens suggests that other non-public benefits are disregarded in any assessment of the overall benefit.

**Is the Bias against Commercial Research Rational?**

Surveys find that "a majority of consumers are positive about health research and, if asked in general terms, support their medical information being made available for research" (Institute of Medicine 2009; Topol 2015; Kish and Topol 2015; Health Data Exploration Project 2014). Despite this general support for research, a majority of surveyed individuals express a desire to be consulted before their information is used in research, and many express concern about sharing of identifiable or re-identifiable information (Institute of Medicine 2009). Another common theme is that there is less public support for activities that are for commercial purposes (ibid., 84; Damschroder et al. 2007; Williston et al. 2007).

This negative view of commercial uses of biospecimens may reflect a misunderstanding of the role commercial entities play in biomedical research and clinical translation of discoveries. Among surveyed individuals, there is "much ambiguity in who 'researchers' are [and] what kind of 'health research' is involved" (Institute of Medicine 2009, citing Westin 2007).

"Some feared that researchers would sell information to drug companies" (Institute of Medicine 2009, 83), which seemingly are the archetype of an exploitive user of biospecimens in the minds of many who contemplate donating their biospecimens to science. Yet commercial drug and medical device companies conduct research that is overtly translational in nature— that is, their research frequently aims to develop useful new therapies and obtain regulatory clearances and approvals to move the products into the clinic to help patients. Commercial success is a rain that sprinkles on useful therapeutic products, more than on idle discoveries that offer no practical benefit to anyone. The public's rejection of commercial research, if carried to its logical conclusion, seemingly would destine their specimens for use only in upstream, basic science that may be decades away from improving the health of any patient.

Another point is that in modern times there is very little upstream, basic biomedical research that can be characterized as wholly free of commercial motives. Courts consider research, even grant-funded research carried out at academic institutions and non-profit foundations, to be a commercial activity unless it is "solely for amusement, to satisfy idle curiosity, or for strictly philosophical inquiry," a standard that almost no modern biomedical research meets or should aspire to meet (*Madey v. Duke Univ.*, 207 F.3d 1351, 1362 (Fed. Cir. 2002)). Commercial motives of academic researchers may be as modest as luring additional research grants and tuition-paying students, but these things "unmistakably further the institution's legitimate business objectives" (id.). Moreover, many academic institutions reap direct, substantial commercial benefits by patenting and out-licensing discoveries from their federally funded research under the Bayh-Dole amendments enacted in the 1980s (35 U.S.C. §§ 200–212). In contrast to some academic disciplines, such as history or philosophy, the distinction between academic and commercial research is blurred in biomedicine. Virtually all biomedical research has commercial aspects.

If the distinction between commercial and non-commercial uses of biospecimens is illusory, why are so many members of the public uncomfortable with commercial research? In part, this may reflect the basic human desire not to be "made fools of" by being the only actor with donative instincts in a sea of profit-seekers. It is disconcerting to watch others profit at one's own expense. If this were, in fact, the underlying concern, then a possible solution would be to let specimen donors share in the profits derived from studies that use their specimens. Profit-sharing of this sort draws various objections, including ethical concerns about potential coercion and commodification of human tissue as well as practical concerns about developing a workable profit-sharing formula. Heller and Eisenberg have noted various cognitive biases that can cause persons who contribute to a collective effort to overvalue their own contribution while disparaging contributions of others (Heller and Eisenberg 1998). A dispassionate valuation of the contribution of a single biospecimen to a discovery—which may have required the use of multiple specimens, capital equipment, and skilled scientific insight of current and past researchers—might prove so small as to insult specimen contributors, if indeed it were worth more than the transaction cost of administering payments. This chapter does not reject the notion of profit-sharing. Instead, it simply elects not to focus on that topic and accepts at face value the representations of persons who claim donative or altruistic motives when contributing their biospecimens for scientific study. Their expressed aversion to commercial uses must, therefore, reflect a different concern.

Survey data suggest that people who contribute their informational resources for research want their contributions to produce tangible benefits for future patients and the public (Health Data Exploration Project 2014). Specimen contributors' distaste for commercial research may reflect erroneous assumptions that academic and non-profit institutions outperform commercial entities in producing clinically useful innovations, or that academic and non-profit institutions place their discoveries into the public domain and make them available to patients for free. These assumptions are at odds with the realities of the post-Bayh-Dole academic research environment, discussed above. Even in the pre-Bayh-Dole era, commercial pharmaceutical and device companies typically led the process of translating academic discoveries into FDA-reviewed therapies available to help patients—a process that requires significant up-front investment. The concept of a publicly beneficial medical technology has never been synonymous with "free."

Negative attitudes about commercial research may simply be a misguided attempt to enunciate a public benefit standard. Survey data lend some credence to this possibility. A recent survey of attitudes about sharing of personal health data (PHD) found that individuals were very willing to share "if they knew the data would advance knowledge in the fields related to PHD such as public health, health care, computer science and social and behavioral science," although most expressed a desire for their data to be shared in anonymous form (Health Data Exploration Project 2014). Topol reports that multiple global consumer surveys have found that over 80 percent of people would be willing to share their medical data if the data were anonymized and subject to strict privacy protections (Topol 2015).

A potential problem with biospecimens is that the rich stores of data that can be extracted from them—including, for example, genomic data—make them susceptible to re-identification even if overt identifiers such as names have been removed (Department of Homeland Security et al. 2015). Biospecimens are not unique in this respect, however. Concerns about re-identification do not justify "biospecimen exceptionalism"—that is, a policy of treating biospecimens differently from other types of health data. With any rich, multiparametric personal dataset, there may be only one individual in the world for whom all the parameters simultaneously fit. Re-identification is also a concern, for example, with consumer sensor data, such as data from fitness tracking devices, given that people can be uniquely identified merely by their gait (Peppet 2014). This implies that biospecimens as well as many other types of health data—particularly the rich, longitudinal datasets that have high utility in biomedical research—all need to be viewed as potentially re-identifiable. It is no longer safe for ethical and privacy policies to assume that overt anonymization will satisfactorily resolve all privacy concerns. Instead, policies must be sufficiently strong to remain ethically robust even under the assumption that biospecimens and data may be re-identifiable.

**Ethics of Access to Identifiable Biospecimens**

Past efforts to explore the ethical boundaries of access to identifiable biospecimens are taking on new relevance in the modern research context, where all biospecimens are potentially re-identifiable. The ethics of sharing identifiable medical information received close attention in the 1977 report of the Privacy Protection Study Commission ("PPSC") formed under Section 5 of the Privacy Act of 1974. (US Department of Justice 2015; US Privacy Protection Study Commission 1977) The PPSC recommended that "no medical care provider should disclose, or be required to disclose, in individually identifiable form, any information" about an individual who has not consented to the disclosure, but the PPSC recognized several ethically justified exceptions to this general rule. These included a research exception that would allow unconsented disclosures of identifiable information:

for use in conducting a biomedical or epidemiological research project, provided that the medical-care provider maintaining the medical record:

(i) determines that such use or disclosure does not violate any limitations under which the record or information was collected;

(ii) ascertains that use or disclosure in individually identifiable form is necessary to accomplish the research or statistical purpose for which use or disclosure is to be made;
(iii) determines that the importance of the research or statistical purpose for which any use or disclosure is to be made is such as to warrant the risk to the individual from additional exposure of the record or information contained therein;
(iv) requires that adequate safeguards to protect the record or information from unauthorized disclosure be established and maintained by the user or recipient, including a program for removal or destruction of identifiers; and
(v) consents in writing before any further use or redisclosure of the record or information in individually identifiable form is permitted (US Privacy Protection Study Commission 1977, at ch. 7, rec. 10(c), *emphasis added*).

Our motive in citing this passage is not to advocate for nonconsensual use of biospecimens. Rather, the PPSC's recommendations are relevant as an attempt to enunciate conditions under which the balance of public benefits and individual harms is so compelling that it would be ethically acceptable to proceed with a use of identifiable information, even if the affected individual were reluctant to share. As has already been discussed, surveys show people are generally willing to let their anonymized data and biospecimens be used for socially beneficial purposes, but less willing to allow identifiable information to be used. Concerns about re-identifiability thus may cause people to hesitate to share biospecimens, even for socially beneficial purposes. Yet the public—including every individual whose specimens are used—stands to gain by having socially beneficial uses go forward. The PPSC's recommendations are informative as an attempt to balance individual autonomy with the public interest in a context—that is, the use of identifiable information—where the two are potentially in conflict.

The PPSC's recommendations also are important because of their historical role in the development of the Common Rule (45 C.F.R. pt. 46, subpt. A) (see Evans 2013 for a discussion of this history). The National Research Act of 1974 (Pub. L. No. 93–348) established a National Commission for the Protection of Human Subjects of Biomedical and Behavioral Research (the "National Commission"), which was charged with developing guidelines for the protection of human subjects, and these guidelines evolved over time to become the Common Rule. The National Commission, in its own guidelines, cited and adopted the PPSC's recommendations on nonconsensual use of identifiable information (US Department of Health, Education, and Welfare 1978). The National Commission concluded: "In studies of documents, records, or pathological specimens, where the subjects are identified, informed consent may be deemed unnecessary but the IRB must assure that subject's interests are protected" and then concluded that such uses are ethically justified if the five conditions identified by the PPSC, as outlined above, are met (ibid., 56, 181).

The National Commission called for the "IRB to assure that such conditions exist before approving proposed research in which documents, records, or pathology specimens are used for research purposes without explicit consent, and that the importance of the research justifies such use" (US Department of Health, Education, and Welfare 1978, 56, 181). In other words, the founders of the Common Rule saw the use of identifiable biospecimens as ethically acceptable, even without consent, but only if the uses were held to a utilitarian public benefit standard—that is, if "the importance of the research or statistical purpose for which any use or disclosure is to be made is such as to warrant the risk to the individual from additional exposure of the record or information contained therein" (US Privacy Protection Study Commission 1977; US Department of Health, Education, and Welfare 1978). The National Commission thus recognized that individual autonomy is bounded by considerations of the public's interest.

Subsequent ethical analyses also have called for a public benefit standard for assessing when it is ethically acceptable to use data and specimens without individual consent. Pharmacoepidemiological research is notorious for requiring access to identifiable health information in order to compile longitudinal records that incorporate data from multiple health-care settings (Weiss 2011; US Department of Health and Human Services/Food and Drug Administration 2011). In this context, Casarett et al. (2005) noted that the "central ethical issue in pharmacoepidemiologic research is deciding what kinds of projects will generate generalizable knowledge that is widely available and highly valued, and do this in a manner that protects individuals' right to privacy and confidentiality." This statement is a call for a public benefit standard: the prospect of highly valued results that will be widely shared. Jacobson also has commented that the most important issue to resolve is which public health objectives are sufficiently important to override the individual's interest in nondisclosure (Jacobson 2002). The National Bioethics Advisory Commission also called for IRBs to determine that "the benefits from the knowledge to be gained from the research study outweigh any dignitary harm associated with not seeking informed consent" for the use of data and specimens pursuant to consent waivers, which can include uses in identifiable form (National Bioethics Advisory Commission 2001). These examples display a fairly broad agreement it can be ethically justifiable to allow access to identifiable/re-identifiable data, even without individual consent, *but only if* the proposed use satisfies a public benefit standard.

## The Missing Public Benefit Standard in Current Regulations

Given such agreement, it is surprising that important federal regulations that govern access to biospecimens lack a public benefit standard. The history of how this omission occurred has been traced elsewhere (Evans 2011) and is only briefly summarized here.

The Federal Policy for the Protection of Human Subjects, or Common Rule (45 C.F.R. pt. 46, supt A), first included a waiver provision in 1981 (US Department of Health and Human Services 1981). This waiver provision, which allows IRBs to waive individual informed consent, was not originally designed for the purpose of waiving consent for research uses of data and biospecimens; rather, it was directed at clinical studies of the optimal design of federal benefit programs such as Medicare and Medicaid (Evans 2011, 122). This explains why the waiver provision did not incorporate the National Commission's recommendations regarding a public benefit standard for unconsented uses of medical records and biospecimens (US Department of Health, Education, and Welfare 1978, 56,181). Only later was the Common Rule's waiver provision embraced as a way to facilitate access to data and biospecimens (Evans 2011). The Common Rule lacks a public benefit standard in its waiver provision, which historically has appeared at 45 C.F.R. § 46.116(d) but which will be replaced by a new provision at 45 C.F.R. § 46.116(f) of the final rule published in January 2017, to become effective January 2018 (Department of Homeland Security et al. 2017, 7149, 7267). Under the current and future Common Rule waiver provisions, IRBs can approve unconsented access to people's data and biospecimens—including, under some circumstances, information in identifiable form—without first having to make a determination that the proposed research use satisfies a public benefit standard. When the Department of Health and Human Services and a group of other federal agencies initiated proceedings to amend the Common Rule on September 8, 2015 (Department of Homeland Security et al. 2015), they considered a requirement for IRBs to apply a rudimentary public benefit standard when approving waivers for biospecimens, but not for data, as will be discussed below. The final rule published in January 2017 rejected this requirement, however, thus leaving the Common Rule without a public benefit standard in its waiver criteria (Department of Homeland Security et al. 2017, 7267).

The Health Insurance Portability and Accountability Act of 1996 ("HIPAA," Pub. L. No. 104-191 1996) Privacy Rule (45 C.F.R. pts. 160, 164) also lacks a public benefit standard in its waiver provisions, which have remained unchanged since the Privacy Rule was promulgated in 2002 (US Department of Health and Human Services 2002). This omission, however, was more deliberate than was the case with the Common Rule. When developing the Privacy Rule, HHS was aware that its waiver provisions—unlike those of the Common Rule—are specifically directed at informational research. It proposed a new set of waiver criteria for the Privacy Rule, and these would have required an IRB or Privacy Board (together, "IRB") to determine that "the research is of sufficient importance so as to outweigh the intrusion of the privacy of the individual whose information is subject to the disclosure" (See Evans 2011, 123–124). This proposed criterion, which amounted to a public benefit standard, attracted a large number of negative comments during the public comment period for the proposed Privacy Rule. Commenters felt the criterion was too subjective, relied on conflicting value judgments about whether research is important, and would be inconsistently applied by IRBs (US Department of Health and Human Services 2000, 82,698).

In response, the next (December 2000) version of the HIPAA Privacy Rule substituted a different public benefit standard: that the risks of research must be reasonable in relation to the anticipated benefits of the research—if any—to the individual and the importance of the knowledge that may reasonably be expected to result from the research (US Department of Health and Human Services 2000, 82,698; Evans 2011, 123). This mirrored the familiar test at 45 C.F.R. section 46.111(a)(2) of the Common Rule, which IRBs apply when approving *any* research, whether consented or unconsented. This change in HIPAA's waiver provisions misconstrued the purpose of a public benefit standard. As applied at section 46.111(a)(2) of the Common Rule, this criterion sets a minimum threshold for ethically acceptable research (Evans 2011). Research that does not meet this threshold is so lacking in scientific merit that it would be unethical to proceed with it, even if participants were willing to consent (US Department of Health, Education and Welfare 1978, 56,180). The section 46.111(a)(2) test answers the question: Is this research so meritless that people should not be allowed to consent to it? That is a different question from the one a public benefit standard must answer: Is this research of such immense value that it is ethical to use people's data and specimens even if they do not consent? The latter question is what the PPSC and the National

Commission had in mind when they proposed the public benefit standard in the 1970s (Evans 2016).

Two years later, in 2002, HHS adjusted the Privacy Rule's waiver provision (US Department of Health and Human Services 2002). The adjustment consisted of eliminating the troublesome public benefit standard. The currently effective HIPAA waiver provision, like that of the Common Rule, does not require research to offer any public benefit as a condition of granting waivers. (See 45 C.F.R. § 46.116(d) (current Common Rule); id. § 164.512(i) (HIPAA Privacy Rule).)

### Clarifying the Public Benefit Standard for Research

During the first HIPAA rulemaking, when commenters questioned whether IRBs are capable of applying a public benefit standard, they raised a valid concern. Balancing public good against individual rights is conceptually at odds with the autonomy-based bioethical principles that the Privacy Rule and Common Rule seek to uphold. Miller notes that some ethicists feel that even if research has high social value, and even if a consent requirement might bias the research results, and even if obtaining consent may be impracticable, these facts "do not in themselves constitute valid ethical reasons for waiving a requirement of informed consent" (Miller 2008). Many of the foundational works of bioethics rely on an atomistic concept of autonomy and never explain how the principle of autonomy competes with other important social interests (Tauber 2005). Some recent bioethical works recognize that public interests also are important and suggest that individuals may have certain duties to undertake low-risk forms of informational research that contribute to a learning health-care system and other public health objectives (Faden et al. 2013). On the latter view, individual autonomy does not entirely override public interests. Yet the question of *how* to balance public and privacy interests is difficult because the interests are quite often incommensurable.

In property law, such tradeoffs have been equally difficult to navigate in takings cases, in which the question is "When can a person's property legitimately be taken for a beneficial public use?" Takings law is often described as a "muddle" (Rose 1984; Halper 1995; Claeys 2003). Subjecting biospecimen access to a public benefit standard has the potential to plunge bioethics into a similar muddle. The Privacy Rule and Common Rule avoid this muddle: they do not require IRBs to balance the individual's and public's interests when approving waivers that allow unconsented research use of data and biospecimens. Ignoring the issue implicitly presumes that all research uses of data and specimens do offer public benefits. This presumption often is correct, because many research uses—even those directed at abstract scientific questions—do produce real, practical benefits that ultimately are enjoyed by many members of the public. Yet some research—even research that is scientifically meritorious—may serve narrower or even private aims, such as an investigator's desire to answer a question that will impress colleagues but offers little prospect of translating into useful treatments for patients in a foreseeable time frame. Perhaps because it is so devilishly hard to predict which scientific advances will genuinely benefit the public, the Common Rule calls on IRBs to ponder "the importance of the knowledge that may reasonably be expected to result from research" and "anticipated benefits" (45 C.F.R. § 46.111(a)(2)) before approving any research, but it does not call on them to parse whether the anticipated benefits are "public" as opposed to serving more private aims (such as career advancement for an individual scientist, intellectual satisfaction within a narrow scientific community, or commercial benefits of a principally pecuniary nature). Moreover, the Common Rule does not include a public benefit requirement in its waiver criteria for biospecimen access at 45 C.F.R. § 46.116(d)) of the current regulations or at 45 C.F.R. § 46.116(f) of the amended final rule. By avoiding this issue, these regulations fail to vindicate individuals' desire—expressed in many surveys—for reassurance that their specimens will only be used in ways that benefit other patients and the public.

In its September 2015 proposal to amend the Common Rule, HHS considered the possibility of imposing a public benefit standard on biospecimen research. It proposed a waiver provision that would require IRBs to "find and document" that "[t]here are compelling scientific reasons for the research use of biospecimens" before granting a waiver of informed consent (proposed 45 C.F.R. § __.116(f)(2)(i); Department of Homeland Security et al. 2015, 54,054). HHS indicated its intent that "waiver of consent for research involving biospecimens (regardless of identifiability) will occur only in very rare circumstances" under the proposed standard (Department of Homeland Security et al. 2015, 53,937). This proposal differed in a number of respects from the public benefit standard enunciated by the PPSC and National Commission (US Privacy Protection Study Commission 1977; US Department of Health, Education and Welfare 1978). For example, the 2015 HHS proposal only would have applied its public benefit standard to biospecimen research, whereas PPSC and National Commission called for a public benefit standard for health data as well as biospecimens. From a privacy standpoint, data and biospecimens both are subject to re-identification, so there appeared to be no rational basis for distinguishing the two if the goal was to protect privacy. Approximately sixty public comments on the 2015 HHS proposal stated that no justification exists to treat data and biospecimens differently, and HHS ultimately abandoned the proposed public benefit standard for biospecimens (Department of Homeland Security et al. 2017, 7225).

From a practical standpoint, the "compelling scientific reason" standard HHS proposed in 2015 and the earlier PPSC/National Commission standard both raise serious implementation concerns. These are the same concerns that ultimately caused HHS, in 2002, to abandon the public benefit standard it had tried to include in the HIPAA waiver provisions: IRBs may not be ideal decision makers to assess what is a "compelling scientific reason" and what is not, making the standard subjective and vulnerable to inconsistent application. IRBs, by design, have a local perspective; a national oversight body or legislature may be in a better position to judge how a specific project fits into the "big picture" of national public health objectives.

There also was something worrisome about the assertion that waivers of consent for biospecimen research would occur "only in very rare circumstances" under the public benefit standard HHS proposed in 2015 (Department of Homeland Security et al. 2015, 53,937). The Common Rule applies to federally funded research. It should not be a "very rare" occurrence that federally funded research meets the criterion of serving compelling public interests; if this is rare, then research funding criteria need to be revised with the aim of reducing waste of taxpayers' funds. The waiver provision HHS proposed in 2015 would potentially have placed non-expert, private IRBs in the position of second-guessing research funding decisions made through the extensive peer-review processes of expert federal funding agencies (Evans 2016).

The idea of subjecting biospecimen consent waivers to a public benefit standard is, however, a good one and should not be abandoned, even though further work is needed to address these and other implementation issues. It may be possible to glean useful ideas from other contexts that subject decisions to a public-benefit standard. The legal literature on takings of private property is a potentially rich resource. It warns that there is no easy solution, but offers ideas that may be worth exploring.

One such approach would be to form a publicly accountable body to identify—and update—a list of general categories of research that offer sufficient public benefit to warrant access to biospecimens (Evans 2011, 2016). Patient advocacy groups could petition this body to approve biospecimen access for their diseases of concern, just as they petition Congress to fund research into those diseases. In situations where Congress or a state legislature has authorized funding for a specific program of research, this could be treated as a Blackstonian "consent of the people" to the research, and the use of biospecimens could be presumed to serve the public interest.

An alternative approach is to identify attributes of research that are inconsistent with the public interest, and restrict the use of waivers to facilitate biospecimen access for these presumptively "private" lines of research. Merrill suggests a similar approach for takings of private property (Merrill 1986), recognizing that it sometimes is easier to spot what *is not* a public-serving use of resources than to define what *is*. For example, it could be seen as presumptive evidence that a commercial actor is pursuing private (non-public) aims, if it uses people's biospecimens to develop a medical product that it then sells it at a high, whatever-the-market-of-desperate-people-will-bear price that has no relation to the cost of developing the product. Companies wishing access to the public's biospecimens might be required, in return, to undertake to price the resulting discoveries at a reasonable price, which might draw on utility rate principles to ensure an adequate allowance for operating and capital costs as well as a rate of return aimed at sustaining incentives for medical innovation (Evans 2016). Another possibility would be to establish a registry of biospecimen research, similar to the ClinicalTrials.gov for clinical trials. If a researcher is unwilling to disclose information about the research on this registry, this could be treated as presumptive evidence that the researcher is motivated by private rather than public aims and should not be allowed to use the public's biospecimens under a consent waiver (Evans 2011). Private-purpose uses of biospecimens would be allowed, but they would require individual informed consent. The point of this example is that biospecimens are a valuable input for research. Policy makers sell the public short if they fail to leverage that value by conditioning commercial researchers' biospecimen access on voluntary adoption of policies that could benefit the people who contributed those specimens.

A final strand of takings theory is potentially rich with insights relevant to biospecimen access. Claeys (2003) has described a line of nineteenth-century cases decided under natural law principles rather than under the more modern approach of balancing competing public and private interests. These cases explored projects that burdened individuals (for example, by making them install sidewalks on their land), but offered them a benefit of enjoying the improvements that others were similarly forced to make

(*Palmyra v. Morton*, 25 Mo. 593, 593 (1857)). Such projects were found to be permissible intrusions on autonomy if each individual, although burdened, received "reciprocity of advantage" from other members of the public. When this was true, the projects were treated as instances of the state using its coercive powers to broker mutually advantageous exchanges that individuals would not have been to arrange for themselves. A similar argument using reciprocity has been suggested for personalized medicine (Meslin and Cho 2010).

## Conclusion

There are strong reasons to subject biospecimen research to a public benefit standard. Such a standard presents complex implementation issues, but there are various options for designing a workable public benefit standard. These should be studied and adapted for possible use in biospecimen research. When resolving conflicts between individual and public interests in biospecimen use, however, it is important to remember that not all uses of biospecimens present a direct clash between individual and public interests. Many research uses of biospecimens have the character of mutually advantageous exchanges: individuals bear the burdens of biospecimen research, but they also stand to benefit from discoveries that can only be made by studying large samples of specimens reflective of the larger community.

When weighing competing uses of specimens, a worthy goal is to facilitate flows of biospecimens to those activities that offer the strongest prospect of producing benefits to the public. Whether an activity is commercial or non-commercial may be a distraction. Commercial research that produces an effective treatment for a troublesome disease may offer far more public benefit than an abstract academic study that serves only to burnish an individual investigator's scientific reputation. However hard it may be to develop a workable public benefit standard, it is time for ethicists and regulators to grapple with this challenge. Persons whose specimens are used in research deserve assurance that the burdens they bear further an important public purpose.

## Note

1. The views expressed in this chapter are those of the authors and should not be attributed to their employers.

## References

Casarett, David, Jason Karlawish, Elizabeth Andrews, and Arthur Caplan. 2005. Bioethical issues in pharmacoepidemiologic research. In *Pharmacoepidemiology*, fourth edition, ed. Brian L. Strom. Wiley.

Claeys, Eric R. 2003. Takings, regulations, and natural property rights. *Cornell Law Review* 88: 1549–1671.

Damschroder, L. J., J. L. Pritts, M. A. Neblo, R. J. Kalarickal, J. W. Creswell, and R. A. Hayward. 2007. Patients, privacy, and trust: Patients' willingness to allow researchers to access their medical records. *Social Science & Medicine* 64 (1): 223–225.

Department of Homeland Security et al. 2015. Federal policy for the protection of human subjects. *Federal Register* 80: 53,933–54,061.

Department of Homeland Security et al. 2017. Federal policy for the protection of human subjects. *Federal Register* 82: 7,149–7,274.

Evans, Barbara J. 2011. Much ado about data ownership. *Harvard Journal of Law & Technology* 25 (1): 69–130.

Evans, Barbara J. 2013. Why the common rule is hard to amend. *Indiana Health Law Review* 10: 365–414.

Evans, Barbara J. 2016. Comments on US Department of Health and Human Services (HHS) Proposed Rule: Federal Policy for the Protection of Human Subjects (Docket HHS-OPHS-2015–0008). http://www.regulations.gov/#!documentDetail;D=HHS-OPHS-2015-0008-1424.

Faden, Ruth R., Nancy E. Kass, Steven N. Goodman, Peter Provonost, Sean Tunis, and Thom L. Beauchamp. 2013. An Ethics Framework for a Learning Health Care System: A Departure from Traditional Research Ethics and Clinical Ethics. Hastings Center Report (Special Report: Ethical Oversight of Learning Health Care Systems): S16–S24.

Halper, Louise A. 1995. Why the nuisance knot can't undo the takings muddle. *Indiana Law Review* 28: 329–352.

Health Data Exploration Project. 2014. Personal Health Data for the Public Good: New Opportunities to Enrich Understanding of Individual and Population Health. http://hdexplore.calit2.net/wp-content/uploads/2015/08/hdx_final_report_small.pdf.

Health Insurance Portability and Accountability Act of 1996, Pub. L. No. 104–91, 110 Stat. 1936 (codified as amended in scattered sections of 18, 26, 29 and 42 U.S.C.).

Heller, Michael A., and Rebecca S. Eisenberg. 1998. Can patents deter innovation? The anticommons in biomedical research. *Science* 280 (5364): 698–701. doi:10.1126/science.280.5364.698.

Institute of Medicine. 2009. *Beyond the HIPAA Privacy Rule: Enhancing Privacy, Improving Health Through Research*, ed. Sharyl J. Nass, Laura A. Levit, and Lawrence O. Gostin.

Jacobson, Peter D. 2002. Medical records and HIPAA: Is it too late to protect privacy? *Minnesota Law Review* 86: 1497–1514.

Kish, Leonard J., and Eric J. Topol. 2015. Unpatients—Why patients should own their medical data. *Nature Biotechnology* 11 (9): 921–924.

*Madey v. Duke Univ.*, 307 F.3d 1351, 1362 (Fed. Cir. 2002).

Merrill, Thomas W. 1986. The economics of public use. *Cornell Law Review* 72: 61–116.

Meslin, E. M., and M. K. Cho. 2010. Research ethics in the era of personalized medicine: Updating science's contract with society. *Public Health Genomics* 13: 378–384.

Miller, Franklin G. 2008. Research on medical records without informed consent. *Journal of Law, Medicine & Ethics* 36: 560–566.

National Bioethics Advisory Commission. 2001. *Ethical and Policy Issues in Research Involving Human Participants*, volume 1. https://bioethics.georgetown.edu/nbac/human/overvol1.pdf

National Research Act of 1974 (National Research Service Award Act of 1974), Pub. L. No. 93–348, 88 Stat. 342 (codified as amended in scattered sections of 42 U.S.C.).

*Palmyra v. Morton*, 25 Mo. 593, 593 (1857).

Peppet, Scott R. 2014. Regulating the Internet of Things: First steps toward managing discrimination, privacy, security, and consent. *Texas Law Review* 93: 85–166.

Rose, Carol M. 1984. Mahon reconstructed: Why the takings issue is still a muddle. *Southern California Law Review* 57: 561–599.

Tauber, Alfred I. 2005. *Patient Autonomy and the Ethics of Responsibility*. MIT Press.

Topol, Eric. 2015. The big medical data miss: Challenges in establishing an open medical resource. *Nature Reviews Genetics* 16: 253–254.

US Department of Health, Education and Welfare. 1978. Protection of human subjects: Institutional Review Boards: Report and recommendations of the National Commission for the Protection of Human Subjects of Biomedical and Behavioral Research. *Federal Register* 43: 56,174–56,198.

US Department of Health and Human Services. 1981. Final regulations amending basic HHS policy for the protection of human research subjects. *Federal Register* 46: 8366–8391.

US Department of Health and Human Services. 2000. Standards for privacy of individually identifiable health information. *Federal Register* 65: 82,462–829.

US Department of Health and Human Services. 2002. Standards for privacy of individually identifiable health information. *Federal Register* 67: 53,182–53,273.

US Department of Health and Human Services/Food and Drug Administration. 2011. Report to Congress: The Sentinel Initiative—A National Strategy for Monitoring Medical Product Safety. http://www.fda.gov/downloads/Safety/FDAsSentinelInitiative/UCM274548.pdf.

US Department of Justice. 2015. Overview of the Privacy Act of 1974. https://www.justice.gov/opcl/role-privacy-protection-study-commission.

US Privacy Protection Study Commission. 1977. Personal Privacy in an Information Society. https://epic.org/privacy/ppsc1977report.

Weiss, Stanley H. 2011. Letter to Jerry Menikoff, Dir., Office for Human Subject Prots. 4–5. http://www.regulations.gov/#!documentDetail;D=HHS-OPHS-2011-0005-1066.

Westin, A. 2007. How the public views privacy and health research, as cited in *Institute of Medicine*, 2009.

Williston, D. J., L. Schwartz, J. Abelson, C. Charles, M. Swinton, D. Northrup, and L. Thabane. 2007. Alternatives to project-specific consent for access to personal information for health research: What do Canadians think? Presented at Conference on Data Protection and Privacy Commissioners, Montreal.

# 6   What Specimen Donors Want (and Considerations That May Sometimes Matter More)

Suzanne M. Rivera and Heide Aungst

In an era of increased interest in precision medicine, it is important to develop an adequate supply of biospecimens and data for research. Although it is estimated that there are more than 300 million stored samples in the United States (Baker 2012), the number must continue to grow in order for researchers to answer important scientific questions about human health.

Typically, scientists obtain samples for research in one of two ways. People can donate tissue prospectively for research (in the context of a single, hypotheses-driven study or for the deliberate creation of a tissue bank or "biospecimen repository"), or medical facilities can retain specimens (originally collected for diagnostic purposes or during treatment) to be used for research. In the latter case, donors may be asked for permission to retain specimens for research purposes or—if the specimens are de-identified—current regulations permit the specimens to be considered discarded medical waste and retained without the knowledge or permission of the person from whom they were removed. This approach to using de-identified specimens is a matter of some controversy within research circles, as is discussed throughout this volume.

The practice of retaining medical waste for research is a controversial one that we will not focus on here. In this chapter, we will examine what is known about why people do and do not donate specimens for research, discuss the ethical and logistical challenges of trying to honor donors' wishes, and explore other important ethical considerations.

## Why Donors[1] Give Specimens

Many empirical studies have looked both hypothetically and retrospectively at individuals' motivations for voluntarily donating specimens for research.

**Altruism**
Overwhelmingly, people cite altruistic reasons, such as "it's my obligation" and "to benefit future patients" for donating specimens (Porteri et al. 2014; Shin et al. 2011; Ma et al. 2012; van Schalkwyk et al. 2012; Manegold et al. 2010; Dang et al. 2014). One study showed that almost 98 percent of actual donors "agreed or strongly agreed" that there is widespread benefit to sharing genetic information. Similarly high numbers of people cite "advancing science" (Neidich et al. 2008; Lee et al. 2012) while others understand it as a "gift" to society (Kerath et al. 2013).

**Self-Interest**
In addition to altruism, for many the motivation to donate comes from self-interest: people want something in return. Donors may expect a direct, short-term benefit, such as personal health information provided through return of results (Bovenberg et al. 2009; Abou-Zeid et al. 2010; McMurter et al. 2011; Ma et al. 2012) or a long-term contribution to science that benefits them, a relative, a particular disease group, or a specific population. In one qualitative study, some donors expressed the view that researchers have an "ethical obligation" to give the donor any health information generated (Murphy et al. 2008).

Donor willingness increases when researchers can promise that the knowledge gained from using specimens will benefit a specific population that impacts the donor directly. Such populations can be as varied as disease groups or geographical boundaries. For example, studies in Africa (Moodley et al. 2014), Canada (McMurter et al. 2011), Finland (Tupasela et al. 2010), and Egypt (Abou-Zeid et al. 2010) found the majority of donors wanted assurance that their samples would benefit residents of their own country exclusively. Accordingly, some patient advocacy organizations, such as Skin of Steel (which backs melanoma research), have developed their own disease-specific biospecimen repositories, recruiting melanoma survivors and family members for tissue donation (http://skinofsteel.org/melanoma-tissue-bank.aspx).

In some cases, willingness to donate based on "membership" in a disease community actually depends on the type and state of disease. In almost all cases, cancer patients are willing to consent to research with residual tissue (Vermeulen et al. 2008). But one study found that people were less likely to give specimens for "stigmatizing" diseases, such as hypertension, diabetes, hepatitis B, or depression (Ma et al. 2012). Long's study on patients with irritable bowel disease (IBD) found that they were slightly more likely to give when they felt well, rather than when they had a "flare" (Long et al. 2015). When donating to dementia research—a disease with a strong genetic component—donors were more likely to contribute citing "advancement of knowledge for possibility of a cure," but they also expressed more fears of finding out results that might impact their own futures (Porteri et al. 2014).

The more concrete a potential therapy may result from the research, the more likely people are to donate. Murphy gave subjects four hypothetical scenarios and asked if they would donate. People were most willing to donate when research was both actionable and accurate. Research on diseases that had no treatment interested potential donors the least (Murphy et al. 2008).

At least one study found that patients often feel that they will receive better treatment if they donate (Peterson et al. 2014). And, as with other types of research, therapeutic misconception can be an issue in biospecimen donation. Donors may think their participation will lead to a cure if they are donating because of a specific disease or diagnosis (Lee et al. 2012).

**Trust**

Another important factor is trust. Individuals are more willing to donate specimens to researchers or within medical systems where they have established a trusting relationship. Trust emerges from transparency and honesty. Donors must have the confidence that specimens will be used appropriately, as well as that any personal genetic information will be kept confidential (Rahm et al. 2013; Neidich et al. 2008; Ma et al. 2012). Murphy et al. (2008) found that participants wanted to be ensured there would be "consequences for researchers who violated" an agreement about how the specimen would be used. Gaskell et al. (2013) found that both knowledge and trust contribute to a willingness to participate in biospecimen repositories throughout Europe. His research shows "a strong association between a country's level of engagement and the intention to participate in [biospecimen repositories]," as well as "an association between trust, participation and consent at a country level." More specifically, those who were better informed were in Northern European countries, including Sweden, Finland and Iceland. And, those living in countries such as Finland and The Netherlands, who reported "trust of key actors," reported a higher willingness to participate. In the United States, Dang et al. (2014) found, trust is "an essential component" for recruiting a diverse pool of participants.

## Barriers to Donation

One of the largest barriers to specimen donation is the public's lack of understanding about biospecimen repositories and their value to society (Rahm et al. 2013; Cervo et al. 2013; Igbe and Adebamowo 2012; Tupasela et al. 2010; Gaskell et al. 2013). While lay people often are unaware of the existence of—or the purpose for—biospecimen repositories, once informed, most will donate (Igbe and Adebamowo 2012; Dang et al. 2014; Partridge 2014).

Educational attainment also has a large impact on donation. While the influence of gender and age of donors varies greatly across studies, willingness to donate is correlated consistently with higher education levels (Porteri et al. 2014; Ma et al. 2012).

Cultural background influences willingness to donate, too. As with participation in other types of biomedical research, African Americans donate specimens for research in disproportionately low numbers, citing lack of trust most often as the reason (Dang et al. 2014). When African Americans do agree to donate, they overwhelmingly want to be able to consent for each new study, rather than give the specimen generally and trust the system to decide the circumstances under which it may be used (Murphy et al. 2009). In addition, Goldenberg et al. (2009) found that 61 percent of subjects would be influenced by helping others "learn about diseases that might affect people of your race or ethnicity"; and, of those, African Americans are more likely than whites (83 percent to 56 percent) to want to donate to learn about diseases that may affect people of their race.

Dang et al. (2014) found a barrier to donation among those with Eastern cultural values and backgrounds: Chinese, Vietnamese, and Hmong living in the United States. These populations expressed concern over donating blood because of its significance in representing life and vitality. If people identify donated tissue or blood as a sacred part of themselves, then getting them to agree to donation is more difficult. Morrell et al. (2011) found that Australian Aboriginals often felt that "something had been taken from them."

In reviewing the literature to understand and reduce cancer disparities by increasing participation of minority and underserved populations in biospecimen repositories, Partridge (2014) found multiple studies showing that minorities, including African Americans, Hispanics and Asians, are more willing to donate following education through community-based participatory research (CPBR). "It is encouraging," he concludes (897), "to find that at least with a community-based participatory approach, minority populations are willing to provide biospecimens for research and to participate in clinical trials."

Religion is another barrier to specimen donation. Although some studies show that people may think that donating specimens is against their religion (Al-Jumah et al. 2011; Ahram et al. 2013), once they are educated about their doctrine's teachings or receive "religious permission" (Ahram et al. 2013) they often will donate.

Other reasons given for refusing to donate run along a spectrum from plausible, such as the ability to be identified and exposed for a particular health issue, to deeply cynical, such as that research specimens can be used by the police to "commit suspected criminals and exonerate the innocent" (Godard et al. 2009). More fantastical reasons given for not donating specimens stretch the imagination into science fiction territory, including concerns about human cloning (Melas et al. 2010), the use of samples for "bioweapons development" (Luque et al. 2011), for "monitoring people through DNA" (Melas et al. 2010), and for "satanic rituals" (Moodley et al. 2014).

Consent plays a critical role in donation. Proponents of a standardized, "broad" consent[2] for biospecimen collection argue it would simplify and improve the process, but the data suggest that it may actually push potential donors away. Trinidad et al. (2012, 7) found participants unwilling to sign broad consent, quoting one who stated, "I probably wouldn't sign it, because I don't know what 'anything' means." One study found only 24 percent of adults would be willing to give broad consent when presented with a hypothetical situation for donating to a biospecimen repository (Gaskell et al. 2013). Another concluded that the majority of donors want to re-consent for additional research beyond the original purpose; 43 percent for every new study and 27 percent "at certain points" (Lewis et al. 2013). With broad consent, participants want the option to withdraw at any point (Gaskell et al. 2013). Kaufman et al. (2009) found that 81 percent of participants would feel "respected and involved" if asked for consent each time and 74 percent would feel that they "had control." When presented with the possibility that the researcher or system might even profit from a specimen donation, donors balk (Gaskell et al. 2013; Chan, Mackey, and Hegney 2012; Ma et al. 2012; Nilstun and Hermeren 2006; Moodley et al. 2014; Porteri et al. 2014). Despite such evidence suggesting broad consent may not be a panacea, the revisions to the Federal Policy for the Protection of Human Subjects finalized in January 2017 (and described in chapter 1 of this volume) nonetheless encourage its use (Department of Homeland Security et al. 2017).

### Must (Can) We Always Honor Donors' Wishes?

This review of the literature demonstrates that donors want control over their own specimens and results, while being ensured privacy and confidentiality (Gaskell et al. 2013; Kaufman et al. 2009; Luque et al. 2011; Trinidad et al. 2012). They either want to dictate up-front the specific type(s) of research that can use their specimens, or they expect to be contacted and re-consented every time another study or researcher wants to use their samples. They want results (individual or study-level), and they want the option to withdraw their specimens from use at any point. In sum, they want to be able to exercise their autonomy, even when it comes to materials that are no longer connected to them physically.

The demand for autonomy appears to be greater in the United States than in other countries where the sense of connection to a collective national identity is stronger. But the challenge remains the same worldwide: how can we create biospecimen repositories with both breadth and depth in gender, age, race and ethnicity, sample types, and diseases to meet the growing demands of precision medicine when donors want to maintain control?

### Challenges

Providing donors all that they want is difficult. First, large academic medical centers (the keepers of most of the world's scientific specimens) do many things very well, but they are administrative behemoths. The size and complexity of each biospecimen repository varies, but the activities involved in collecting, cataloging, and preserving large volumes of specimens and their associated data are highly technical, very demanding, and require financial resources. To maintain value, quality controls of freezer temperatures, data dictionaries, even disaster-recovery plans are essential. As Scott et al. (2012) succinctly summarize, "Logistics loom large." Thus, the more specific our promises about how to handle, store, share, and dispose of specimens, the more complex the systems required and the greater the likelihood we will inadvertently fail to honor donor choices.

Furthermore, we cannot predict all potentially beneficial future uses. Maximizing the benefit of specimens means leaving the door open to the unknowable. Researchers and institutions face an ethical and administrative conundrum with the matter of consent when we cannot foresee the future. Is it better to insist on broad consent, explaining clearly that donation entails relinquishing the possibility of knowing and approving of all future specimens uses (pro: honest, con: not exactly "informed" consent in the fashion we have come to expect, and may result in lower donation rates), or is it better to offer a menu of options regarding future uses, re-contacting, and eventual disposition even though we know that it is highly likely that some donor's wishes will not be honored due to bureaucratic complexity and human error (pro: closer to "informed" consent ideal; con: likely to be violated)?

Eventually, the courts may weigh in on this. So far, all available legal precedent supports the notion that a specimen, once it leaves the body, is not the property of the donor (see chapter 2). In three cases involving academic researchers (*Washington University v. Catalona*, *Moore v. Regents of the University of California*, and *Greenberg v. Miami Children's Hospital Research Institute*), the court or a settlement declared biospecimens to be the property of the institution at which the collection took place and not of the person who donated them. Accordingly, despite donors' wishes, they could not re-claim, direct, or share the profits generated by their former tissues.

In addition to the ethical and legal issues, the idiosyncrasies of donors' lives raise practical challenges. What happens if a donor moves or dies and the informed consent document did not explicitly address this? Donors' mobility makes re-contacting them notoriously difficult. And, decedents' specimens pose special concerns. Few people feel comfortable giving permission for a loved one's sample to be used after death (Al-Jumah et al. 2011). In addition, most parents believe that their children should be re-consented at 18 and given the option to pull their specimens if they do not agree with their parents' consent (McMurter et al. 2011). Regardless of the perception of an ethical mandate, a requirement to either find donors up to 18 years after the initial tissue collection or toss out their biospecimens will surely result in the loss of valuable resources.

Finally, specimen sharing dramatically increases the value of specimens. Aggregating large numbers of them allows us to answer questions more quickly, more definitively, and with a greater ability to search for patterns within subgroups. However, given what we know about the high degree of heterogeneity in specimen-sharing practices among academic research institutions (Goldenberg et al. 2015; Rothwell et al. 2015), it seems impossible to create a single menu of items in a consent form that would take into account all possible future permutations. This difficulty increases when sharing specimens internationally, with differences in both the legal frameworks (see chapter 1) and socio-cultural practices between countries (and their various subpopulations) (Hewitt and Hainaut 2011; Harris et al. 2012; H3Africa Consortium 2014).

**Are Donors' Wishes Binding?**
If biospecimen donation is a "social contract" that "explicitly links the ethic of donation with the goal of medical research (Mitchell 2010)," then to what should donors be entitled—morally—in return for their specimens? Ethical principles must be weighed when considering donors' wishes.

The principle of autonomy—or, as expressed more broadly in the Belmont Report (National Commission for the Protection of Human Subjects of Biomedical and Behavioral Research 1979)—"respect for persons"), considers options that maximize transparency, choice, and control for donors. But, given the limitations on our ability to honor donors' wishes combined with a pressing need to use specimens to answer important scientific questions for the advancement of human health, is there an ethical obligation to provide each donor the ability to mandate limitations of future uses?

Or, would a one-size-fits-all, take-it-or-leave-it protocol for biospecimen donation—with the presumption that most will opt-in—be more just? Such an approach would presume that all specimens are equal, that all donors have the same rights, and that they share an understanding about the importance of donation for research. The ethos behind this could be an understanding that all who benefit from science are expected to participate through a trusting system of donation. However, certain "special" cases, such as pediatric donors, donations from discrete and insular populations, and persons who object on religious grounds may pose unique challenges to this approach.

Consider all the ways that people relinquish control of their biological material, personal data, and images. An implicit understanding already exists in many non-research contexts about what people will trade away to get something desirable. For example, in exchange for real-time traffic news, I surrender information about my whereabouts through my smartphone. In exchange for feeling connected with friends and family, I can post a sonogram on Facebook and understand that a stranger may see it. When I visit a barber, I understand that my hair will be swept up and disposed of, leaving me unable to get it back a week later.

So, if we need specimens to advance important science, and if providing a menu of options to donors is logistically difficult to manage, and if the courts have ruled that former pieces of ourselves are not our property, and if people trade-away pieces of themselves routinely in non-research settings without the ability to make future claims on them, must we collect specimens for research in a manner that privileges the notion of autonomy so thoroughly?

## Discussion

As is discussed throughout this volume, there was a proposal in 2015 to change the Federal Policy for the Protection of Human Subjects (i.e., the Common Rule), which would have expanded the definition of "human subject" to cover all research using human biospecimens, even those that have had all identifying information removed. Although the revised regulation published in January 2017, did not expand the definition of "human subject" in this way, there was strong support by the Director of NIH and others for doing so (Hudson and Collins 2015).

Instead, the new rule offers the option of utilizing "broad" informed consent for prospective collection of biospecimens, for their storage, and for future unspecified research uses. The new policy does not provide a template for broad consent, as had been proposed originally, but instead gives institutions the flexibility to create their own broad consent forms. Still, it remains to be seen whether the availability of a broad consent approach will encourage more participants to donate samples that can be used for more than one study.

Many question the appropriateness of so-called "broad" consent and whether it will accomplish something better than what the longstanding human research protection rules already allow (Cadigan et al. 2015; Williams and Wolf 2013). In one of the most infamous cases alleging misuse of biological specimens, Arizona State University compensated the Havasupai Indian tribe for conducting genetic research on their samples without sufficient consent. The tribe members thought they had donated blood for diabetes research, but researchers also used samples to study the heritability of mental illness, among other things. The vaguely worded informed consent document said samples would be used to "study the causes of medical/behavioral disorders" (Mello and Wolf 2010). It is unlikely that a "broad" consent (now permitted under the revised Common Rule) would be much more specific than the one used by researchers at Arizona State University, and eventually objected to by the Havasupai.

## Ethical Considerations

In addition to the many practical challenges, numerous ethical considerations must be taken into account when regulating biospecimen research. Notably, respecting donors by honoring their wishes is only one of them.

Three fundamental principles guide human research ethics: respect for persons ("autonomy"), beneficence, and justice (National Commission for the Protection of Human Subjects of Biomedical and Behavioral Research 1979). Treating biospecimens as if they have the moral significance of people gives greater weight to the principle of autonomy than to the equally important principles of beneficence and justice. It also privileges privacy over the utility of specimens (Oliver et al. 2012).

Obtaining consent before using existing biospecimens for research purposes may prevent potential dignitary harms. But, when such consent cannot be obtained, the destruction of existing specimens will require new/more people to provide specimens to answer important questions. This violates the principle of beneficence because it would expose new donors to unnecessary risks, and would delay beneficial advances in human health, hurting us all. A careful examination of violations of the justice principle is provided in chapter 4 of this volume.

Given scarce resources, the expense of collecting new specimens also must be considered. In addition, delays caused by collecting new specimens despite the availability of existing ones means lives could be lost waiting to answer questions we otherwise would have addressed more quickly. Are the potential dignitary harms of the specimen donors more important than the actual material harms (in dollars and lives) caused by study delays? A more communitarian approach would recognize the societal value of specimens and would maximize their safe and respectful use, even when informed consent is not practicable.

Regulatory precedent exists for prospectively waiving the need for informed consent when the benefits of the knowledge to be gained outweigh the potential harms caused by infringing on individual autonomy. The US Food and Drug Administration allows an exception from informed consent for studies conducted in certain emergency settings. This is morally justified because it is not possible to prospectively identify all the individuals who might someday be eligible for an emergency study but unable to consent. Because requiring consent would relegate emergency patients to a standard of care that is not based on empirical evidence, the regulations permit exceptions to consent when extensive community education and public notification have taken place for the people most likely to be enrolled without their knowledge and when people are given a way to "opt out," such as by means of a bracelet or a wallet card.

Of course, studies with biospecimens typically are not done to answer questions arising in emergency settings. But, given this ethical precedent for allowing research on an unconscious or cognitively impaired patient in an emergency room, it follows that one could ethically justify the use of existing biospecimens when there is no obvious way to get informed consent. As Adams and Wegener (1999, 221) state, "If therapeutic advances could proceed no other way, if the risks are minimized, and if other protections are maximized, it would sometimes be unethical not to conduct research that could ultimately benefit those most in need."

## Conclusion

Empirical evidence indicates that donors want to control the research use of their biospecimens to a degree that exceeds expectations of control over data, images or specimens in other aspects of everyday life. It is extremely hard to honor those choices and still advance important science. Must we always honor donor preferences, even if they impede important scientific breakthroughs?

A different ethos, in which all people who could benefit from medical advances willingly contribute biospecimens for research, could change how we think about informed consent for specimen collection.

Rather than offering a menu of choices from which donors could choose, an atmosphere of trust and collaboration between scientists and donors could support a default position that all will donate and that, only under unusual circumstances (pediatric donors, discrete and insular populations, religious exceptions), would special provisions be made to offer additional protections of privacy. In this sense, we would create the expectation that only by exception would people "opt out" of contributing their specimens to science.

It would take a national public health campaign directed at changing attitudes to bring about this communitarian view of specimen donation—not unlike those we have seen successfully change attitudes about smoking, seat belts, and use of bicycle helmets. And, of course, such a campaign will require resources. Although the revised Policy for the Protection of Human Subjects does not include provisions for such a campaign, patient advocacy groups and other health promotion organizations could play an important role in shaping public opinion about specimen donation for research.

## Notes

1. We use the word "donor" here because we are describing motivations to give willingly.

2. Broad consent is described in more detail in chapter 8 of this volume.

## References

Abou-Zeid, A., H. Silverman, M. Shehata, M. Shams, M. Elshabrawy, T. Hifnawy, S. A. Rahman, et al. 2010. Collection, storage and use of blood samples for future research: Views of Egyptian patients expressed in a cross-sectional survey. *Journal of Medical Ethics* 36 (9): 539–547.

Adams, James G., and Joel Wegener. 1999. Acting without asking: An ethical analysis of the Food and Drug Administration Waiver of Informed Consent for Emergency Research. *Annals of Emergency Medicine* 33 (2): 218–223.

Ahram, Mamoun, Areej Othman, Manal Shahrouri, and Ebtihal Mustafa. 2013. Factors influencing public participation in biobanking. *European Journal of Human Genetics* 22 (4): 445–451.

Al-Jumah, M., M. A. Abolfotouh, I. B. Alabdulkareem, H. H. Balkhy, M. I. Al-Jeraisy, A. F. Al-Swaid, E. M. Al-Mussaaed, and B. Al-Knawy. 2011. Public attitude towards biomedical research at outpatient clinics of King Abdulaziz Medical City, Riyadh, Saudi Arabia. *Eastern Mediterranean Health Journal* 17 (6): 536–545.

Baker, M. 2012. Biorepositories: Building better biobanks. *Nature* 486: 141–146.

Bovenberg, Jasper, Tineke Meulenkamp, Sjef Gevers, and Ellen Smets. 2009. Biobank research: Reporting results to individual participants. *European Journal of Health Law* 16 (3): 229–247.

Cadigan, R. J., D. K. Nelson, G. E. Henderson, et al. 2015. Public comments on proposed regulatory reforms that would impact biospecimen research: The good, the bad, and the puzzling. *IRB: Ethics & Human Research* 37 (5): 1–10.

Cervo, Silvia, Jane Rovina, Renato Talamini, Tiziana Perin, Vincenzo Canzonieri, Paolo De Paoli, and Agostino Steffan. 2013. An effective multisource informed consent procedure for research and clinical practice: An observational study of patient understanding and awareness of their roles as research stakeholders in a cancer biobank. *BioMed Central Medical Ethics* 14 (1): 30.

Chan, Tuck Wai, Sandra Mackey, and Desley Gail Hegney. 2012. Patients' experiences on donation of their residual biological samples and the impact of these experiences on the type of consent given for the future research use of the tissue: A systematic review. *International Journal of Evidence-Based Healthcare* 10 (1): 9–26.

Dang, Julie H. T., Elisa M. Rodriguez, John S. Luque, Deborah O. Erwin, Cathy D. Meade, and Moon S. Chen. 2014. Engaging diverse populations about biospecimen donation for cancer research. *Journal of Community Genetics* 5 (4): 313–327.

Department of Health and Human Services and Food and Drug Administration. 2011. Human subjects research protections: Enhancing protections for research subjects and reducing burden, delay, and ambiguity for investigators. Advance notice of proposed rulemaking. *Federal Register* 76 (143): 44512–44531.

Department of Homeland Security et al. 2017. Final rule. Federal policy for the protection of human subjects. *Federal Register* 82 (12): 7149–7274.

Federal Policy for the Protection of Human Subjects. Notice of Proposed Rulemaking 2015. https://www.hhs.gov/ohrp/humansubjects/regulations/nprmhome.html.

Gaskell, George, Herbert Gottweis, Johannes Starkbaum, Monica M. Gerber, Jacqueline Broerse, Ursula Gottweis, Abbi Hobbs, et al. 2013. Publics and biobanks: Pan-European diversity and the challenge of responsible innovation. *European Journal of Human Genetics* 21 (1): 121.

Godard, B., V. Ozdemir, M. Fortin, and N. Egalite. 2009. Ethnocultural community leaders' views and perceptions on biobanks and population specific genomic research: A qualitative research study. *Public Understanding of Science* 19 (4): 469–485.

Goldenberg, Aaron J., Sara Chandros Hull, Jeffrey Botkin, and Benjamin S. Wilfond. 2009. Pediatric biobanks: Approaching informed consent for continuing research after children grow up. *Journal of Pediatrics* 155: 578–583.

Goldenberg, A. J., S. C. Hull, B. S. Wilfond, and R. R. Sharp. 2011. Patient perspectives on group benefits and harms in genetic research. *Public Health Genomics* 14 (3): 135–142.

Goldenberg, Aaron J., Karen J. Maschke, Steven Joffe, Jeffrey R. Botkin, Erin Rothwell, Thomas H. Murray, Rebecca Anderson, Nicole Deming, Beth F. Rosenthal, and Suzanne M. Rivera. 2015. IRB practices and policies regarding the secondary research use of biospecimens. *BioMed Central Medical Ethics* 16 (1): 32.

*Greenberg v. Miami Children's Hospital Research Institute, Inc.*, 264 F. Supp. 2d 1064 (S.D. Fla. 2003).

H3Africa Consortium. 2014. Enabling the genomic revolution in Africa: H3Africa is developing capacity for health-related genomics in Africa. *Science* 344: 1346–1348.

Harris, Jennifer R., Paul Burton, Bartha Maria Knoppers, Klaus Lindpaintner, Marianna Bledsoe, Anthony J. Brookes, Isabelle Budin-Ljøsne, et al. 2012. Toward a roadmap in global biobanking for health. *European Journal of Human Genetics* 20 (11): 1105–1111.

Hewitt, Robert, and Pierre Hainaut. 2011. Biobanking in a fast moving world: An international perspective. *Journal of the National Cancer Institute Monographs* (42): 50–51.

Hudson, Kathy L., and Francis S. Collins. 2015. Bringing the common rule into the 21st century. *New England Journal of Medicine* 373 (24): 2293–2296.

Igbe, Michael A., and Clement A. Adebamowo. 2012. Qualitative study of knowledge and attitudes to biobanking among lay persons in Nigeria. *BioMed Central Medical Ethics* 13 (1): 27.

Kaufman, David J., Juli Murphy-Bollinger, Joan Scott, and Kathy L. Hudson. 2009. Public opinion about the importance of privacy in biobank research. *American Journal of Human Genetics* 85 (5): 643–654.

Kerath, Samantha M., Gila Klein, Marlena Kern, Iuliana Shapira, and Jennifer Witthuhn. 2013. Beliefs and attitudes towards participating in genetic research—a population based cross-sectional study. *BMC Public Health* 13 (114): 1–9.

Lee, Christoph I., Lawrence W. Bassett, Mei Leng, Sally L. Maliski, Bryan B. Pezeshki, Colin J. Wells, Carol M. Mangione, and Arash Naeim. 2012. Patients' willingness to participate in a breast cancer biobank at screening mammogram. *Breast Cancer Research and Treatment* 136 (3): 899–906.

Lewis, C., M. Clotworthy, S. Hilton, C. Magee, M. J. Robertson, L. J. Stubbins, J. Corfield, et al. 2013. Consent for the use of human biological samples for biomedical research: A mixed methods study exploring the UK public's preferences. *BMJ Open* 3 (8).

Long, Millie D., R. Jean Cadigan, Suzanne F. Cook, Kaaren Haldeman, Kriste Kuczynski, Robert S. Sandler, Christopher F. Martin, Wenli Chen, and Michael D. Kappelman. 2015. Perceptions of patients with inflammatory bowel diseases on biobanking. *Inflammatory Bowel Diseases* 21 (1): 132–138.

Luque, John S., Gwendolyn P. Quinn, Francisco A. Montel-Ishino, Mariana Arevalo, Shalanda A. Bynum, Shalewa Noel-Thomas, Kristen J. Wells, Clement K. Gwede, and Cathy D. Meade. 2011. Formative research on perceptions of biobanking: What community members think. *Journal of Cancer Education* 27 (1): 91–99.

Ma, Yi, Huili Dai, Limin Wang, Lijun Zhu, Hanbing Zou, and Xianming Kong. 2012. Consent for use of clinical leftover biosample: A survey among Chinese patients and the general public. *PLoS One* 7 (4): 1–7.

Manegold, Gwendolin, Sandrine Meyer-Monard, André Tichelli, Christina Granado, Irene Hösli, and Carolyn Troeger. 2010. Controversies in hybrid banking: Attitudes of Swiss public umbilical cord blood donors toward private and public banking. *Archives of Gynecology and Obstetrics* 284 (1): 99–104.

McMurter, Britney, Louise Parker, Robert B. Fraser, J. Fergall Magee, Christa Kozancyzn, and Conrad V. Fernandez. 2011. Parental views on tissue banking in pediatric oncology patients. *Pediatric Blood & Cancer* 57 (7): 1217–1221.

Melas, P. A., L. K. Sjoholm, T. Forsner, M. Edhborg, N. Juth, Y. Forsell, and C. Lavebratt. 2010. Examining the public refusal to consent to DNA biobanking: Empirical

DATA from a Swedish population-based study. *Journal of Medical Ethics* 36 (2): 93–98.

Mello, Michelle M., and Leslie E. Wolf. 2010. The Havasupai Indian Tribe case—Lessons for research involving stored biologic samples. *New England Journal of Medicine* 363: 204–207.

Mitchell, Robert. 2010. Blood banks, biobanks, and the ethics of donation. *Transfusion* 50 (9): 1866–1869.

Moodley, Keymanthri, Nomathemba Sibanda, Kelsey February, and Theresa Rossouw. 2014. "It's my blood": Ethical complexities in the use, storage and export of biological samples: Perspectives from South African research participants. *BioMed Central Medical Ethics* 15 (1): 4.

*Moore v. Regents of the University of California*, 51 Cal. 3d 120, 271 Cal. Rptr. 146, 793 P.2d 479, cert. denied 499 U.S. 936 (1991).

Morrell, B., W. Lipworth, R. Axler, I. Kerridge, and M. Little. 2011. Cancer as rubbish: Donation of tumor tissue for research. *Qualitative Health Research* 21 (1): 75–84.

Murphy, Juli, Joan Scott, David Kaufman, Gail Geller, Lisa Leroy, and Kathy Hudson. 2008. Public expectations for return of results from large-cohort genetic research. *American Journal of Bioethics* 8 (11): 36–43.

Murphy, Juli, Joan Scott, David Kaufman, Gail Geller, Lisa Leroy, and Kathy Hudson. 2009. Public perspectives on informed consent for biobanking. *American Journal of Public Health* 99 (12): 2128–2134.

National Commission for the Protection of Human Subjects of Biomedical and Behavioral Research. 1979. The Belmont Report: Ethical Principles and Guidelines for the Protection of Human Subjects of Research.

Neidich, Alon B., Josh W. Joseph, Carole Ober, and Lainie Friedman Ross. 2008. Empirical data about women's attitudes towards a hypothetical pediatric biobank. *American Journal of Medical Genetics* 146A (3): 297–304.

Nilstun, Tore, and Göran Hermerén. 2006. Human tissue samples and ethics. *Medicine, Health Care, and Philosophy* 9 (1): 81–86.

Norohna, Nicole, Myriam Kline, Farisha Baksh, Peter Gregersen, and Emanuela Taioli. 2013. Beliefs and attitudes towards participating in genetic research—A population based cross-sectional study. *BioMed Central Public Health* 13 (1): 114.

Oliver, J. M., M. J. Slashinski, T. Wang, P. A. Kelly, S. G. Hilsenbeck, and A. L. McGuire. 2012. Balancing the risks and benefits of genomic data sharing: Genome research participants' perspectives. *Public Health Genomics* 15 (2): 106–114.

Partridge, E. E. 2014. Yes, minority and underserved populations will participate in biospecimen collection. *Cancer Epidemiology, Biomarkers & Prevention* 23 (6): 895–897.

Peterson, Imme, C. Desmedt, A. Harris, F. Buffa, and R. Kollek. 2014. Informed consent, biobank research and locality: Perceptions of breast cancer patients in three European countries. *Journal of Empirical Research on Human Research Ethics* 9 (3): 48–55.

Porteri, Corinna, Patrizio Pasqualetti, Elena Togni, and Michael Parker. 2014. Public's attitudes on participation in a biobank for research: An Italian survey. *BioMed Central Medical Ethics* 15 (1): 81.

Rahm, Alanna Kulchak, Michelle Wrenn, Nikki M. Carroll, and Heather Spencer Feigelson. 2013. Biobanking for research: A survey of patient population attitudes and understanding. *Journal of Community Genetics* 4 (4): 445–450.

Rothwell, Erin, Karen Maschke, Jeffrey R. Botkin, Aaron Goldenberg, Thomas H. Murray, and Suzanne M. Rivera. 2015. Biobanking research and human subjects protections: Perspectives of IRB leaders. *IRB: Ethics & Human Research* 37 (2): 8–13.

Scott, Christopher Thomas, Timothy Caulfield, Emily Borgelt, and Judy Illes. 2012. Personal medicine—The new banking crisis. *Nature Biotechnology* 30 (2): 141–147.

Shin, Sue, Jong Hyun Yoon, Hye Ryun Lee, Byoung Jae Kim, and Eun Youn Roh. 2011. Perspectives of potential donors on cord blood and cord blood cryopreservation: A survey of highly educated, pregnant Korean women receiving active prenatal care. *Transfusion* 51 (2): 277–283.

The H3Africa Consortium. 2014. Enabling the genomic revolution in Africa: H3Africa is developing capacity for health-related genomics in Africa. *Science* 344: 1346–48.

Trinidad, Susan Brown, Stephanie M. Fullerton, Julie M. Bares, Gail P. Jarvik, Eric B. Larson, and Wylie Burke. 2012. Informed consent in genome-scale research: What do prospective participants think? *American Journal of Bioethics Primary Research* 3 (3): 3–11.

Tupasela, A., S. Sihvo, K. Snell, P. Jallinoja, A. R. Aro, and E. Hemminki. 2010. Attitudes towards biomedical use of tissue sample collections, consent, and biobanks among Finns. *Scandinavian Journal of Public Health* 38 (1): 46–52.

van Schalkwyk, Gerrit, Jantina De Vries, and Keymanthri Moodley. 2012. "It's for a good cause, isn't it?"—Exploring views of South African TB research participants on sample storage and re-use. *BioMed Central Medical Ethics* 13 (1): 19.

Vermeulen, E., M. K. Schmidt, N. K. Aaronson, M. Kuenen, P. van Der Valk, C. Sietses, P. van Den Tol, and F. E. van Leeuwen. 2008. Opt-out plus, the patients' choice: Preferences of cancer patients concerning information and consent regimen for future research with biological samples archived in the context of treatment. *Journal of Clinical Pathology* 62 (3): 275–278.

*Washington University v. Catalona*, No. 4:03CV1065 (E. Dist. Mo. April 14, 2006), on appeal Nos. 06–2286 & 06–2301 (8th Cir.).

Williams, Brett A., and Leslie E. Wolf. 2013. Biobanking, consent, and certificates of confidentiality: Does the ANPRM muddy the water? *Journal of Law, Medicine & Ethics* 41 (2): 440–453.

# 7 Assessing Risks to Privacy in Biospecimen Research

Ellen Wright Clayton and Bradley A. Malin[1]

The last two decades have seen enormous attention devoted to worries about the use of biospecimens generally and DNA in particular for research. One of the most prominent of these concerns is that people whose DNA is used in research will be re-identified and harmed. (Hazell 2013; Kolata 2013; Meyer 2013). Despite its current fate, the recent Notice of Proposed Rule Making (NPRM) (US Department of Homeland Security et al. 2015) regarding changes to the Common Rule, published in September 2015, merits discussion as an example of how these perceived threats have been cited to affect policy. The NPRM's drafters asserted that "it is acknowledged that a time when investigators will be able [sic] readily ascertain the identity of individuals from their genetic information may not be far away" (id. at 53943), apparently implying that if re-identification can be accomplished from research samples, it *will* be accomplished, and that harms will follow. Although the regulations for the protection of human research participants existing at the time the changes were proposed did not mandate extensive consent for a great deal of research using biospecimens and clinical data from which identifiers have been removed, the drafters of the NPRM proposed to require express, elaborate informed consent for all uses of biospecimens regardless of their identifiability—a major change in policy.

In this chapter, we undertake a two-part analysis to address the question of whether this level of "biospecimen exceptionalism" (Clayton 2014) and the drafters' concern are warranted empirically, legally, and as a matter of policy. We agree with arguments, such as those made by McEwen et al. (2013), that the proposed recommendations regarding the use of biospecimens, by focusing on the *possibility* of re-identification, would have unduly limited research by placing too emphasis on remote risk and that they should instead have reflected the *probability* that a person will be identified from genomic or other data. Defining the likelihood of identification is a complex calculus, which requires looking at the circumstances in which attempts to do so occur and weighing incentives and disincentives for multiple actors in the ecosystem. In this regard, we will discuss the implications of recent research that demonstrates that under some conditions, efforts will *not* be undertaken to identify an individual from whom genomic information was obtained. We will also consider the probability that individuals could more easily be identified from phenotypic research data, medical records, and even information in the public domain than from genomic data. While further investigation into this area is needed, the research completed to date allows us to define areas where policy changes and other interventions can be invoked to decrease the likelihood that someone will turn to research biospecimens in order to try to identify someone.

Assuming that at least some attempts to use biospecimens for re-identification will take place, we also consider the likelihood that re-identification will cause harms, weighing the legal, policy, and contractual protections that are or could be put into place to decrease the risk of re-identification without the individual's permission. Our primary focus will be on assessing the risk of harms to individuals, since that is a relevant consideration in the Belmont Report (National Commission for the Protection of Human Subjects of Biomedical and Behavioral Research 1979), which underlies the Common Rule (Office for Human Research Protection 2009b). We will also consider potential harms to institutions and to the research enterprise as well, looking at both the consequences of re-identification of individuals as well as those that could attend the implementation of these proposed regulations.

## What Is Involved in Re-identifying Biospecimens from Which Traditional Identifiers Have Been Removed?

As was noted by the drafters of the NPRM, a few articles have been published in which genomic sequence data have been leveraged to distinguish some individuals. A number of authors have reviewed such efforts to re-identify individuals (re-ID attempts), often referred to as "attacks"[2] (Akgun et al. 2015; Erlich and Narayanan 2014; Naveed et al. 2015). Here we highlight several for the reader to make clear the problem at hand. The simplest and most direct re-ID attempt is based on the fact that genomic data is unique to each individual and is inherited. Notably, a whole genome sequence is not necessary to distinguish an individual. Rather, as few as 100 independent single nucleotide polymorphisms (SNPs) can uniquely characterize a single person (Lin, Owen, and Altman 2004). Thus, when a person who has such data also has access to an identified sequence, the DNA can serve as the way to identify the person from whom the target sample was obtained. Additionally, any inherited variation in the genome, such as short tandem repeats (STRs)—particularly on sex chromosomes—can be used to track and discover family members, a strategy that has been invoked by law-enforcement organizations (Bieber, Brenner, and Lazer 2006) and by individuals looking for unknown sperm donors (Sample 2005).

Beyond such direct re-ID attempts, a growing number of investigations illustrate how summary statistics about the genomic data involved in specific association studies may be sufficient to detect the presence of an individual and thus provide some information about his or her phenotype. Homer et al. (2008) were the first to determine whether an individual in a study was a case or a control, basing their methodology on forensic techniques for discovering the presence of an individual in a mixture of biospecimens (e.g., a mass grave). This re-ID attempt is particularly notable because it led to policy changes at the National Institutes of Health (NIH) and the Wellcome Trust in which such summary statistics were pulled out of the public domain (Zerhouni and Nabel 2008). In a subsequent study, Gymrek et al. (2013) were able to identify a small percentage of the males from the Utah CEPH who had contributed genomic data to the 1000 Genomics Project by extracting STRs from Y-chromosome sequence data, which they used to predict surnames using publicly available genealogical data that summarized rates of STRs by surname. The inferred surname, along with other information, such as age and geographic locale of residence, were then linked with public records to determine an individual's identity. We highlight this investigation because it illustrates that an adversary does not necessarily need to be in possession of a corresponding identified sequence genome to be successful in an identification attack but can rely in part on non-genomic information. These studies led the drafters of the NPRM to conclude that biospecimens cannot be guaranteed to be, in their terms, nonidentifiable.

The aforementioned studies show that efforts to re-identify can succeed, but risk assessment involves more than simply identifying what is possible. Rather, it requires (at a minimum) understanding why and how genomic data would be made available, understanding why someone with access to the data would attempt to re-identify a person in the dataset, and understanding the likelihood and the consequences of success of such an attack.

One reason why identifiability has become an issue is that federal research funding agencies now require that genomic data, usually with some phenotypic or other data, be shared with others in order to speed up research, while decreasing the need to collect additional biospecimens and data, and so lessening burdens on individuals (National Institutes of Health 2014). While data sharing is important for advancing our understanding of the causes of health and disease, increased availability of data may create opportunities for pinpointing individuals from whom they came. This expanded opportunity raises the question of whether there are conditions under which data sharing can occur without provoking an attempt by an adversary to identify those from whom the data were obtained. For purposes of this analysis, we will focus on the setting of controlled access to genomic and other data from which traditional social identifiers, such as those enumerated in the Health Insurance Portability and Accountability Act's Safe Harbor implementation of de-identification (e.g., personal name, Social Security numbers, address, phone numbers) (Health Insurance Portability and Accountability Act Privacy Rule 2016 at §164.514(b)(2)), have been removed since this is the most common research scenario. We do recognize, however, that allowing open access to data increases the possibility of re-identification with less ability to identify who made the re-ID attempt.

**Understanding the Likelihood of Re-identification**

Assessing the probability of re-identification necessitates considering the actions and responses of multiple stakeholders. Requirements for data sharing as a condition of funding strongly incentivize data collectors and data holders to share. At the same time, these entities also have powerful motivations to protect the identity of individuals. These incentives include the possibility of penalties, including fines, under the Health Insurance Portability and Accountability Act of 1996 (HIPAA) when such data is derived from the health-care setting (Office for Civil Rights 2015), penalties imposed by the Office of Human Research Protections (OHRP) (Office for Human Research Protection 2009a) which include shutting down federally funded research at the entire institution, impaired ability to obtain grant funding (National Institutes of Health 2014), as well as damage to the reputation of investigators and institutions (Solove 2013).

Despite institutional efforts to protect data, there are a number of reasons why an adversary might seek to re-identify individuals within a de-identified research dataset. One might simply wish to demonstrate that it is possible to do so, which the adversary often characterizes as "white hat hacking," undertaken to clarify the need for greater efforts to protect the data. For these purposes, it typically does not matter for these "proof of concept attacks" who is identified so long as someone is. It may not even matter whether that individual's identity is made public as was true for the members of the Utah CEPH in the attack conducted by Gymrek and colleagues (Gymrek et al. 2013). To the best of our knowledge, the identification demonstrations that have been reported to date all take this form. Such efforts may not be entirely altruistic on the part of the people effecting the re-identification since success can enhance the prestige or grant or job getting ability of the adversary. And in some ways, these demonstrations may have been counterproductive by raising fears unduly about the likelihood of serious economic harms.

In certain cases, an adversary might seek to identify a particular person in a data set either to demonstrate that this is possible or to cause that individual harm. Potential harms can range from the re-identification itself or revelation of the individual's genomic information, to impairing the person's access to other goods, such as life- or long-term care insurance if the person is not yet symptomatic (Solove 2006; Solove and Schwartz 2011). Once again, to the best of our knowledge, such "targeted attacks" to harm individuals whose genomic data have been de-identified have not been reported in the research context. In addition, both proof of concept and targeted attacks may also be relied upon to harm the data publisher, exposing the entity that shares data to a variety of civil and criminal penalties as well as damaging their reputation for protecting data.

Finally, an attacker may seek to re-identify individuals in a data set for personal economic gain, either to market their own products or services, as in use of prescription data for direct pharmaceutical marketing (*Sorrell v. IMS* 2011) or to sell the information gained to a third party. Whether a market exists for such information is difficult to predict, but it will depend in part on what uses are legally permissible. Since attackers pursuing these aims may not necessarily publicize these activities, another factor is the likelihood that the person(s) seeking to sell the re-identified data will be detected.

Importantly, it should be recognized that attempting to re-identify one or more people in a dataset is not cost-free. These attempts often require significant expertise, computational resources, and human capital, which may not be available to everyone who seeks to re-identify individuals who have participated in genomics research. At least at present, these efforts also require access to identified information that overlaps at least in part with variables in the targeted dataset, which is often difficult for either the genomic data itself or the associated phenotypic or demographic data needed for most genomics research.

Currently, the sources of identified genomic data are relatively few in the public domain. Some people post their own data on online genealogy or other websites (Telenti et al. 2014). Other people, such as Craig Venter, James Watson, and members of the Personal Genome Project, have placed their identified genomic data on the Internet for all to use. Forensic DNA databases, such as the CODIS database managed by the Federal Bureau of Investigation, by contrast, are much less widely available. Identified data in those resources are not kept in one place, and access is very tightly controlled (Federal Bureau of Investigation 2015). Progress is being made in the prediction of physical features from a genomic sequence (Fagertun et al. 2015). As this research develops, it may be increasingly possible to identify an individual by matching these predictions with photographs available on the Internet (Acquisti et al. 2011), an approach that also requires an identified dataset, a more readily available but messy one. Whether this approach will actually be used to re-identify someone from a research data set will probably depend on the motives for such efforts.

Most genomics research requires access to phenotypic, demographic, or social data. Identified versions of these types of data are often much more available than genomic data. Commercial data brokers, such as Intelius.com, InfoUSA.com, and Lexis-Nexis, sell a range of personal information associated with demographic and geographic details. Even health data are, at times, made available. Many people post information on the Internet about their own health and that of their families, and some have even disclosed results from 23andMe on social media (Yin et al. 2015; Asiri, Asiri and Househ 2014; Househ 2011; Lee et al. 2013). We note that, to date, there has been only one reported instance of a HIPAA breach involving research data (Office for Civil Rights 2016); all other HIPAA breaches have concerned clinical health information (Office for Civil Rights 2015). None of the breaches to date involve genomic data. Thus, one might reason that de-identified demographic and phenotypic data in a research dataset warrant special concern since they are likely to have more accessible sources of matches (Sweeney et al. 2013) but the NPRM actually would have expanded the circumstances in which such data can used for research with limited or no consent (NPRM § ___.101(a)(2)(C)(ii)).

Other costs to the person seeking to re-identify individuals from genomic data include possible penalties if laws or data use agreements are violated as well as limiting access to grant funding or genomic data. As will be discussed below, the likelihood of these penalties depends on the probability that the attempts at re-identification are detected and that civil damages or criminal penalties will be imposed for these efforts.

**The Contributions of Modeling**

Economic modeling is beginning to shed light on the issue of whether and how much data can probably be shared without incurring the risk that a third party will attempt to learn hidden knowledge about the individuals to whom data corresponds. A certain portion of these investigations assume that explicit identifiers, such as name, address, phone number, and social security number, are removed, obscured, or generalized, but also that some attributes of the individual that could be matched with identified external resources are retained. The question is under what conditions attempts to re-identify individuals will be undertaken.

A recent study reported on the results of a Stackelberg game (also known as a leader-follower game) with two entities—a data publisher and a person with access to the data (and potential attacker)—that specifically accounted for the value to the data holder of sharing as well as the cost to the party attempting to make the identification of mounting the attack and the benefits that would accrue to the attacker if the efforts were successful (Wan et al. 2015). The authors examined a case study in which (1) the benefit of sharing a record was $1200 (the average money paid to a grant recipient for each DNA sequence generated and shared), (2) the cost of obtaining identified records was $4 (based on the price of buying a record from a broker such as Intelius), and (3) the loss to the publisher of the data was approximately $300 (the average fine paid per person for a HIPAA Security Rule violation). This analysis revealed that in many instances the data holder would have the incentive to share quite a bit of data (at times even more than HIPAA Safe Harbor model of de-identification) while the third party would not attempt to re-identify because the benefits of success to the attacker were outweighed by the costs of mounting the effort. At the same time, the authors illustrated that the results were relatively stable to orders of magnitude change in the benefit and loss values, suggesting that even if the exact values of the case study were inaccurate, the results probably would hold true in practice.

A subsequent study (Xia et al. 2015) sheds further light on the importance of deterrence. These investigators considered the case in which those attempting to re-identify have to make a series of decisions to commit a re-identification. Specifically, these people have to decide whether it is worth purchasing access to a resource with identified information and then having done so, must decide whether the data are sufficient for mounting an attack. In this situation, the risk may depend on the extent to which the attackers can actually discern from the outset what information is available in the resource with identified data. They may, for instance, initially overestimate the chance that they can make a profit from accessing such data and attempting to re-identify an individual. Once they purchase access to the resource, they may discover that the resource does not contain sufficient information about the individuals in the research dataset. Had such knowledge been available from the outset, the attackers rationally may have chosen not to buy the identified data in the first place. Yet once the investment has been made, rational adversaries may view the situation through the perspective of "sunk cost," reasoning that it may be possible to recoup some of the money that has been spent by continuing to try to re-identify even if they ultimately cannot make a profit. To illustrate this point, let us consider the following example: The value of re-identifying a person in the dataset is $10, but there is only a 40 percent chance that that person is in the identified data, so that the expected value of an attack on that person is $4. If the cost to the attacker of another effort to re-identify the person in the research data is only $1, the attacker may continue trying three more times since there could be at least some net gain. Notably, however, the attacker may choose not to proceed if he or she knows that there is a 50 percent chance that efforts will be detected and, if detected, carry a penalty of $8. Thus, oversight of data use and assessment of penalties become crucial.

While additional research is needed to create more complete models (e.g., including more actors and identified datasets for comparison, as well as varying the potential gains and losses or costs to the various actors), the findings that have been generated to date provide a number of important lessons. One is that it is not the case that efforts to re-identify individuals from genomic data will always occur. This point may appear obvious since so few such efforts have been undertaken to date, but it needs to be reiterated because these findings suggest that simple disincentives may play a significant role in mitigating risk. Admittedly, it is likely that the technical cost of trying to re-identify a record will decrease over time as knowledge increases. Yet it is still possible to design conditions, some of which could be incorporated into policy, that would encourage data holders to share genomic and other associated data for research purposes while incurring little risk that a third party would attempt to re-identify individuals whose data are contained therein by increasing the cost of making the effort.

## What Needs to Happen to Achieve a More Appropriate Balance of Risk and Benefit for Research Using Biospecimens and Associated Data

A central ethical concern in research is protecting participants from harm. While acknowledging that some have argued that simply being observed or identified constitutes a harm from which research participants should be protected (Rothstein 2010), for this analysis, we will focus on the nondignitary injuries that law typically attempts to deter and compensate. Informed by evidence from modeling, we suggest that developing standards for research using biospecimens should begin by assessing the realistic likelihood that such harms will occur. This requires considering what data, genomic as well as phenotypic, demographic, and social, most of which will have explicit identifiers removed, will be available in the research dataset. Understanding where identified information that includes some or all of the remaining variables that can be used for matching the research data exist is essential. The costs of accessing both the research and identified data need to be factored in as well as the skills and resources needed for computation.

Assessing how hard it is to access research data is an essential first step. The NPRM provisions required the same level of protection for biospecimens, regardless of whether they are identified or not, and *identifiable* health information, even though they do not pose equivalent risk to individuals. This differential treatment of specimens and data was rejected in the Final Rule. Moreover, the drafters provided little guidance about how the reasonableness of these protections is to be assessed. Happily, there is remarkably little evidence that genomic data has been inappropriately revealed. As of July 1, 2015, the NIH reported that only approximately 0.1 percent of the groups that have requested access to dbGaP (Database of Genotypes and Phenotypes) have violated their Data Use Certification requirements (National Institutes of Health 2015). Although more than half of these involved issues with data access or security, it appears that none of these breaches caused harm to individuals (personal communication, eMERGE meeting, September 2015). Penalties for breach were relatively light, typically denying access to dbGaP for no more than 6 months.

More, however, can and should be done to prevent efforts to prevent unauthorized re-identification efforts. More emphasis needs to be placed on monitoring those who use these data to ensure biospecimens and data are not misused. Neither the NPRM nor the Final Rule addressed this issue at all even though it is hard to deter and impossible to penalize efforts to re-identify individuals from research data about them if attempts to re-identify are not detected. Indeed, it is widely acknowledged that detection and penalties need to be strengthened. (Gutmann 2013) The Working Group of the President's Precision Medicine Initiative, for example, recently recommended that "unauthorized re-identification or recontacting of participants should be expressly prohibited in agreements for the use of specimens and data, and NIH should pursue legislation penalizing such actions" (Precision Medicine Initiative (PMI) Working Group Report to the Advisory Committee to the Director 2015). The Secretary's Advisory Committee for Human Research Protections (2016) recommended that such penalties should extend to all uses of biospecimens and data for research, with enhanced penalties for those who reveal identities. At a minimum, individuals who violate these agreements should not be valorized, and journals should not publish their results.

Providing compensation to individuals who are harmed by re-identification would also act as a deterrent. Here, the law that protects privacy outside the research context provides additional, but incomplete, support. As noted above, the law tends to provide damages only when the identification leads loss of personal opportunities or other types of economic injuries. A number of laws address the use of genomic and other data, including the Genetic Information Nondiscrimination Act (GINA) (2008b), the Americans with Disabilities Act (ADA)(2008a), HIPAA, and tort laws (Solove 2006), creating a patchwork of protections that vary in their efficacy. Importantly, many of these laws depend for their enforcement on other entities such as government agencies, which has led some to call for the creation of private rights of action. (McGraw 2009) While room for improvement clearly exists, there nonetheless is remarkably little evidence that people have suffered harm on the basis of the use of genetic information for research (Green, Lautenbach and McGuire 2015; Clayton 2015).

Finally, it is worth asking whether the NPRM's requirements would have increased individuals' autonomy and trust in research, both in their own right and in ways that actually increase the availability of biospecimens for research, which are stated goals of the NPRM. While these potential outcomes are beyond the direct scope of this analysis, there is much reason to question whether these happy outcomes would come to pass. First, numerous studies show that complex consent forms do not always enhance understanding and decision making (Henderson et al. 2014; Robinson et al. 2013; Nishimura et al. 2013; Kass et al. 2011), essential elements of honoring autonomy or respect for persons. Nor is it clear that disclosure will enhance trust and participation, particularly among those who are already skeptical (Eyal 2012; O'Neill 2004). Ameliorating mistrust requires governance, transparency, and engagement, not overstating the risk and then asking participants to assume it.

Indeed, there is real reason to worry that these proposals would actually have decreased access to biospecimens for research. In particular, healthcare institutions might have chosen not to incorporate complex consent forms for the use of biospecimens into their admission, consent to treat, and surgical consent forms. If they failed to do so, many biospecimens would no longer be available for research. These are empirical questions that can be assessed. The better solution will almost surely be to reject "biospecimen exceptionalism" and instead to focus efforts on creating incentives and oversight that decrease the likelihood of re-identification by increasing the probability of detection and enhancing penalties.

The final changes to the Common Rule were issued on January 19, 2017 (US Department of Homeland Security et al., 2017). Fortunately, the drafters ultimately elected not to include de-identified biospecimens within the definition of human subjects. They did, however, adopt a requirement that the definition of identifiability be reexamined routinely, (§ __.102(7)) and suggested that examination should start with whole genome sequencing (Id. at 7169). They apparently expect that biospecimens and the data derived from them will become more identifiable over time and, as such, would increasingly become subject to the Common Rule on that basis alone. This approach can be understood as at best "biospecimen exceptionalism delayed." What is really needed is to assess the real risks of re-identification and to develop strategies to mitigate those risks.

## Notes

1. Neither author has any conflicts of interest to disclose. This work was supported in part by grants 1 U01 HG008672-01, 1 R01 HG006844-01, and 1RM1 HG009034-01. No part of this work has been published previously.

2. In many circles, all efforts to re-identify are referred to as attacks. In this chapter, however, because we address issues of harm, we reserve the terms attack and attacker for efforts to inflict injury in accordance with the common understanding of those terms. We use the term adversary when intent to inflict harm is less clear.

## References

Acquisti, A., R. Gross, and F. Stutzman. 2011. Face recognition and privacy in the age of augmented reality. *Journal of Privacy and Confidentiality* 6 (2): 1–20.

Akgun, M., A. O. Bayrak, B. Ozer, and M. S. Sagiroglu. 2015. Privacy preserving processing of genomic data: A survey. *Journal of Biomedical Informatics* 56: 103–111. doi:10.1016/j.jbi.2015.05.022.

Americans with Disabilities Amendments Act of 2008. 2008. Vol. 42 USCA §§ 12101 et seq.

Asiri, E., H. Asiri, and M. Househ. 2014. Exploring the concepts of privacy and the sharing of sensitive health information. *Studies in Health Technology and Informatics* 202: 161–164.

Bell, E. A., L. Ohno-Machado, and M. A. Grando. 2014. Sharing my health data: A survey of data sharing preferences of healthy individuals. Presented at American Medical Informatics Association Annual Symposium.

Bieber, F. R., C. H. Brenner, and D. Lazer. 2006. Human genetics: Finding criminals through DNA of their relatives. *Science* 312 (5778): 1315–1316. doi:10.1126/science.1122655.

Clayton, E. W. 2014. Biospecimen exceptionalism in the ANPRM. In *Human Subjects Research Regulation: Perspectives on the Future*, ed. I. Glenn Cohen and Holly Fernandez Lynch. MIT Press.

Clayton, E. W. 2015. Why the Americans with Disabilities Act matters for genetics. *Journal of the American Medical Association* 313 (22): 2225–2226. doi:10.1001/jama.2015.3419.

Erlich, Y., and A. Narayanan. 2014. Routes for breaching and protecting genetic privacy. *Nature Reviews Genetics* 15 (6): 409–421. doi:10.1038/nrg3723.

Eyal, N. 2012. Using informed consent to save trust. *Journal of Medical Ethics* 40: 437–444. doi:10.1136/medethics-2012-100490.

Fagertun, J., K. Wolffhechel, T. H. Pers, H. B. Nielsen, D. Gudbjartsson, H. Stefansson, K. Stefansson, R. R. Paulsen, and H. Jarmer. 2015. Predicting facial characteristics from complex polygenic variations. *Forensic Science International Genetics* 19: 263–268. doi:10.1016/j.fsigen.2015.08.004.

Federal Bureau of Investigation. 2015. Combined DNA Index System (CODIS). https://www.fbi.gov/about-us/lab/biometric-analysis/codis.

Genetic Information Nondiscrimination Act, Vol. 42 USC §§ 2000ff et seq. 122 Stat. 881.

Green, R. C., D. Lautenbach, and A. L. McGuire. 2015. GINA, genetic discrimination, and genomic medicine. *New England Journal of Medicine* 372 (5): 397–399. doi:10.1056/NEJMp1404776.

Gutmann, A. 2013. Data re-identification: Prioritize privacy. *Science* 339 (6123): 1032.

Gymrek, M., A. L. McGuire, D. Golan, E. Halperin, and Y. Erlich. 2013. Identifying personal genomes by surname inference. *Science* 339 (6117): 321–324. doi:10.1126/science.1229566.

Hazell, A. J. 2013. An education in "re-identification": Learning from the Personal Genome Project. The DNA Exchange. https://thednaexchange.com/2013/06/07/guest-post-an-education-in-re-identification-learning-from-the-personal-genome-project/.

Health Insurance Portability and Accountability Act Privacy Rule, 45 C.F.R. Part 160 and Part 164 Subparts A and E. 2016. http://www.hhs.gov/hipaa/for-professionals/privacy/.

Henderson, G. E., S. M. Wolf, K. J. Kuczynski, S. Joffe, R. R. Sharp, D. W. Parsons, B. M. Knoppers, J. H. Yu, and P. S. Appelbaum. 2014. The challenge of informed consent and return of results in translational genomics: Empirical analysis and recommendations. *Journal of Law, Medicine & Ethics* 42 (3): 344–355. doi:10.1111/jlme.12151.

Homer, N., S. Szelinger, M. Redman, D. Duggan, W. Tembe, J. Muehling, J. V. Pearson, D. A. Stephan, S. F. Nelson, and D. W. Craig. 2008. Resolving individuals contributing trace amounts of DNA to highly complex mixtures using high-density SNP genotyping microarrays. *PLOS Genetics* 4 (8): e1000167. doi:10.1371/journal.pgen.1000167.

Househ, M. 2011. Sharing sensitive personal health information through Facebook: The unintended consequences. *Studies in Health Technology and Informatics* 169: 616–620.

Kass, N. E., L. Chaisson, H. A. Taylor, and J. Lohse. 2011. Length and complexity of US and international HIV consent forms from federal HIV network trials. *Journal of General Internal Medicine* 26 (11): 1324–1328. doi:10.1007/s11606-011-1778-6.

Kolata, G. 2013. 2013. Web Hunt for DNA Sequences Leaves Privacy Compromised. *New York Times*, January 17, 2013. http://www.nytimes.com/2013/01/18/health/search-of-dna-sequences-reveals-full-identities.html.

Lee, S. S., S. L. Vernez, K. E. Ormond, and M. Granovetter. 2013. Attitudes towards social networking and sharing behaviors among consumers of direct-to-consumer personal genomics. *Journal of Personalized Medicine* 3 (4): 275–287. doi:10.3390/jpm3040275.

Lin, Z., A. B. Owen, and R. B. Altman. 2004. Genomic research and human subject privacy. *Science* 305 (5681): 183. doi:10.1126/science.1095019.

McEwen, J. E., J. T. Boyer, and K. Y. Sun. 2013. Evolving approaches to the ethical management of genomic data. *Trends in Genetics* 29 (6): 375–382.

McGraw, D. 2009. Privacy and health information technology. *Journal of Law, Medicine & Ethics* 37 (suppl. 2): 121–149.

Meyer, M. 2013. Ethical Concerns, Conduct and Public Policy for Re-Identification and De-identification Practice: Part 3 (Re-Identification Symposium). Bill of Health. October 2, 2013. http://blogs.harvard.edu/billofhealth/2013/10/02/ethical-concerns-conduct-and-public-policy-for-re-identification-and-de-identification-practice-part-3-re-identification-symposium.

National Commission for the Protection of Human Subjects of Biomedical and Behavioral Research. 1979. Ethical Principles and Guidelines for the Protection of Human Subjects of Research. http://www.hhs.gov/ohrp/humansubjects/guidance/belmont.htm.

National Institutes of Health. 2014. Genomic Data Sharing. https://gds.nih.gov/03policy2.html

National Institutes of Health. 2015. Genomic Data Sharing, Compliance Statistics for Policies that Govern Data Submission, Access, and Use of Genomic Data. https://gds.nih.gov/20ComplianceStatistics_dbGap.html#violation_access.

Naveed, M., E. Ayday, E. W. Clayton, J. Fellay, C. Gunter, J. P. Hubaux, B. Malin, and X. Wang. 2015. Privacy in the genomic era. *ACM Computing Surveys* 48 (1): 6.

Nishimura, A., J. Carey, P. J. Erwin, J. C. Tilburt, M. H. Murad, and J. B. McCormick. 2013. Improving understanding in the research informed consent process: A systematic review of 54 interventions tested in randomized control trials. *BioMed Central Medical Ethics* 14: 28. doi:10.1186/1472-6939-14-28.

Office for Civil Rights, US Department of Health and Human Services. 2015. Breaches affecting 500 or more individuals. https://ocrportal.hhs.gov/ocr/breach/breach_report.jsf.

Office for Civil Rights, US Department of Health and Human Services. 2016. Improper disclosure of research participants' protected health information results in $3.9 million HIPAA settlement. http://www.hhs.gov/hipaa/for-professionals/compliance-enforcement/agreements/feinstein/index.html.

Office for Human Research Protection, Department of Health and Human Services. 2009a. Compliance Oversight Procedures for Evaluating Institutions. http://www.hhs.gov/ohrp/compliance-and-reporting/.

Office for Human Research Protection. 2009b. Protection of Human Subjects. In 45 C.F.R. Part 46.

Office of the Secretary, Department of Health and Human Services. 2011. Human subjects research protections: Enhancing protections for research subjects and reducing burden, delay, and ambiguity for investigators. *Federal Register* 76 (143): 44512–44531.

O'Neill, O. 2004. Accountability, trust and informed consent in medical practice and research. *Clinical Medicine* 4 (3): 269–276.

Precision Medicine Initiative (PMI) Working Group Report to the Advisory Committee to the Director, NIH. 2015. The Precision Medicine Initiative Cohort—Building a Research Foundation for 21st Century Medicine. http://acd.od.nih.gov/reports/DRAFT-PMI-WG-Report-9-11-2015-508.pdf.

Robinson, J. O., M. J. Slashinski, T. Wang, S. G. Hilsenbeck, and A. L. McGuire. 2013. Participants' recall and understanding of genomic research and large-scale data sharing. *Journal of Empirical Research on Human Research Ethics* 8 (4): 42–52. doi:10.1525/jer.2013.8.4.42.

Rothstein, M. A. 2010. Is deidentification sufficient to protect health privacy in research? *American Journal of Bioethics* 10 (9): 3–11. http://www.ncbi.nlm.nih.gov/pmc/articles/PMC3032399/pdf/nihms-264889.pdf.

Sample, I. 2005. Teenage finds sperm donor data on internet. *The Guardian*, November 2, 2005.

Schwartz, P. H., K. Caine, S. A. Alpert, E. M. Meslin, A. E. Carroll, and W. M. Tierney. 2015. Patient preferences in controlling access to their electronic health records: A prospective cohort study in primary care. *Journal of General Internal Medicine* 30 (Suppl. 1): S25–S30. doi:10.1007/s11606-014-3054-z.

Secretary's Advisory Committee for Human Research Protections. 2016. Recommendations on the Notice of Proposed Rulemaking entitled "Federal Policy for the Protection of Human Subjects" http://www.hhs.gov/ohrp/sachrp-committee/recommendations/2016-january-5-recommendation-nprm-attachment-a/index.html.

Solove, D. J. 2006. A taxonomy of privacy. *University of Pennsylvania Law Review* 154 (3): 477–464.

Solove, D. J. 2013. HIPAA turns 10. *Journal of American Health Information Management Association* 84 (4): 22–29.

Solove, D. J., and P. M. Schwartz. 2011. *Privacy Law Fundamentals*, second edition. International Association of Privacy Professionals.

*Sorrell v. IMS*, 131 S.Ct. 2653 (2011).

Sweeney, L., A. Abu, and J. Winn. 2013. Identifying Participants in the Personal Genome Project by Name (A Re-identification Experiment). *arXiv:1304.7605*.

Telenti, A., E. Ayday, and J. P. Hubaux. 2014. On genomics, kin, and privacy. *F1000 Research* 3: 80.

US Department of Health and Human Services (HHS). 2016. Improper disclosure of research participants' protected health information results in $3.9 million HIPAA settlement. http://www.hhs.gov/about/news/2016/03/17/improper-disclosure-research-participants-protected-health-information-results-in-hipaa-settlement.html.

US Department of Homeland Security et al. 2015. Notice of proposed rule making, federal policy for the protection of human subjects. *Federal Register* 80:53933–54061.

US Department of Homeland Security et al. 2017. Federal policy for the protection of human subjects. *Federal Register* 82:7149–7274.

Wan, Z., Y. Vorobeychik, W. Xia, E. W. Clayton, M. Kantarcioglu, R. Ganta, R. Heatherly, and B. A. Malin. 2015. A game theoretic framework for analyzing re-identification risk. *PLoS One* 10 (3): e0120592. doi:10.1371/journal.pone.0120592.

Xia, W., Z. Wan, Y. Vorobeychik, M. Kantarcioglu, R. Heatherly, and B. Malin. 2015. Process-driven data privacy. In *CIKM'15*. ACM.

Yin, Z., D. Fabbri, S. T. Rosenbloom, and B. Malin. 2015. A scalable framework to detect personal health mentions on Twitter. *Journal of Medical Internet Research* 17 (6): e138. doi:10.2196/jmir.4305.

Zerhouni, E. A., and E. G. Nabel. 2008. Protecting aggregate genomic data. *Science* 322 (5898): 44. doi:10.1126/science.1165490.

# III  Consent and Its Implications

# Introduction

P. Pearl O'Rourke

A researcher recently began working with a multi-site team that boasts an expansive research portfolio for which it is important that biospecimens be eligible for broad research and sharing. The researcher collects biospecimens for her own projects with consent, but has limited funding and does not want to maintain identifiers and contact information, as would be necessary to obtain specific consent from each specimen source for each future use. Broad consent is critical to the success of her research partnership.

A potential research participant is excited about the prospect of providing a specimen for research on a disease that runs in her family. She may be willing to let her specimen be used in other research, but she would first like to know what that other research would entail. Without a promise to be specifically asked for additional uses, she refuses to provide any specimen at all.

A different potential research participant is also excited about providing a specimen for research on a disease that runs in her family and wants to be certain her specimen will be made available for other areas of research. Unlike the aforementioned participant, she does not want to be contacted by the research team again in the future.

These scenarios illustrate the basic tension between maximizing access to and use of biospecimens and the autonomy interests of participants. The chapters in this part begin to address this tension by examining broad and package consent.

Broad consent for biospecimen research is exactly what it sounds like: consent to use a specimen for broad, but not necessarily all, future uses. Broad consent, unlike blanket consent, allows for some limitations, but depending on the extent of limitations, it is easy to imagine blanket and broad as neighbors along a consent continuum. Package consent—which can be blanket, broad, or very limited—describes the situation in which agreement to secondary uses of collected biospecimens is a condition of participating in the primary research. As noted by Pritchard and Kaneshiro in their chapter, such linking may raise questions of undue influence and/or coercion. Although broad and package consent differ, they share both the goal of maximizing the use of specimens collected for research, as well as the concern regarding individual control over future/secondary use.

What are the advantages of broad consent? The resources needed for recruiting and obtaining consent from potential specimen donors can be significant, and research resources are of course finite. In large population studies, inadequate resources resulting in inadequate numbers of enrolled participants can cripple the research. Broad consent maximizes the use of each specimen collected and provides a better return on investment for the process of recruitment and enrollment. Additionally, the pace of scientific discovery makes it difficult to accurately predict potential future uses of existing specimens: limiting consent to specific uses at enrollment can preclude valuable unanticipated applications in the future.

But, does broad consent truly respect an individual's autonomy? Should not each individual have the right to decide/determine how his or her specimens will be used? Specific limited consent with options of re-contact and re-consent for additional uses would be one approach. But it is expensive, logistically difficult (due to the requirement of tracking individuals for the duration of the research or until their death), and may impede certain research.

We might think these issues could be solved just by asking what potential specimen donors want. The problem, however, is that their views are mixed, as highlighted both by Grady et al. and Garrison in their respective chapters. While some want to tightly control the use of their specimens with detailed and frequent re-consent, others are comfortable with notification of secondary uses rather than consent, and still others prefer to give broad, if not blanket, consent and do not want to be re-contacted.

The chapters in this part describe the problem, the goals, and possible approaches that bridge the extremes of complete participant control versus open and unrestricted access. I add to this discussion by highlighting a few related issues.

First, why focus singularly on biospecimens? It is not clear if or why participant concern for the use of their biospecimens versus their data is dramatically different. In fact, if data storage fees were reasonable, some have proposed that rather than storing specimens, the whole genome sequence could be stored. Have big data and genomics blurred the distinction between data and biospecimens? Do participants truly perceive

# Introduction to part III

the need for a different level of autonomy for each? Ultimately, it seems that what is reasonable for biospecimens should be reasonable for data. The chapters in this part should therefore be read with an eye toward their broader implications for consent to the future research use of data as well.

Second, context matters. There is a huge difference between an investigator who maintains specimens from a single study in his or her own laboratory versus specimen collection for a formal central repository. Specimen collection and storage come in a variety of models. Too often this heterogeneity is not considered. Discussions should perhaps initially focus on approaches to formal biospecimen repositories and then see how these best practices can be applied, or not, to individual collections. A formal repository should have business rules that address details such as: the type of specimens that will be accepted pursuant to what consent process; how specimens will be maintained; and the processes and rules for accessing specimens. Such repositories should already fulfill many of the key components for broad consent described by Grady et al. in their chapter, but the required infrastructure may not be scalable for every one-off collection. A number of options could then be considered, for example: should access to specimens for secondary uses always be arbitrated through a formal repository? Should there be tiered requirements for different types of collections?

Third, it is notable that this increasing focus on individual autonomy when biospecimens are used for secondary research is happening at the same time that mandates are being proposed and implemented for broad sharing of research results, both in aggregate and at the level of de-identified individual data. There is a growing international call for transparency through broad sharing of results and data. Proponents also suggest that ensuring broad use best respects research participants by maximizing the application of their contribution; once available, data could be used in secondary analysis and study by a variety of researchers.

Even those who believe that data and biospecimens merit different types of consideration may see a conflict here. This push for making research results and data publicly available seems to be in conflict with the increasing push for autonomy and control by limiting uses and avoiding blanket consent when it comes to biospecimens. Mandated sharing diminishes an individual's control regarding secondary use of their data and perhaps the only control that is left is to refuse to participate in the study to begin with.

Finally, there are general issues concerning the consent form and process. Well-intended, meticulously developed broad consents with detailed limitations are drafted with the goal of protecting autonomy by allowing each individual to make his or her own informed decision. But setting limitations is challenging. Are the limitations meaningful to the potential participant? Do they make scientific sense to researchers? How will the limits be respected? Each specimen must be clearly labeled regarding what uses are and are not allowable; access to those specimens by secondary (and possibly tertiary and beyond) users should be conditioned on certification that boundaries of use will be respected. And as novel areas of research emerge, how should we proceed if there is honest disagreement regarding whether or not the new use can be assigned to the "allowable" category? Should the novel uses be disallowed, or should individuals be re-contacted to determine acceptability of this new area of research?

In this context, as in others, we must remain cognizant of the fact that, in general, the effectiveness of informed consent is suspect. Even in interventional studies with very specific research procedures, significant risks, and benefits, participants may fail to remember not only the details of consent, but giving consent at all. Given that donation of a biospecimen is minimal risk, often for future, less-defined research, is biospecimen consent doomed to be less memorable and hence even more poorly remembered?

How to proceed? Assuming that the value of genetic and population research is a given, we must find ways to respectfully access large numbers of biospecimens, with the ability to re-access those specimens over time. This will only be possible with active and ongoing engagement and partnership with the public that goes well beyond recruitment and informed consent; the (understandable) focus on the initial donation is not adequate on its own. Rather than focusing on the consent itself, there should be more attention, education, and communication that remind individuals about the storage of their specimens, how those specimens are used, what a participant agreed to, and reminders of their ability to withdraw.

Equally important is making sure that the public is aware of why biospecimens are so critical to research, and the fact that the value of a specific specimen may not be realized for many years. The research community must collaborate with the public to explore strategies for recruiting, engaging, and enrolling participants. Different consent models should be discussed with honest dialogue regarding the benefits and downsides of each.

For example, a model that requires tracking and frequent re-consent will decrease the funds available for the research itself.

It is an interesting time. The promises of today's science will only be realized if large numbers of people agree to make their biospecimens (and data) available for study. We must find a way to meaningfully bridge the conflict between total personal control and autonomy and maximize the use of biospecimens. Broad consent as described in these chapters, now an option under the revised regulations published in January 2017, and perhaps even package consent, can serve as a foundation for that bridge.

# 8 Broad Consent for Research on Biospecimens[1]

Christine Grady, Lisa Eckstein, Benjamin Berkman, Dan Brock, Sara Chandros Hull, Bernard Lo, Rebecca Pentz, Carol Weil, Benjamin S. Wilfond, and David Wendler

Biological samples or biospecimens have been collected and stored in clinical and research settings for decades. Billions of samples are in storage (SACHRP 2011). Valuable research has been conducted with these biospecimens, including, for example, identifying prevalence estimates and clinical outcomes for Hepatitis C (Alter et al. 1997; Seeff et al. 1992), characterization of dengue virus types (Lewis et al. 1993), relative efficacy estimations of tamoxifen chemoprevention for *BRCA1* versus *BRCA2* breast cancer (King et al. 2001), identifying possible viral causes of post-transplant idiopathic pneumonia syndrome (Seo et al. 2015), and others.

Variable processes and practices exist for obtaining consent for the future research use of biospecimens. These include consent obtained at the time of specimen collection for each specific use with re-consent for subsequent uses, consent involving choices on a checklist, and in some cases no consent at all (Edwards et al. 2014). Some confusion and uncertainty remain about appropriate and ethically permissible types of consent for biospecimen research. Further, reliance on different approaches necessitates keeping track of the type of consent used for particular biospecimens and handling them accordingly, potentially increasing the costs of research, decreasing its scientific value, and increasing the possibility of violating the terms of consent. Confusion and uncertainty about consent can also result in decisions to not use certain specimens for research and consequent loss in research-related public benefit.

In September 2013, the NIH Clinical Center Department of Bioethics convened a group of subject-matter experts with diverse perspectives to debate the merits of broad consent for biospecimen collection for future research use. We defined "broad consent" as consent for an unspecified range of future research conditional on a few content and/or process restrictions and coupled with governance. We understood broad consent as less specific than consent for each use, but narrower than open-ended permission without any limitations (i.e., "blanket" consent). Participants discussed and debated ethical justifications and permissibility of broad consent for biospecimen research compared to alternative approaches, identified an approach that could be adopted across diverse sites and studies, and discussed optimal implementation. The specific focus was informed consent at the time of biospecimen collection—in clinical or research settings—and not research with previously collected samples, community consent, return of results and incidental findings, or other important and related issues. A summary of workshop deliberations was published in September 2015 in the *American Journal of Bioethic*s (*AJOB*) as a target article with fifteen concurrent open peer commentaries (Grady et al. 2015).

This chapter, adapted from the *AJOB* publication, describes ethical justifications for an approach to broad consent envisioned by workshop participants with areas of agreement and disagreement, and considers issues and objections expressed in accompanying *AJOB* commentaries, current literature, and recently proposed and final regulatory changes.

## Changing Regulations and Guidance

Several recent and projected changes to research regulations and guidance have endorsed a broad-consent approach for research with biospecimens and data. Workshop participants considered the US Department of Health and Human Services' Office of Human Research Protections (OHRP) 2011 Advanced Notice of Proposed Rulemaking (DHHS 2011). In writing this chapter, we considered the 2015 Notice of Proposed Rule Making (NPRM), which proposed defining research with biospecimens as human subjects research and requiring written informed consent for collecting biospecimens and identifiable data for future research, including those collected through clinical encounters, with very few exceptions. The NPRM proposal allowed specific or broad consent, but stipulated the use of a broad-consent template that was to be developed by the DHHS Secretary, and would have required no institutional review board or other oversight for research with biospecimens collected using DHHS broad-consent templates (DHS et al. 2015). The final rule, published in the Federal Register on January 19, 2017 and described in detail in chapter 1 of this volume, did not adopt the proposal to define research with unidentified biospecimens as human subjects research and therefore does not require informed consent for research with unidentified biospecimens. The final rule, however, permits researchers to seek broad consent that allows participants to consent to the unspecified future use of their identifiable private information and biospecimens (DHS et al. 2017; Menikoff et al. 2017).

The 2013 Privacy Rule amendments to the Health Insurance Portability and Accountability Act allow research authorization for use and disclosure of protected health information for future research purposes, as long as participants are provided with sufficient information to make a reasonably informed decision (DHHS 2013). The NIH Genomic Data Sharing Policy (NIH 2014) expects investigators submitting genomic data to NIH to provide documentation of participants' informed consent to broad sharing of genomic and phenotypic data for future research purposes. The Secretary's Advisory Committee on Human Research Protections (SACHRP), responding to a requirement for explicit parental consent for research using newborn dried blood spots in the Newborn Screening Saves Lives Reauthorization Act of 2014, recommended that OHRP develop guidance emphasizing simplified one-time broad permission for future research and a sample broad-consent form (SACHRP 2014). The Newborn Screening Saves Lives Reauthorization Act of 2014 will no longer be effective following the effective date of the final Common Rule. Secondary research with unidentified newborn dried blood spots, therefore, will be treated similarly to secondary research with any other type of nonidentified biospecimen (DHS et al. 2017).

Each of these proposals supports the idea of broad consent for future research use of biospecimens, an idea previously recommended by scholars (Wendler 2006; Hansson et al. 2006). However, these proposals do not necessarily agree on what broad consent is and how it should be implemented.

## Ethical Reasons in Favor of Consent for Research with Biospecimens

Multiple options exist for obtaining consent for the future research use of biospecimens, a range defined by the extent to which donors are informed about and able to decide whether their samples are used for research purposes (see figure 8.1). Identifying the best approach involves first considering the reasons to obtain consent at all.

At least five positive reasons support obtaining donors' consent for biospecimen research:

- It shows respect for donors.
- It allows them to control whether their samples are used for research.
- It allows them to decide whether research risks and burdens are acceptable to them.

|  | TYPE OF CONSENT | DESCRIPTION |
|---|---|---|
| Less burden, less control | No consent | No donor consent |
|  | Blanket | Consent to future research without limitations |
| ↕ | Broad* | Consent to future research with specified limitations |
| More burden, more control | Checklist | Donors choose types of allowable future studies |
|  | Study specific | Consent for each specific future study |

*The framework proposed here couples initial broad consent with oversight and ongoing communication when feasible.

**Figure 8.1**
Approaches to consent for future research with biospecimens.

- It allows donors to decide whether contributing to research is consistent with their fundamental values and non-welfare interests,
- It makes transparent decisions about donating biospecimens for research, which can promote public trust, and the ongoing viability of research with biospecimens.

Identifying an ethically appropriate approach to obtain consent for biospecimens also requires estimating the costs and burdens of obtaining consent. Costs and burdens include donors' and investigators' time, resources needed to obtain consent, and costs of maintaining systems that record and honor individual choices, or later seeking donor re-consent. Further, requiring consent raises the possibility that donors may decline, possibly diminishing the potential for future valuable research.

**Empirical Support for Consent for Research with Biospecimens**

The ethical reasons to obtain consent when collecting biospecimens for future research described above support a presumption in favor of consent, but provide no clear reason to prefer any particular type of consent. A reasonable person standard, which holds that information provided to donors should be based on what a reasonable person would want to know to decide whether to donate samples, could help to inform the choice of consent. Empirical studies have surveyed more than 100,000 patients, research participants, family members, religious leaders, and the general public regarding views on future research use of stored biospecimens (Brothers et al. 2011; Chen et al. 2005; Mezuk et al. 2008; Murphy et al. 2009; Simon et al. 2011; Valle-Mansilla et al. 2010), although some groups are insufficiently represented. Across different populations and countries and diverse survey methodologies, respondents indicate that they want to decide whether their biospecimens are used for research (see also chapter 6 of this volume). However, the majority's willingness to donate specimens seems largely unaffected by specific details of future research, such as the disease being studied, technology used, study target, or product (Hoeyer et al. 2004).

Existing data further suggest that after initial consent, the majority are willing to have their samples used for research without being asked about each use, with the possible exception of a few types of research, for example, research involving human cloning and commercial or for-profit research (Stegmayr and Asplund 2002; Tupasela et al. 2010; Gaskell et al. 2013; Brothers et al. 2012; McCarty et al. 2008; Ma et al. 2012). Further, many are willing to provide a type of broad consent for future research with biospecimens. Simon et al. found, for example, that US survey and focus group participants wanted to give initial consent, but commonly preferred broad consent because, among other reasons, "the research would help others," "I would only have to sign the paper or be asked about the research once," and "broad consent allows for research in the future that might not have been considered yet" (Simon et al. 2011). Respondents to an online public survey favored options labeled as blanket consent with a caution, with limits, and with an option for withdrawing, which are similar to how we describe broad consent, and rated specific consent for each research use and blanket consent as the worst options for future research with biospecimens (Tomlinson et al. 2015). In a recent qualitative study in a clinical setting, women supported the idea of broad consent and opposed the idea of notice without consent (Brown et al. 2015). Studies suggest older individuals are more comfortable than younger ones with broad consent (Trinidad et al. 2012), and certain populations are less accepting of broad consent (Moodley et al. 2014; Murphy et al. 2009). Studies also show that a minority (up to 40 percent) would not accept broad consent for future unspecified research use of their biospecimens (Kettis-Lindblad et al. 2006; McQuillan et al. 2006; Tomlinson et al. 2015; Treweek et al. 2009; Wendler 2006).

Commentators urge caution in relying on public perception data such as these. Botkin (2015) notes that many members of the public have a poor understanding of biomedical research. Rivera and Aungst (chapter 6 in this volume) and others (Ma et al. 2012) show that education, culture, and trust affect public opinion about consent for research use of biospecimens. Master cautions that public perception data are unreliable, inconsistent, and not alone determinative, yet "public opinion certainly matters on issues of research ethics policy" (Master 2015, 63).

## Considering Costs and Burdens of Obtaining Consent

Ethical analysis and available empirical data support obtaining broad or more specific consent in research settings. This raises the question of costs and burdens and whether consent approaches that offer more information, specificity of choice, and control to donors are better than broad consent.

As is indicated in figure 8.1, increasing the level of control offered to donors generally increases the costs and burdens of the consent approach. The costs of checklist, tiered, and study-specific consent exceed the costs and burdens of broad consent. Costs associated with requiring consent for each subsequent study or following a complex menu of choices include the need to track and monitor compliance for any re-use. These methods may also preclude subsequent use of biospecimens because of restrictions or ambiguity in the initial consent, especially if the limitations are vaguely worded or wide-ranging. For example, if donors limit their samples for "HIV-related" research, researchers and review bodies might disagree as to whether this allows or precludes research on common HIV co-morbidities such as weight loss or cancer, or studies of white cell dynamics or other retroviruses. In contrast, the costs of maintaining a system of broad consent should be relatively low, although there may be significant infrastructure and start-up costs.

Although the costs would be higher, the added benefits of other approaches to consent compared to broad consent seem minor, at least for a majority of donors, especially if sufficient oversight ensures that subsequent research is for purposes that do not conflict with donors' consent or known values. Although allowing donors to consent to specific studies for biospecimen research appears to give them some increased control, at the time of sample collection neither the donor nor the researcher know the range of possibilities for future research, including research that could have substantial social value (Eriksson et al. 2011; Korn and Sachs, chapter

4 in this volume). Little is gained by increased control given the low risks to donor welfare and uncommon circumstances in which research might conflict with donor values. In contrast, robustly engaging communities to provide input on methods of consent, and provide ongoing guidance or governance over the use of biospecimens could be respectful and valuable (Allyse et al. 2015; Bardill and Garrison 2015; Garrett et al. 2015).

Ethical reasons in favor of consent for research with biospecimens, the lower burden of broad consent over more specific approaches, and empirical data supporting its acceptability suggest that broad consent is a reasonable approach. Broad consent allows donors some control and can protect their interests, especially when coupled with oversight. Studies show that individuals are reassured that their interests will be protected when oversight mechanisms exist to review proposed research (Botkin et al. 2014) and protect their privacy (Ma et al. 2012). Finally, attending to public views shows respect for these views, and helps to ensure public acceptability and long-term viability of research. Broad consent may be problematic for those who want their samples used for limited types of studies. However, when collected with broad consent, those who do not want their biospecimens used for a broad range of unspecified future research can exercise their right not to donate. Spellecy (2015) notes that broad consent respects donors and is better than no consent, but could also facilitate autonomy because it is likely to be more acceptable and understandable than long, complex, and difficult to understand consent forms. In sum, this analysis suggests consent is preferable to no consent and broad consent is preferable to other approaches to consent for future biospecimen research in most circumstances, assuming that obtaining consent at the time of biospecimen collection is reasonable and not overly burdensome.

**Objections to Broad Consent**

Some commentators, however, question whether the rationale for any kind of donor consent weakens when recognizing the low risks and the great potential for biospecimen research to lead to useful scientific advances. Rhodes (2015) argues, for example, that contributing biospecimens is a social responsibility and policies should support maximal use of samples. Ballantyne (2015) says because biospecimens are important co-constructed social resources, consent may not be needed, especially for de-identified clinically derived samples stored in publicly available databases. Elsewhere in this volume, authors reason that because biospecimen research benefits all, considerations of justice (Korn and Sachs, chapter 4 in this volume) and communitarian and logistical concerns (Rivera and Aungst, chapter 6 in this volume) argue for creating public expectations that all should donate biospecimens, which would override the need to respect individual donor's autonomy and contrary wishes. Others point out that in a busy clinical environment it is naive to think that meaningful consent that shows respect for donors is even possible (Botkin 2015; SACHRP 2015). These arguments point to the need to evaluate feasible methods for consent for biospecimens.

Further, some object to the use of broad consent for research with biospecimens, and suggest other methods. Some suggest that public disclosure of information coupled with an opt-out option for donating clinical biospecimens for research would suffice (Botkin 2015; Ballantyne 2015; SACHRP 2015). Hornstein et al. (2015) suggest designating a power of attorney to protect donors' interests, since obtaining consent for future research is challenging. Loe et al. (2015) recommend "nudging" donors toward blanket consent, but if they are unwilling, offering, in a cascading fashion, other types of consent that offer more control. Plough and Holm (2015) argue that a more flexible model they call meta-consent is better for respecting donors, allowing individual donors to design their own consent, which could be broad or specific or in-between. Bardill and Garrison (2015), noting that broad consent is better than no consent, urge us to consider biospecimens as "loans in trust" rather than strict donations, and to allow more ongoing control to loaners and their communities. Overall, as others have noted, we believe that broad consent is better than no consent and there is ethical, practical, and empirical support for favoring it over other approaches to consent.

**Proposal for Broad Consent**

Workshop participants agreed that broad consent for collecting biospecimens for future research use is ethically permissible and, in many cases, optimal, when it includes the following three components:

- initial broad consent
- an oversight process for future research activities
- wherever feasible, an ongoing process of providing information or communicating with donors.

Together, these features promote the ethical acceptability and scientific value of future research with biospecimens and demonstrate respect for donors' contributions. Participants also agreed that broad consent is not appropriate in all cases, including circumstances where it might be ethically appropriate and consistent with governing regulations to use samples without consent, and circumstances where a population of donors should be able to limit future research use. An example of the former might involve a national pandemic or institutional outbreak that requires obtaining the widest number of samples possible. An example where specific consent might be appropriate is when populations have specific concerns regarding future uses, such as certain indigenous or rare disease populations (discussed in chapter 9).

Importantly, this proposal entails a greater degree of consent than required under US regulations, under which biospecimens can be used for secondary research purposes without consent or oversight as long as identifying information is removed and donor identity is unavailable to researchers (OHRP 2008, DHS et al. 2017).

**Initial Consent**

Consent alerts persons considering donation about the broad spectrum of research that could occur and promotes individual reflection on risks and benefits. To facilitate prospective donors' decisions, initial broad-consent forms should advise about possible future uses of biospecimens, any limitations, and existing processes of oversight that apply to specific studies. Workshop participants' opinions differed regarding what information should be included in initial consent. Most agreed that consent forms should briefly describe that samples will be stored and shared with a wide range of researchers and institutions and conditions under which sharing would be allowed, that general health information accompanies biospecimens, the possibility of commercial or therapeutic applications, the oversight process for reviewing proposed research, any potential for re-contact or ongoing communication, and the possibility of opting out of further future research on stored biospecimens.

Some workshop participants felt that initial consent should inform prospective donors that any research was possible unless specifically limited in the consent form or overruled by oversight bodies. Others thought consent should include a broad but non-exhaustive description of possible research topics, including the possibility of genetic analyses and keeping cells for indefinite periods, as well as other techniques. Some participants felt strongly that donors should be informed that certain kinds of sensitive or controversial research might be conducted and examples provided. Others felt that specimens from donors who gave broad consent should simply not be used for controversial research without further safeguards, such as oversight and sometimes re-consent. Botkin (2015) notes that concerns about controversial research may be exaggerated because biospecimen research objectionable to the public is exceedingly rare. Empirical research with current and potential future donors could play a pivotal role in identifying appropriate content for initial consent forms.

Broad consent should be sufficiently flexible to allow specific limitations based on the site circumstances or donor population. Specific limitations in initial consent should be based on evidence about research that a large number of people or certain populations find objectionable, such as certain types of reproductive research, e.g., human cloning or developing human embryonic stem cells from frozen embryos (Shepherd et al. 2007; McCarty et al. 2008). Additional limitations might be appropriate for particular donor groups, for example, rare disease groups may want to specifically limit their biospecimens for studies related to their disease. For other groups, including preferences about long-term sample disposition after death might appropriately respect culturally grounded values (Bardill and Garrison 2015). Certain groups might want to limit other specific research topics, for example studies of human evolution or genetic ancestry. Attention to formulating sufficiently clear and implementable descriptions is important. Individuals who are uncomfortable with the future research possibilities or feel that the limitations are insufficient can choose not to donate their specimens.

**Independent Oversight**
A process for approving and overseeing future biospecimen research will help ensure its ethical acceptability and scientific value, especially given the limitations of prospective consent for achieving these goals. Oversight adds further protections since future uses cannot all be explained, predicted, or are not known at the time specimens are collected, and donors entrust research institutions and biospecimen repositories to make reasonable decisions about future research on their behalf (Mongoven and Solomon 2012). Such oversight exceeds the scope of review required by the US Common Rule with respect to de-identified or coded biospecimens, but has been recommended by others (SACHRP 2015). Some might worry that such oversight will be too onerous. Others feel that the process of ethical oversight should be the emphasis for ethical research with biospecimens and initial consent should be consent for future governance (Boers et al. 2015).

Workshop participants envisaged a possible two-step oversight process to minimize burden. In the first step, an investigator briefly describes the proposed study and applies for release of samples. In the second, the oversight body designee reviews the application and either approves it or refers it for further review. Further review would only be prompted if the initial reviewer has concerns about the scientific value or rationale of proposed research, whether risks are more than minimal, whether research is inconsistent with specified consent limitations, or whether research might conflict with known donor values (Tomlinson 2013). Others recommend similar oversight processes to assess the scientific merits of each proposed study, justification for the use of specimens and data, and minimization of any possible harm (Rhodes 2015; SACHRP 2015). Oversight processes could be tailored to the specific research and governance characteristics of individual institutions or biospecimen repositories.

Where feasible, existing oversight bodies, such as institutional review boards or data access committees, could be used or adapted to provide oversight (Pulley et al. 2010), especially for research use of samples retained by investigators or institutions. Large biospecimen repositories might establish oversight bodies, and instructive lessons can be drawn from presently operational biospecimen repository review mechanisms (Bédard et al. 2009). Community representation on oversight bodies is important, including Community Advisory Boards (Mongoven and Solomon 2012; Lemke et al. 2010). Allyse et al. (2015) highlight the value of community engagement and community advisory board participation in the long-term ethical and socially responsible stewardship of biospecimens.

**Ongoing Communication with Donors**
Workshop participants recommended a commitment when feasible to periodically informing donors about research activities and emphasizing donors' right to withdraw from further distribution of their biospecimens. The structure and processes for such communication will differ according to technological capacity, donor characteristics, and so forth. One approach is a regularly updated website that describes research projects with stored samples and seeks donor comments (Kaye et al. 2012). Websites or IT systems could also integrate mechanisms for donors to withdraw consent for future use of their biospecimens, if they disagree with research topics or practices for which samples have been used. A robust system for ongoing communication mirrors proposals for "dynamic consent," a mechanism for ongoing communication with donors with an openness to revision of consent that can be built on either broad or specific initial consent (Wee et al. 2013; Kaye et al. 2015).

### Need for Future Research and Debate

Workshop participants acknowledged the need for further research on the adoption and implementation of broad consent for future biospecimen research. Comments and controversy about broad consent have demonstrated the need for more evidence of what is acceptable to the public, researchers, and institutions, as well as what is practically implementable in diverse settings. Research could help us better understand donor attitudes, the contours of the oversight process, and applicability of this proposal to international sample collection or certain donor groups, e.g., donors with rare or highly stigmatized disorders. Designing initial consent and oversight processes depends on understanding potential donors' views on research topics and practices that affect willingness to donate or are objectionable, and specifying information that donors want about possible future research and how they regard oversight processes. Data on experience in developing oversight principles and criteria, and circumstances that trigger wider review or modification of requests would also be useful.

There has also been significant debate about the practical challenges of implementing broad consent, especially in clinical settings (Edwards et al. 2014; Rivera and Aungst, chapter 6 in this volume). Further research is needed to explore the strategies, costs and burdens, and effects of implementation in a variety of settings. As many assume that broad consent would utilize an opt-in method, research to evaluate the ethics and practicality of opt-out methods of broad consent are needed. Research is also needed to help identify any ethical, practical, or policy grounds for distinguishing consent for health data research from consent for biospecimens research.

### Conclusion

Broad consent allows donors some control over the use of their biospecimens while minimizing costs and burdens on donors and researchers. Further, broad consent is consistent with majority views in surveys about research use of biospecimens. Workshop participants agreed that broad consent is ethically appropriate and preferable to no consent for the majority of biospecimen collections for research. The proposed framework for acceptable broad consent includes initial consent for a broad range of future research with some limitations, oversight of future research projects, and, when feasible, mechanisms for communicating with donors.

## Note

1. This chapter was adapted from Grady et al. 2015. The authors acknowledge workshop participants and *AJOB* authors who chose not to participate, as well as authors of the *AJOB* open peer commentaries. Views are those of the authors and do not reflect the official policies or positions of the NIH, the Department of Health and Human Services, or other institutions.

## References

Allyse, M., J. McCormick, and R. Sharp. 2015. Prudentia populo: Involving the community in biobank governance. *American Journal of Bioethics* 15 (9): 1–3.

Alter, H. J., Y. Nakatsuji, J. Melpolder, J. Wages, R. Wesley, J. W. Shih, and J. P. Kim. 1997. The incidence of transfusion-associated hepatitis G virus infection and its relation to liver disease. *New England Journal of Medicine* 13 (11): 747–754.

Ballantyne, A. 2015. In favor of a no-consent/opt-out model of research with clinical samples. *American Journal of Bioethics* 15 (9): 65–67.

Bardill, J., and N. Garrison. 2015. Naming indigenous concerns, framing considerations for stored biospecimens. *American Journal of Bioethics* 15 (9): 73–75.

Bédard, K., S. Wallace, S. Lazor, and B. Knoppers. 2009. Potential conflicts in governance mechanisms used in population biobanks. In *Principles and Practice in Biobank Governance*, ed. Jane Kaye and Mark Stranger. Ashgate.

Boers, S., J. van Delden, and A. Bredenoord. 2015. Broad consent is consent for governance. *American Journal of Bioethics* 15 (9): 53–55.

Botkin, J. 2015. Crushing consent under the weight of expectations. *American Journal of Bioethics* 15 (9): 47–49.

Botkin, J., E. Rothwell, R. Anderson, L. Stark, and J. Mitchell. 2014. Public attitudes regarding the use of electronic health information and residual clinical tissues for research. *Journal of Community Genetics* 5 (3): 205–213.

Brothers, K. B., and E. W. Clayton. 2012. Parental perspectives on a pediatric human non-subjects biobank. *American Journal of Bioethics Primary Research* 3 (3): 21–29.

Brothers, K. B., D. R. Morrison, and E. W. Clayton. 2011. Two large-scale surveys on community attitudes toward an opt-out biobank. *American Journal of Medical Genetics* 155A (12): 2982–2990.

Brown, K., B. Drake, S. Gehlert, L. Wolf, J. DuBois, J. Seo, K. Woodward, H. Perkins, M. Goodman, and K. Kaphings. 2016. Differences in preferences for models of consent for biobanks between black and white women. *Journal of Community Genetics* 7 (1): 41–49.

Chen, D. T., D. L. Rosenstein, P. G. Muthappan, S. G. Hilsenbeck, F. G. Miller, E. J. Emanuel, and D. Wendler. 2005. Research with stored biological samples: What do research participants want? *Archives of Internal Medicine* 165: 652–655.

DHHS (Department of Health and Human Services). 2011. Advance notice of proposed rulemaking (ANPRM). Human subjects research protections: Enhancing protections for research subjects and reducing burden, delay, and ambiguity for investigators. *Federal Register* 76 (143): 44512–44531.

DHHS. 2013. Modifications to the HIPAA Privacy, Security, Enforcement, and Breach Notification Rules under the Health Information Technology for Economic and Clinical Health Act and the Genetic Information Nondiscrimination Act; other modifications to the HIPAA rules. *Federal Register* 78 (17): 5566–5702. https://federalregister.gov/articles/2013/01/25/2013-01073/modifications-to-the-hipaa-privacy-security-enforcement-and-breach-notification-rules-under-the.

DHS (Department of Homeland Security) et al. 2015. Notice of proposed rulemaking (NPRM). Federal policy for the protection of human subjects. *Federal Register* 80 (173): 53933–54061. http://www.hhs.gov/ohrp/regulations-and-policy/regulations/nprm-home/.

DHS et al. 2017. Final rule. Federal policy for the protection of human subjects. *Federal Register* 82 (12): 7149–7274. https://www.gpo.gov/fdsys/pkg/FR-2017-01-19/pdf/2017-01058.pdf.

Edwards, T., R. J. Cadigan, J. P. Evans, and G. E. Henderson. 2014. Biobanks containing clinical specimens: Defining characteristics, policies, and practices. *Clinical Biochemistry* 47 (4–5): 245–251.

Eriksson, C., H. Kokkonen, M. Johansson, G. Hallmans, G. Wadell, and S. Rantapää-Dahlqvist. 2011. Autoantibodies predate the onset of systemic lupus erythematosus in northern Sweden. *Arthritis Research & Therapy* 13 (1): R30.

Garrett, S., D. Dohan, and B. Koenig. 2015. Linking broad consent to biobank governance: Support from a deliberative public engagement in California. *American Journal of Bioethics* 15 (9): 56–57.

Gaskell, G., H. Gottweis, J. Starkbaum, M. M. Gerber, J. Broerse, U. Gottweis, A. Hobbs, et al. 2013. Publics and biobanks: Pan-European diversity and the challenge of responsible innovation. *European Journal of Human Genetics* 21 (1): 14–20.

Grady, C., L. Eckstein, B. Berkman, D. Brock, R. Cook-Deegan, S. M. Fullerton, H. Greely, S. Hull, S., B. Lo, and D. Wendler. 2015. Broad consent for research with biological samples: Workshop conclusions. *American Journal of Bioethics* 15 (9): 34–42.

Hansson, M., J. Dillner, C. Bartram, J. Carlsson, and G. Helgesson. 2006. Should donors be allowed to give broad consent to future biobank research? *Lancet Oncology* 7: 266–269.

Hoeyer, K., B. Olofsson, T. Mjundal, and N. Lynoe. 2004. Informed consent and biobanks: A population based study of attitudes towards tissue donation for genetic research. *Scandinavian Journal of Public Health* 32: 224–229.

Hornstein, D., S. Nakar, S. Weinberger, and D. Greenbaum. 2015. More nuanced informed consent is not necessarily better informed consent. *American Journal of Bioethics* 15 (9): 51–53.

Kaye, J., L. Curren, N. Anderson, K. Edwards, S. M. Fullerton, N. Kanellopoulou, D. Lund, et al. 2012. From patients to partners: Participant-centric initiatives in biomedical research. *Nature Reviews Genetics* 13 (5): 371–376.

Kaye, J., E. A. Whitley, D. Lund, M. Morrison, H. Teare, and K. Melham. 2015. Dynamic consent: A patient interface for twenty-first century research networks. *European Journal of Human Genetics* 23 (2): 141–146.

Kettis-Lindblad, A., L. Ring, E. Viberth, and M. G. Hansson. 2006. Genetic research and donation of tissue samples to biobanks: What do potential sample donors in the Swedish general public think? *European Journal of Public Health* 16 (4): 433–440.

King, M. C., S. Wieand, K. Hale, M. Lee, T. Walsh, K. Owens, J. Tait, et al. 2001. Tamoxifen and breast cancer incidence among women with inherited mutations in BRCA1 and BRCA2. *Journal of the American Medical Association* 286 (18): 2251–2256.

Lemke, A. A., J. T. Wu, C. Waudby, J. Pulley, C. P. Somkin, and S. B. Trinidad. 2010. Community engagement in biobanking: Experiences from the eMERGE network. *Genomics, Society, and Policy* 6 (3): 35–52.

Lewis, L. A., G. J. Chang, R. S. Lanciotti, R. M. Kinney, L. W. Mayer, and D. W. Trent. 1993. Phylogenetic relationships of Dengue-2 viruses. *Virology* 197 (1): 216–224.

Loe, J., C. Robertson, and D. A. Winkelman. 2015. Cascading consent for research on biobank specimens. *American Journal of Bioethics* 15 (9): 68–70.

Ma, Y,H. Dai, L. Wang, L. Zhu, H. Zou, and X. Kong . 2012. Consent for use of clinical leftover biosamples: A Survey among Chinese Patients and the General Public. *PLoS One* 7.4.

Master, Z. 2015. The US National Biobank and consensus on informed consent. *American Journal of Bioethics* 15 (9): 63–65.

McCarty, C. A., D. Chapman-Stone, T. Derfus, P. F. Giampietro, N. Fost, and the Marshfield Clinic PMRP Community Advisory Group. 2008. Community consultation and communication for a population-based DNA biobank: The Marshfield

Clinic Personalized Medicine Research Project. *American Journal of Medical Genetics* 146A (23): 3026–3033.

McQuillan, G. M., Q. Pan, and K. Porter. 2006. Consent for genetic research in a general population: An update on the National Health and Nutrition Examination Survey experience. *Genetics in Medicine* 8 (6): 354–360.

Menikoff, J., J. Kaneshiro, and I. Pritchard. 2017. The Common Rule, updated. *New England Journal of Medicine.* On line first 10.1056/NEJMp1700736.

Mezuk, B., W. Eaton, and P. Zandi. 2008. Participant characteristics that influence consent for genetic research in a population-based survey: The Baltimore Epidemiologic Catchment Area follow-up. *Community Genetics* 11 (3): 171–178.

Mongoven, A., and S. Solomon. 2012. Biobanking: Shifting the analogy from consent to surrogacy. *Genetics in Medicine* 14 (2): 183–188.

Moodley, K., N. Sibanda, K. February, and T. Rossouw. 2014. "It's my blood": Ethical complexities in the use, storage and export of biological samples: Perspectives from South African research participants. *BioMed Central Medical Ethics* 15 (4).

Murphy, J., J. Scott, D. Kaufman, G. Geller, L. Leroy, and K. Hudson. 2009. Public perspectives on informed consent for biobanking. *American Journal of Public Health* 99 (12): 2128–2134.

NIH. 2014. Genomic Data Sharing Policy. https://gds.nih.gov/PDF/NIH_GDS_Policy .pdfhttp://gds.nih.gov/PDF/NIH_GDS_Policy.pdf.

Office for Human Research Protections (OHRP). 2008. Guidance on Research Involving Coded Private Information or Biological Specimen. http://www.hhs.gov/ohrp/ regulations-and-policy/guidance/research-involving-coded-private-information/.

Plough, T., and S. Holm. 2015. Going beyond the false dichotomy of broad or specific consent: A meta-perspective on participant choice in research using human tissue. *American Journal of Bioethics* 15 (9): 44–46.

Pulley, J., E. Clayton, G. R. Bernard, D. M. Roden, and D. R. Masys. 2010. principles of human subjects protections applied in an opt-out, de-identified biobank. *Clinical and Translational Science* 3 (1): 42–48.

Rhodes, R. 2015. Love thy neighbor: Replacing paternalistic protection as the grounds for research ethics. *American Journal of Bioethics* 15 (9): 49–51.

SACHRP (Secretary's Advisory Committee on Human Research Protections. 2011. FAQs, Terms and Recommendations on Informed Consent and Research Use of Biospecimens. *Secretarial Communications.* http://www.hhs.gov/ohrp/sachrp-committee/ recommendations/2011-october-13-letter-attachment-d/index.html.

SACHRP. 2014. Recommendations Regarding Research Uses of Newborn Dried Bloodspots and the Newborn Screening Saves Lives Reauthorization Act. http://

www.hhs.gov/ohrp/sachrp-committee/recommendations/2015-april-24-attachment-e/index.html.

SACHRP. 2015. Recommendations on the Notice of Proposed Rulemaking. http://www.hhs.gov/ohrp/sachrp-committee/recommendations/2016-january-5-recommendation-nprm-attachment-a/index.html.

Seeff, L. B., Z. Buskell-Bales, E. C. Wright, S. J. Durako, H. J. Alter, F. L. Iber, F. B. Hollinger, et al. 1992. Long-term mortality after transfusion-associated Non-A, Non-B Hepatitis. *New England Journal of Medicine* 327 (27): 1906–1911.

Seo, S., C. Renaud, J. M. Kuypers, C. Y. Chiu, M. L. Huang, E. Samayoa, H. Xie, et al. 2015. Idiopathic Pneumonia Syndrome after hematopoietic cell transplantation: Evidence of occult infectious etiologies. *Blood* 125 (24): 3789–3797.

Shepherd, R., J. Barnett, H. Cooper, A. Coyle, J. Moran-Ellis, V. Senior, and C. Walton. 2007. Towards an understanding of British public attitudes concerning human cloning. *Social Science & Medicine* 65 (2): 377–392.

Simon, C., J. L'Heureux, J. Murray, P. Winokur, G. Weiner, E. Newbury, L. Shinkunas, and B. Zimmerman. 2011. Active choice but not too active: Public perspectives on biobank consent models. *Genetics in Medicine* 13 (9): 821–831.

Spellecy, R. 2015. Facilitating autonomy with broad consent. *American Journal of Bioethics* 15 (9): 43–44.

Stegmayr, B., and K. Asplund. 2002. Informed consent for genetic research on blood stored for more than a decade: A population based study. *BMJ (Clinical Research Ed.)* 325 (7365): 634–635.

Tomlinson, T. 2013. Respecting donors to biobank research. *Hastings Center Report* 43 (1): 41–47.

Tomlinson, T., R. De Vries, K. Ryan, H. M. Kim, N. Lehpamer, and S. Y. Kim. 2015. Moral concerns and the willingness to donate to a research biobank. *Journal of the American Medical Association* 313 (4): 417–419.

Treweek, S., A. Donley, and D. Lieman. 2009. Public attitudes to storage of blood left over from routine general practice tests and its use in research. *Journal of Health Services Research & Policy* 14 (1): 13–19.

Trinidad, S., S. M. Fullerton, J. Bares, G. Jarvik, E. Larson, and W. Burke. 2012. Informed consent in genome-scale research: What do prospective participants think? *American Journal of Bioethics Primary Research* 3 (3): 3–11.

Tupasela, A., S. Sihvo, K. Snell, P. Jallinoja, A. R. Aro, and E. Hemminki. 2010. Attitudes towards biomedical use of tissue sample collections, consent, and biobanks among Finns. *Scandinavian Journal of Public Health* 38 (1): 46–52.

Valle-Mansilla, J., M. Ruiz-Canela, and D. P. Sulmasy. 2010. Patients' attitudes to informed consent for genomic research with donated samples. *Cancer Investigation* 28: 726–734.

Wee, R., M. Henaghan, and I. Winship. 2013. Dynamic consent in the digital age of biology: Online initiatives and regulatory considerations. *Journal of Primary Health Care* 5 (4): 341–347.

Wendler, D. 2006. One-time general consent for research on biological samples. *BMJ (Clinical Research Ed.)* 332 (7540): 544–547.

# 9 Evolving Consent: Insights from Researchers and Participants in the Age of Broad Consent and Data Sharing

Nanibaa' A. Garrison

In an era of rapid advancements in genomics research, huge amounts of data are being generated from biospecimens (coupled with electronic health information) and stored in large repositories for other researchers to access. As is described in chapter 8 of this volume, there is a trend among some ethicists and regulators to encourage researchers to obtain broad, rather than study-specific, consent from participants for future research with their biological materials collected in both research and clinical contexts—and the option for broad consent for future research use of identifiable biospecimens has now been formally adopted in the U.S. Common Rule, as amended in January 2017. However, some research participants may be uncomfortable with such broad consent. This chapter explores perspectives from several distinct groups of people, including certain indigenous tribes involved in recent litigation or repatriation regarding consent to genomics research, researchers and institutional review boards (IRBs) responding to changing regulations in light of that litigation, and more generally, biospecimen repository participants in the United States. Overall, there are a wide range of opinions and perspectives on the propriety of broad consent, but this chapter focuses on the cautious voices to further explore the types of reservations that some people may have about broad consent.

## Genetic Research with Tribes

Indigenous people have long been subjects of genetic research, from population-based studies to determine the biological basis of genetic diseases to research on human migration and ancestry. While some indigenous groups have become active participants in research, others have had negative experiences that have raised questions about the risks and benefits of participation, and raised awareness regarding what participants might want or expect from their participation in research. This section describes two cases in which tribes that participated in genetic research raised concerns about informed consent and whether secondary uses of their genetic information were appropriate given their initial consent for research.

### Diabetes Research with the Havasupai Tribe

As noted in chapter 3, the Havasupai Tribe's lawsuit against the Arizona Board of Regents has become the most publicized lawsuit involving biospecimens in recent history. In particular, this case highlighted concerns among tribes, academic researchers, and IRBs related to secondary uses of biological material and derived data that were collected under broad consent.

The Havasupai have high rates of Type 2 diabetes, affecting roughly half of adults in the population. In the early 1990s, they asked anthropologist John Martin from Arizona State University (ASU), with whom they had a long-standing, trusting relationship, to identify and invite a researcher to study the genetic contributions to the diabetes problem. He invited the geneticist Therese Markow to lead the initial investigations on the genetics of diabetes in the Havasupai. Accordingly, members of the Tribe agreed to participate in the Diabetes Project with both educational and research components; between 1990 and 1994, DNA samples were collected from approximately 400 individuals. At recruitment, informed consent was obtained by "making an oral statement to the donors" (Hart and Sobraske 2003, 1). Initially, it was not clear whether consent documents were presented, but later documents were written in English, not the tribe's native language, for participants to sign. The consent documents stated that "the purpose of the research is to study the causes of behavioral/medical disorders" (Hart and Sobraske 2003, 1); no specific mention was made to the Havasupai of future research beyond diabetes, but the lead investigator believed that the consent language authorized additional studies within the scope of behavioral and medical disorders. Dr. Markow and her collaborators failed to find a genetic link to Type 2 diabetes. The research team conducted subsequent studies on schizophrenia using the samples, and sent samples to other collaborators who used them for studies on migration and inbreeding that were not approved by the Tribe and not specifically consented to by the individuals who had provided the samples.

In 2003, Carletta Tilousi, a member of the Havasupai tribe and a participant in the study, learned at a dissertation defense that the samples from the diabetes study were being used for these unconsented studies on human migration and inbreeding. The tribe found these subsequent studies problematic because they challenged their beliefs about their tribe's origins and conflicted with the tribe's strongly held views about relatedness that prohibit inbreeding. Further, the tribe members were not aware that their samples might be used for studies on schizophrenia, a potentially stigmatizing condition, as they claim that intentions of the researchers to search for evidence of this disease were not disclosed at any time during the original study.

The Havasupai Tribe filed a lawsuit in 2004 over lack of informed consent and misuse of genetic materials, which signified an important moment in which research subjects took a stand and initiated legal action against researchers over misuse of DNA samples (Havasupai Tribe of the Havasupai Reservation v. Arizona Board of Regents and Therese Ann Markow 2004). Further interviews conducted for the Hart Report, an investigative report that was jointly sponsored by ASU and the tribe to investigate the claims, suggested that members of the research team failed to disclose their intentions to study schizophrenia to the Havasupai; some junior members of the research team had been asked by the principal investigator to search medical records for evidence of schizophrenia, despite lack of explicit approval from the tribe (Bommersbach 2008; Hart and Sobraske 2003).

The Havasupai Tribe issued a Banishment Order in 2003, effectively banning all academic researchers from entering the Havasupai reservation and halting research (Bommersbach 2008). The Inter Tribal Council of Arizona and the National Congress of American Indians each passed resolutions supporting the Havasupai Tribe. In 2002, the Navajo Nation independently issued a moratorium on genetic research studies based on community discussions that took place before the Havasupai case began, and held off on revising its stance on the moratorium in light of the Havasupai case until such a time that a robust policy could be developed (Navajo Nation Council 2002). Many other tribes took note of the case and restricted or banned certain kinds of research within their borders; some tribes continue to refuse participation in genetic research as a result.

Today, the Havasupai are still wary of genetic researchers although they have allowed a few from other fields to return to their reservation. Their case ultimately settled in April 2010, with the 41 tribe members who were named in the lawsuit receiving a total of $700,000 in monetary compensation from the Arizona Board of Regents, the tribe receiving funds for a clinic and school, and the remaining blood samples returned to the tribe (Harmon 2010). A small group of Havasupai members traveled to ASU to retrieve the DNA samples from the university's freezers and took them home and buried them with a ceremony.

To contextualize the importance of retrieving the samples and conducting a ceremony, consider two examples from members of other tribes that view DNA as sacred. In the 1990s, before the case and at the height of the controversy between Human Genome Diversity Project organizers and indigenous peoples worldwide, Dr. Frank Dukepoo, a Hopi geneticist, tried to help scientists understand indigenous peoples' perceptions of DNA with this explanation: "To us, any part of ourselves is sacred. Scientists say it's just DNA. For an Indian, it is not just DNA, it's part of a person, it is sacred, with deep religious significance. It is part of the essence of a person" (Petit 1998). For other tribes in the Southwest, extraction of DNA samples for research has left some individuals feeling "fragmented" such that return of samples would allow one to become whole again (Sahota 2014). Thus, upon the completion of the study, it would be appropriate for researchers to destroy the samples or return them to the participant to bury them with a ceremony in order to cleanse or protect themselves from not feeling whole. This is particularly important if a research participant passes away; if a participant has DNA in a freezer while his or her spirit transitions to the "next world," without it the person risks not being "whole" (Sahota 2014). This view was shared by some of the Havasupai members who felt that it was important to retrieve the DNA samples and have them returned to their homelands. The return of DNA samples signified closure, allowing the Havasupai to reconnect with the DNA that some viewed as an extension of themselves. Furthermore, the return of samples effectively transferred power and ownership of the DNA samples back to the Havasupai.

The Havasupai case was reported widely in numerous scientific publications including *Nature* magazine (Dalton 2004) and the *New England Journal of Medicine* (Mello and Wolf 2010), in addition to appearing on the front page of *The New York Times* (Harmon 2010) and in *Phoenix Magazine* (Bommersbach 2008). It drew attention to indigenous concerns over genetic material used for secondary studies, prompting discussions about what constitutes broad consent and conditions under which researchers are permitted to share samples and data. The Havasupai case also raised new ethical issues for researchers and regulatory boards. The settlement left no legal precedent surrounding informed consent, however, creating ambiguity for researchers and IRB chairs on best practices, particularly when broad language is used.

### Genetic Research with the Nuu-Chah-Nulth people

Unfortunately, the Havasupai case was not unique. The Nuu-Chah-Nulth people of Canada experienced similar research harms when, in the

1980s, members of the tribe agreed to participate in a genetic study on rheumatoid arthritis. More than 800 blood samples were collected by a genetic researcher, Dr. Ryk Ward, at the University of British Columbia (UBC) (Dalton 2002). When Dr. Ward later moved to the University of Utah and eventually to the University of Oxford, he brought the samples with him. He was unable to demonstrate a genetic basis for arthritis but shared the samples with his collaborators who used them for several subsequent research studies on human migration, HIV/AIDS, and drug abuse research, none of which were studies to which the tribe ever gave specific approval or individuals felt they had given their specific informed consent (Wiwchar 2004).

Although Dr. Ward published more than 200 papers, he did not return or report the results to the tribe. When the tribe found out about the subsequent research in 2000, many tribe members were furious and demanded explanations for why research was conducted with their samples without their permission. After Dr. Ward died suddenly in 2003, many UBC officials and researchers collaborated to retrieve blood samples from UBC and Dr. Ward's collaborators and returned them to the Nuu-Chah-Nulth in 2004, helping to avoid potential litigation or intense media scrutiny. During this process, and to mitigate future research harms, the Nuu-Chah-Nulth formed their own Research Ethics Committee to review and oversee all future research protocols within their community (Wiwchar 2004).

## Indigenous Peoples' Increased Oversight of Genetic Research

The attention garnered by both cases meant that many other tribes took note of the perceived research misconduct, and thus restricted or banned (or publicly supported tribes who banned) many types of research within their borders (National Congress of American Indians Policy Research Center 2006; *American Journal of Medical Genetics Part A* 2010). Other examples have garnered negative attention such as the Human Genome Diversity Project (HGDP), where geneticists were "racing the clock" to collect DNA samples from communities such as the Yanomami Indians of the Amazon rainforest, who were described as "literally becoming extinct" (Roberts 1991). In these cases, the communities viewed their samples as being misused and believed that consent was either uninformed or breached. These actions led to deep distrust of the scientific community among indigenous peoples and created a vivid narrative landscape full of reasons to not participate in future research. Despite the perceived misuses of samples, and even after the Havasupai settlement, Tilousi stated: "I'm not against scientific research. I just want it to be done right. They used our blood for all these studies, people got degrees and grants, and they never asked our permission" (Harmon 2010). Many tribes may support research, but would like to have a voice in dictating how their samples are used; as is explored in chapter 5 of this volume, data suggest that this perspective is not necessarily unique to indigenous peoples, but rather is shared by many Americans.

One major area of concern that resulted from the Havasupai case is the potential danger of seeking broad informed consent without full, specific disclosure of what types of research that might entail in the future, particularly since the original consent for the proposed diabetes research was broad and the research participants were not informed about their samples' specific uses. A broad approach to informed consent can overlook the unique concerns and perspectives of a participating community, which may make members of certain communities less likely to participate in research, and which could ultimately lead to exclusion of minority populations from the development of new scientific knowledge. Although a broad consent form may be logistically easier and may cover a broader range of potential uses of genetic material for future research studies, indigenous peoples—and some other populations—may be far less likely to participate in a study if their concerns are not addressed. The remainder of this chapter further explores concerns related to broad consent through the lens of indigenous peoples and communities.

## Data Sharing and Tribal Sovereignty

The Havasupai case has generated many discussions at the federal funding level, across universities, in professional societies, and in indigenous communities about the implications for tribes who may already be distrustful of research and for researchers who diligently adhere to ethical guidelines while working with indigenous communities. There are also questions more specifically about sharing samples with other researchers and institutions, which is a practice that has become more common due to the time-consuming and expensive nature of sample collection, and has helped to advance scientific discovery at a quicker pace.

Collaborators from one university-tribal partnership held a day-and-a-half-long workshop to discuss data sharing concerns of their tribal partners in order to explore pathways to building long-lasting, trusting relationships to promote ethical and engaged research (James et al. 2014). Some themes that emerged from this workshop were the importance of recognizing tribes as sovereign nations that may have their own research oversight boards, acknowledging that many tribes view knowledge and intellectual property as belonging to the collective group rather than an individual, in contrast to a western framework, and acknowledging that scientific practices such as data sharing may not be in a tribe's best interest. Further, tribal governments have special legal status with the federal government that requires specific accommodations for tribal rights; data-sharing policies may thus fall under a tribe's protection. Additional discussions in the workshop brought attention to the concerns, such as stigmatization or discrimination, that some communities may have regarding how their data are used for broad research purposes.

In genomics, much of the data that are derived from DNA samples are deposited into the online Database of Genotypes and Phenotypes (dbGaP). Researchers who receive federal funding are required to submit a data sharing plan stating when and how their study data will be deposited into dbGaP. Data submitted to the database must be stripped of all personal identifiers such as names, addresses, ZIP codes, birthdates, and social security numbers. Researchers who wish to use the individual-level genotype data in dbGaP must submit a request to the NIH Data Access Committee (DAC) with a description of how they will use the data, which members of their research team will have access, and how they will ensure that the data are used and managed properly. Additionally, the DAC oversees ongoing uses of data in order to ensure appropriate use and to reduce risks to the individuals from whom the data were derived.

Some tribes have raised concerns about the lack of tribal representation on the DAC, which is limited to federal employees, particularly because tribes are sovereign entities in the US. Sovereignty refers to the collective powers of a nation, such as the power to grant access to the population or to negotiate treaties between nations. Federally recognized tribes in the US have the power to self-govern, determine their own membership, and regulate tribal business. Thus, some tribal representatives feel that federal employees should not be the only oversight boards for how data in dbGaP are used if the databases include data derived from their tribes.

Working with indigenous communities also introduces the need for engaging communities in group discussions about research participation (Shelton 2012; Greely 2001). While personal autonomy and respect for individual persons remains important, when working with tribal communities, the additional layer of sovereignty and group consent has to be taken into account. At times, individual autonomy and tribal sovereignty may be in tension with one another, particularly when weighing group decisions with individual consent. Thus, researchers should be attentive to the concerns of tribal leaders as well as the individual tribe members.

Developing and disseminating good models for group consent and community engagement is crucially important for building trusting relationships. Sovereignty can also support tribes who want to work with researchers through the development of policies to guide research partnerships and providing guidance in proposed aspects of a study, such as data sharing, that may not conform to NIH guidelines by allowing for alternatives (James et al. 2008). This additional consent of the group can be time consuming, but helps to build trust with the community as well as ensuring a level of buy-in for the group and its members.

**Changes to the Common Rule**

Efforts to systematize how samples are used for genetic studies include the recent changes to the Common Rule to permit broad consent for secondary research with identifiable specimens (Office of Human Research Protections 2017) and the Genomic Data Sharing (GDS) Policy that requires consent for broad uses of genomic data (National Institutes of Health 2014). These efforts offer many opportunities to explore the concerns of indigenous research participants and researchers who work in these communities to further contextualize these issues and explore ways to maximize tangible benefits. Of note, the GDS Policy allows a data sharing exception for "compelling scientific reasons" (National Institutes of Health 2014). One of these reasons may be that some tribal laws may not permit broad data sharing.

As noted above, broad consent has the benefit of allowing researchers to conduct a wide range of studies without having to return to participants for new consent for each new project. Importantly, however, broad consent would fail to resolve the concerns raised by the Havasupai case, precisely because it does not demand study-specific consent to specific future research uses. In other words, broad consent would have allowed researchers to nonetheless conduct the potentially stigmatizing or offensive research, unless particular uses had been explicitly rejected. One group has reported that many people appreciate being re-contacted and asked for new consent (Ludman et al. 2010). While the practicality of seeking new consent is arguable, given the difficulties in re-contacting the sources of the samples in the future and the need to retain identifiers if such re-contact is desired, asking for specific consent to use existing samples for new studies can be a more cost-effective and less time-consuming option than initiating an entirely new study and sample collection. Moreover, seeking specific consent may be particularly important to preserving the trust of certain populations. Notably, even if the regulations permit options other than study-specific consent, tribes who have sovereign powers to negotiate other commitments regarding the samples could demand specific consent as a condition of engagement in research (National Congress of American Indians 2012; James et al. 2008). Furthermore, although the Common Rule allows institutions to utilize a single IRB (Office of Human Research Protections 2017), tribes may still require investigators to seek additional approval from their tribal or local IRBs.

## Broad Consent, Data Sharing, and Trust

Drawing from the perspectives of researchers, IRB chairs, and patients, this section argues for a need for greater discussion about responsible scientific practices involving broad consent. Some communities that are historically underrepresented in research may also be less willing to participate in research where their data are shared broadly. Greater discussion among these key players would provide an opportunity to address broad data sharing challenges that researchers face while also trying to build community trust and ultimately, greater inclusion of minority communities in research.

First, I will draw on empirical data from semi-structured interviews of 26 IRB chairs and genetic researchers at top NIH-funded research institutions that examines their concerns and perspectives on broad consent approaches, the appropriateness of using data for secondary uses, and sharing with other investigators. Furthermore, additional data are analyzed from published literature on patient perspectives on broad consent and data sharing to incorporate minority viewpoints and concerns that have not garnered much attention but may provide important insights regarding strategies to increase inclusion and build trust for greater participation in research. In general, the IRB chairs and researchers interviewed were concerned that many samples that were previously collected do not have appropriate broad consent for large-scale studies or for deposition into biospecimen repositories and databases (even if that is not a regulatory requirement).

After the Havasupai case was settled (in 2010), I conducted interviews with IRB chairs and genetic researchers to determine their perceptions around broad consent and data sharing in light of the issues raised in the case (Garrison 2012; Garrison and Cho 2013). The purpose of these interviews was to explore the extent to which the Havasupai case has affected practices for obtaining consent, sharing samples, and adapting to regulatory changes in genetics research. Through semi-structured interviews with IRB chairs and researchers, I identified a range of perspectives on the case and opinions about informed consent from IRB chairpersons and biomedical researchers engaged in genetics research involving human subjects. Their insights provide a deeper understanding of how concerns raised by the lawsuit affect decisions made by human research review boards and researchers in the context of genetic research. In particular, discussions about the impact of the case have revealed new awareness of informed consent issues, the importance of recognizing and addressing community concerns, and acceptable uses of biological materials that stem from broad informed consent agreements.

Difficulty may arise when vague wording in older informed consent language leaves the precise terms of the sample donation up to interpretation by researchers and IRBs. For example, the forms used in the collection of DNA from Havasupai members described using the samples for "behavioral/medical disorders," but the participants believed this would only include studies on diabetes. When the Human Genome Diversity Project was proposed, there was a systematic effort to revisit the consent forms from existing collections to determine whether old, legacy collections were appropriately consented for large-scale projects to examine diversity and human migration. Several researchers who were interviewed here reported relying on IRBs to interpret consent documentation and provide permission to use existing samples in a given study. In cases of older consents, some researchers view the IRB as a gatekeeper for gaining access to valuable DNA samples for new uses. One researcher described his experiences working with IRB officials at his institution this way: "I understand why investigators would go back to their IRBs and make impassioned cases to make certain interpretations on the language that might be a stretch. I am comfortable with that as long as they do it with the IRB's full acknowledgement and if that IRB is tuned in to the issues that would have been important to the participants at the time that the consent was obtained." This attribution of a gatekeeper role for the IRB transfers the role of specific consent from research participants to the IRB. If researchers can convince the IRB to accept their interpretation of informed consent documentation they may gain the ability to use these highly valuable existing samples in new studies.

Several researchers and IRB chairs described a tension between collecting and using samples for their studies and the general push toward depositing data into databases that are maintained and governed by the federal government such that researchers would relinquish their control over who will have access to the data and how they will be used. Some researchers have recognized the importance of obtaining broad consent from their research participants in order to feel comfortable with sharing them with the broader scientific community. One researcher stated: "We're part of several studies that have used dbGaP. That's a really important kind of compromise that you have to strike. There's a desire in the scientific community to have broad access to data." Other researchers highlighted concerns that there may be general lack of oversight when samples are shared with collaborative researchers. One researcher had received DNA samples from the Havasupai tribe through a collaborator at a time when there was little oversight and regulation as to how those samples were moved from one institution to another. Further, this researcher and his collaborator did not carefully examine the consent forms to ensure that the new proposed studies aligned with the participants' understanding of the proposed research. "I was really shocked when I realized that I had used that [Havasupai] data, because, you know, I get data sets from people all the time. ... [Anonymized data was] passed around much more freely. ... I need to be really more careful and look at the sources of the data that is sent to me." This statement highlighted the potential issues that may be associated with using DNA samples from collaborators, as researchers often did before the case was settled and as they continue to do today, and led one researcher to pause and think about the sources of data, including the consent forms and original intent of the research.

**Participants' Perspectives on Broad Consent**

The National Institutes of Health (NIH) has emphasized the inclusion of ethnic minorities, women, and children in biomedical research studies (Epstein 2007). However, after the controversies highlighted by cases such as the Havasupai case and the HGDP, many indigenous people have been hesitant to participate in research. On the other hand, by not participating in research, whether by choice or by being excluded from research, individuals and communities may not benefit from research. In recent years, more efforts have been made to engage research participants in research using methods based in community-based participatory research (CBPR) and other engagement practices. Using methodologies like CBPR, tribe members would be viewed as not only as participants in genetic research studies, but also as active contributors to the research process. Recent efforts have been made to bridge the divide between tribes and researchers (Jacobs et al. 2010), particularly in genetic research, such as by providing more educational opportunities for Native American students to pursue genetic research or research in biological sciences. Other examples may include engaging communities in collaborative efforts with researchers where community members become engaged in the research process and are able to provide input by allowing for respect and cultural exchange of ideas.

As is noted in chapters 6 and 8 of this volume, some studies have shown that research participants value being asked for permission for secondary uses of their samples, especially if they were to be stored in a federal repository (Ludman et al. 2010; Trinidad et al. 2011; Garrison et al. 2015). Furthermore, other studies have shown that research participants of diverse ethnic backgrounds have different opinions on informed consent that influence their expectations of the research process; participants are influenced by their cultural backgrounds and communities and by the perceived risks, burdens, and benefits of participation (Lakes et al. 2012). These studies have suggested that not all people are comfortable with broad consent for research. Further exploration into the reasons for not feeling comfortable and solutions for inclusion of diverse populations are important in order to make genomic research more equitable.

## Conclusion

A review of cases, interviews, and the literature suggests that many people do not readily endorse broad consent for secondary research on biospecimens—an important conclusion given the adoption of broad consent as an option for biospecimen research in the 2017 revisions to the Common Rule. The Havasupai and Nuu-Chah-Nulth cases illustrate important instances where communication, trust, and engagement were broken, ultimately resulting in further research on samples collected from these communities being halted. These cases illustrate that, for at least some communities, substantially more engagement is needed to respect the ethical demands of informed consent, sometimes going beyond what may be required as a regulatory matter.

While IRBs and researchers may recognize the value of broad consent (over specific consent for secondary research) and access to samples that allow the conduct of a wide range of studies, they voiced concerns over how to do this in an ethical and respectful manner. These case studies along with the interview data and published findings show that broad consent may not work for all. Indeed, broad consent might lead some potential research participants, regardless of whether they are from indigenous communities or not, to feel uncomfortable with participating in genomic research. Regardless of whether broad consent becomes the norm, it will be valuable to continue to engage research participants in discussions about the range of potential studies that may be conducted with their samples as well as making efforts to increase the general level of engagement with communities about research. Without taking into account the lessons learned from past mistakes, we may actually widen the disparities between those who decide to participate and those who do not, thus lessening the potential for benefit for those communities.

## References

*American Journal of Medical Genetics Part A*. 2010. After Havasupai litigation, Native Americans wary of genetic research. *American Journal of Medical Genetics A* 152A (7): ix.

Bommersbach, J. 2008. Arizona's broken arrow. *Phoenix Magazine*, November: 134. http://www.phoenixmag.com/lifestyle/200811/arizona-s-broken-arrow/.

Dalton, R. 2002. Tribe blasts "exploitation" of blood samples. *Nature* 420 (6912): 111.

Dalton, R. 2004. When two tribes go to war. *Nature* 430 (6999): 500–502.

Epstein, S. 2007. *Inclusion: The Politics of Difference in Medical Research*. University of Chicago Press.

Garrison, N. A. 2013. Genomic Justice for Native Americans: Impact of the Havasupai Case on Genetic Research. *Science, Technology & Human Values* 38 (2): 201–223.

Garrison, N. A., and M. K. Cho. 2013. Awareness and acceptable practices: IRB and researcher reflections on the Havasupai lawsuit. *American Journal of Bioethics Primary Research* 4 (4): 55–63.

Garrison, N. A., N. A. Sathe, A. H. Antommaria, I. A. Holm, S. C. Sanderson, M. E. Smith, M. L. McPheeters, and E. W. Clayton. 2016. A systematic literature review of individuals' perspectives on broad consent and data sharing in the United States. *Genetics in Medicine* 18 (7): 663–71.

Greely, H. T. 2001. Informed consent and other ethical issues in human population genetics. *Annual Review of Genetics* 35: 785–800.

Harmon, A. 2010. Indian Tribe Wins Fight to Limit Research of Its DNA. *The New York Times*, April 22, 2010.

Hart, S., and K. Sobraske. 2003. *Investigative Report Concerning the Medical Genetics Project at Havasupai*. Arizona State University Law Library.

*Havasupai Tribe of the Havasupai Reservation v. Arizona Board of Regents and Therese Ann Markow*. 2004. Coconino County, Superior Court of Arizona.

Jacobs, B., J. Roffenbender, J. Collman, K. Cherry, L. L. Bitsoi, K. Bassett, and C. H. Evans, Jr. 2010. Bridging the divide between genomic science and indigenous peoples. *Journal of Law, Medicine & Ethics* 38 (3): 684–696.

James, R. D., J. H. Yu, N. B. Henrikson, D. J. Bowen, and S. M. Fullerton. 2008. Strategies and stakeholders: Minority recruitment in cancer genetics research. *Community Genetics* 11 (4): 241–249.

James, R., R. Tsosie, P. Sahota, M. Parker, D. Dillard, I. Sylvester, J. Lewis, J. Klejka, L. Muzquiz, P. Olsen, R. Whitener, W. Burke, and Group Kiana. 2014. Exploring pathways to trust: A tribal perspective on data sharing. *Genetics in Medicine* 16 (11): 820–826.

Lakes, K. D., E. Vaughan, M. Jones, W. Burke, D. Baker, and J. M. Swanson. 2012. Diverse perceptions of the informed consent process: Implications for the recruitment and participation of diverse communities in the National Children's Study. *American Journal of Community Psychology* 49 (1–2): 215–232.

Ludman, E. J., S. M. Fullerton, L. Spangler, S. B. Trinidad, M. M. Fujii, G. P. Jarvik, E. B. Larson, and W. Burke. 2010. Glad you asked: Participants' opinions of re-consent for dbGaP data submission. *Journal of Empirical Research on Human Research Ethics* 5 (3): 9–16.

Mello, M. M., and L. E. Wolf. 2010. The Havasupai Indian Tribe case—Lessons for research involving stored biologic samples. *New England Journal of Medicine* 363 (3): 204–207.

National Congress of American Indians. 2012. *Genetics Research and American Indian and Alaska Native Communities*. http://genetics.ncai.org/.

National Congress of American Indians Policy Research Center. 2006. Resolution SAC-06-019. Supporting the Havasupai Indian Tribe in their Claim Against the Arizona Board of Regents Regarding the Unauthorized Use of Blood Samples and Research. National Congress of American Indians.

National Institutes of Health. 2014. Final NIH Genomic Data Sharing Policy. *Federal Register* 79 (167): 51345–51354. https://federalregister.gov/articles/2014/08/28/2014-20385/final-nih-genomic-data-sharing-policy.

Navajo Nation Council. 2002. Approving a moratorium on genetic research studies conducted within the jurisdiction of the Navajo Nation until such time that a Navajo Nation Human Research Code has been amended by the Navajo Nation

Council. In *HSSCAP-20–02*: Health and Social Services Committee of the Navajo Nation Council.

Office of Human Research Protections. 2017. Federal Policy for the Protection of Human Subjects. *Federal Register* 82 (12): 7149–7274. https://www.gpo.gov/fdsys/pkg/FR-2017-01-19/pdf/2017-01058.pdf.

Petit, C. 1998. Trying to study tribes while respecting their cultures: Hopi Indian geneticist can see both sides. *SFGate*. http://www.sfgate.com/news/article/Trying-to-Study-Tribes-While-Respecting-Their-3012825.php.

Roberts, L. 1991. A genetic survey of vanishing peoples. *Science* 252 (5013): 1614–1617.

Sahota, P. C. 2014. Body fragmentation: Native American community members' views on specimen disposition in biomedical/genetics research. *American Journal of Bioethics Empirical Bioethics* 5 (3): 19–30.

Shelton, B. L. 2012. *Consent and Consultation in Genetic Research on American Indians and Alaska Natives*. Indigenous Peoples Council on Biocolonialism 2012. http://www.ipcb.org/publications/briefing_papers/files/consent.html.

Trinidad, S. B., S. M. Fullerton, E. J. Ludman, G. P. Jarvik, E. B. Larson, and W. Burke. 2011. Research practice and participant preferences: The growing gulf. *Science* 331 (6015): 287–288.

Wiwchar, D. 2004. Nuu-chah-nulth blood returns to west coast. *Ha-Shilth-Sa Newsletter* 31 (25): 1–3.

# 10 The Ethics of the Biospecimen Package Deal: Coercive? Undue? Just Wrong? Or Maybe Not?

Ivor Pritchard and Julie Kaneshiro[1]

Is it ethical to require research subjects to agree to donate their biospecimens and associated data for future *companion* studies or for unspecified research as a condition of being eligible to participate in a specific, present clinical research study (the *primary study*)? Applying the ethical principles identified in the Belmont Report, this chapter will consider whether this type of package deal is coercive or unduly influential, and raise the question of whether such package deals should nevertheless be avoided even if they could be ethically justified.[2]

Over the past several years, research involving the use of biospecimens and associated data has become more commonplace, and such research can be expected to increase. Among the reasons for this increase is that advances in technology have made biospecimens a treasure trove for researchers, while generally posing little risk and burden to the individuals from whom the specimens were obtained. This has led sponsors to encourage researchers to make biospecimens and data available for other research studies (NIH Genomic Data Sharing Policy 2014). Package deals offer a means to do this. The question of whether such mandatory companion studies should be permissible is not confined to ethical considerations. The informed consent requirements under the Department of Health and Human Services (HHS) regulations for the protection of human subjects also address this issue through the informed consent requirement that "An investigator shall seek such consent only under circumstances that provide the prospective subject or the representative sufficient opportunity to consider whether or not to participate and that minimize the possibility of coercion or undue influence" (Code of Federal Regulations, title 45 sec. 46.116); this provision was substantively unchanged in the revisions to the rule finalized in January 2017 (Department of Homeland Security et al. 2017).

The archetypal case considered here is where individuals are being asked to participate in a clinical trial of an interventional treatment that offers the prospect of direct benefit, and their participation is conditioned on their agreeing that the identified biospecimens and identified data obtained for the clinical trial will also be made available for a companion study, or for unspecified future use. Furthermore, this discussion assumes that the companion study is indeed a second study distinct from the clinical trial offering the prospect of direct benefit, which is the primary study. In some cases, it may be arguable whether the collection and research use of particular biospecimens or certain data collected as part of a clinical trial should be considered part of the primary study because it is an important element of the primary study, or whether it is really part of a separate but related research study. We will assume that this is clear for the purposes of this analysis. We recognize that in marginal cases a researcher could try to game the system by claiming that the companion study is part of the primary study, and therefore that only one consent is necessary. While this chapter does not focus on the type of package deal where multiple study objectives are bundled into a single primary study, we acknowledge that this other type of package deal may raise some of the same ethical and regulatory questions as the archetypal case considered here.

### The Belmont Report, Current Regulations, and OHRP Guidance

The Belmont Report identifies three salient ethical principles to be considered in research involving human subjects. One of these is Respect for Persons, which implies honoring individuals' autonomy, and the application of which in research includes obtaining their informed consent to participate. The Belmont Report identifies information, comprehension, and voluntariness as being necessary components of informed consent, and describes voluntariness as follows:

**Voluntariness.** An agreement to participate in research constitutes a valid consent only if voluntarily given. This element of informed consent requires conditions free of coercion and undue influence. Coercion occurs when an overt threat of harm is intentionally presented by one person to another in order to obtain compliance. Undue influence, by contrast, occurs through an offer of an excessive, unwarranted, inappropriate or improper reward or other overture in order to obtain compliance. Also, inducements that would ordinarily be acceptable may become undue influences if the subject is especially vulnerable.

Unjustifiable pressures usually occur when persons in positions of authority or commanding influence—especially where possible sanctions are involved—urge a

# The Ethics of the Biospecimen Package Deal

course of action for a subject. A continuum of such influencing factors exists, however, and it is impossible to state precisely where justifiable persuasion ends and undue influence begins. But undue influence would include actions such as manipulating a person's choice through the controlling influence of a close relative and threatening to withdraw health services to which an individual would otherwise be entitled. (National Commission 1979, 9)

As was noted above, the regulations (both before and after the changes finalized in 2017) mandate that researchers seek consent in circumstances that minimize the possibility of coercion or undue influence. However, the regulations provide no further explanation of these terms. The HHS Office for Human Research Protections (OHRP) guidance and assessments reflect the Belmont Report's perspective. OHRP has issued frequently asked questions (FAQs) on the office's interpretation of coercion and undue influence, explaining that "coercion occurs when an overt threat or implicit threat of harm is intentionally presented by one person to another in order to obtain compliance." The FAQ also explains that "Undue influence, by contrast, often occurs through an offer of an excessive or inappropriate reward or other overture in order to obtain compliance" (Office for Human Research Protections 2007). In another FAQ, OHRP clarifies that compensating research subjects for participation in research is a common and acceptable practice, including compensation for the risks of research, so long as the IRB does not find the method and magnitude to constitute "undue influence" (Office for Human Research Protections 2013).

This interpretation of coercion is also reflected in OHRP's responses to specific questions since at least the early 2000s. OHRP's responses have generally addressed the archetypal case of this chapter, and in such circumstances, OHRP has strongly recommended that subjects enrolling in a clinical trial be given the distinct option—separate from consent to that trial itself—of agreeing or declining to have their biospecimens that are collected as part of the clinical trial banked for future unspecified research studies.

## OHRP Compliance Actions

Since 2000, OHRP has made very few determinations of non-compliance related to the requirement that informed consent be sought only under circumstances that minimize the possibility of coercion or undue influence. The most recent determination letter that related to the issue of companion studies occurred in 2012, and involved requiring individuals enrolling in a clinical trial that offered research subjects the prospect of direct benefit to also agree to have their identified biospecimens stored in a repository for future unspecified research (Office for Human Research Protections 2012). In this case, OHRP determined that the consent form for this study may have resulted in subjects being coerced into participating in open-ended future research involving their biospecimens. We are unaware of any OHRP determinations in which OHRP specifically determined that undue influence had occurred in violation of the regulations.

### Does Identifiability of Biospecimens Matter?

The regulatory requirement that informed consent be sought only under circumstances that minimize the possibility of coercion or undue influence does not create a clear and bright line for determining when an individual's voluntariness has been hindered to the point that the regulatory threshold has been violated. The identifiability of biospecimens might be a relevant consideration, for example, if there is reason to believe that individuals have a greater interest in controlling the use of their identified biospecimens than non-identified biospecimens because there is greater potential that their confidentiality will be breached if identified biospecimens are used in research. Alternatively, identifiability could be determined to be irrelevant, or at least not determinative, if consideration of risk is secondary to other factors, such as an interest in autonomy. If this were the case, and if individuals' autonomy interests were to determine the policy, then mandatory companion studies associated with a research study that offered subjects the prospect of direct benefit generally would be impermissible regardless of whether they involved the use of identified or non-identified biospecimens for future unspecified research. In contrast, if the principle of beneficence were to be given greater weight, then mandatory companion studies using non-identified biospecimens might be permissible in the interests of advancing science and the welfare of society.

As a means of respecting individuals' autonomy interest in controlling whether their biospecimens can be used in research, the scope of the regulations could be expanded to cover the secondary research use of all biospecimens regardless of identifiability, as was proposed in a Notice of Proposed Rulemaking (Department of Homeland Security et al. 2015). The proposal suggested that mandatory companion studies involving the use of non-identified biospecimens would not be permissible if the regulations were changed to bring this type of research under the purview of the rule. However, that proposal was not included in the final rule, re-opening the door to the possibility of creating such package deals (Department of Homeland Security et al. 2017).

**Are Package Deals Coercive?**

According to Aristotle, voluntary actions are actions in which the initiative for the action lies within the agent, who knows the circumstances (Aristotle 1962). Coercion either eliminates or limits the voluntariness of the coerced person's actions.

In the first instance, following Aristotle, Wertheimer (1987) points out that coercion may be alleged when someone's actions literally allow someone else no choice at all and the action is something the coerced person should or would not otherwise have chosen to do. In the context of companion studies, coercion would occur if the researcher gives the subject no choice at all about participating in the companion study and the subject would not have chosen to participate in it; note that this could occur if the researcher does not inform the subject of the companion study at all. This would be the case in the circumstance alluded to earlier, for example, where the researcher does not disclose his or her intention to create a biospecimen repository for future unspecified research using specimens collected as part of the primary study, and only seeks single consent for the primary study. Subjects would be similarly coerced if informed consent for the companion study is waived by an IRB, or if the companion study involves the banking of non-identified biospecimens, which is not regulated as human subjects research, and individuals are not informed about researchers' plans to create such a biospecimen repository. (Aristotle would say that the subjects' participation is completely involuntary.)

In the second instance, coercion also affects the voluntariness of an action when the person who is the object of coercion still has some kind of choice. Wertheimer's analysis of this sort of coercion involves the evaluation of both the proposal that creates the alternatives to be chosen from and the evaluation of the choice that is made (Wertheimer 1987). In Wertheimer's two-prong theory, B is under duress if A creates a choice that involves a threat such that B has no choice but to accept the proposal, and A acts wrongfully in creating the proposal. Wertheimer points out that typically coercion is alleged when B still has a choice, but the alternatives would normally be undesirable. (Aristotle would say that such choices are *partly* voluntary.)

In Wertheimer's view, coerced choices occur only when threats are made, not offers, because offers make you better off, and people are not said to be coerced if they are made better off. This implies the need to have a baseline for judging whether someone is coerced, because the evaluation of whether someone is made worse off has to be compared to how well off they were before the proposal is made. If A already has an obligation to do something good for B, and threatens not to do it unless B does X, then A's proposal not to meet the obligation becomes a threat, and is potentially coercive. For example, in the passage from the Belmont Report quoted above, the researcher denies access to needed clinical services the prospective subject is entitled to in order to induce participation.

For companion studies, on this view, there is no coercion involved in a package deal where the primary study involves a potential benefit and the researcher tells the prospective subject that participation in the companion study is mandatory. This is because here the package deal involves an offer, not a threat. The prospective subject is free to decline the offer of the potential benefit of participating. The companion study may involve risks. The risk of the companion study could be so minor as to be insignificant, in which case there is no cost attached to the offer of the package deal. In that case, the package deal is not coercive. If there is a significant risk to the companion study, this makes the offer of the package deal less attractive. But it is still not coercive, because the prospective subject can simply decline the offer, and not have to choose an alternative that makes him or her worse off; he or she would not lose anything to which he or she was otherwise entitled.

The conclusion that the package deal generally is not coercive would be otherwise if the prospective subject had a right to demand being included in a primary study that involves a potential direct benefit. In that case, the baseline for the assessment based on the two-prong theory would be different, because the prospective subject would already be entitled to participation in the primary study, and in that case attaching participation in the companion study would be coercive, because the subject must choose between losing the potential benefit of the study to which the subject was already entitled, or having to agree to the companion study. But under normal circumstances, this is not the case: so long as the recruitment and enrollment procedures are fair, researchers do not have any obligation to enroll any particular individual in a research study. Consequently, individuals do not have a right to participate in any particular research study.

In sum, the package deal is not coercive if the subject can accept or decline it. It is only coercive if the subject is left unaware of the companion study component, and would have objected to it.

## Do Package Deals Create Undue Influence?

The idea of undue influence is more subtle and complex. The Belmont Report, OHRP statements, and the standard bioethics/philosophy literature suggest that there are two general ways of construing undue influence. The first is the converse of coercion, where undue influence involves an offer of a benefit that is somehow untoward or inappropriate when it is attached to a given choice or action, because it entices the individual to decide on the basis of the value of the benefit, and ignore what might be harmful or undesirable about what the offer is attached to. The second way of construing undue influence is in terms of an influence which is inappropriate and somehow overwhelms the individual's will to do what they would otherwise choose to do. The source of this second kind of influence generally derives from the context in which the prospective subject's consent is sought, rather than the terms of the package deal itself. These two interpretations will be considered in turn.

If undue influence is interpreted in terms of inappropriate offers, then this is the interpretation that seems most relevant to the consideration of companion studies. Wertheimer's two-prong theory of coercion, which we have adapted to the circumstances of explaining undue influence offers, is helpful in this analysis. One could say that B (the prospective subject) is under undue influence if A (the researcher) creates a choice of an offer so good that B has no choice but to accept, and A acts wrongfully in creating the proposal. The idea here would be that A should not be attaching an arbitrary condition to B's participation in the primary study of also participating in the companion study, limiting B's choice to the package deal or nothing, and enticing the prospective subject to agree to the companion study through the offer of the potential benefit of the primary study. A is engineering an artificial connection between the two parts of the package deal where the benefit of one creates the incentive to agree to the other, when two distinct decisions about research participation could be made.

The two-prong approach draws attention to the idea that both A's and B's decisions are subject to moral evaluation. A is to be evaluated with respect to creating the constrained choices presented to B, and B may still be evaluated with respect to which of the alternatives is chosen. This suggests further consideration of both A's and B's motivations, because it may be that depending on the specific circumstances, the reasons A has for creating a package deal proposal may be different, and this may affect the evaluation of A's proposal.

Regarding A (the researcher's decision to create the package deal), on the one hand, it could be that this is merely a situation of convenience, where A (the researcher) wants to utilize the package deal because if the subjects of the primary study participate in the companion study as well, A saves a small amount of time and effort, and dismisses entirely the importance of the subject's willingness to agree. This would seem to be an unwarranted basis for utilizing the package deal mechanism. On the other hand, it could be that the companion study—while clearly distinct from the purpose of the primary study—represents an important and unique opportunity to find out something crucial about the subjects in the primary study. Perhaps only the subjects receiving the intervention in the primary study could provide biospecimens or information that would produce the evidence that would illuminate the hypothesis of the companion study, and nearly all of those subjects are needed to accumulate sufficient evidence to test that hypothesis. In such a situation, there appears to be a greater warrant for using the package deal to condition participation in the primary study on consent to participate in the companion study. However, this rationale would not generally apply to companion biospecimen repository studies for future unspecified research, because the special importance of the secondary research has not been set forth. Only if a biospecimen repository of those particular biospecimens represented a unique and scientifically valuable resource might such an argument apply.

At the same time, the second prong—B's (the prospective subject) decision—must also be considered. One issue here concerns whether the prospective subject has a proper appreciation of the value of the offer of the primary study. If the prospective subject overestimates the potential benefit, as is alleged under the notion of the therapeutic misconception, they might be overly inclined to accept the package deal based on that misapprehension. The package deal can only be unduly influential if the prospect of direct benefit is truly significant. Only if the potential benefit of the clinical trial represents a significant chance of being better off in an important way might the package deal represent an undue influence.

And how do we evaluate B's attitude toward the companion study? It would be one thing if the procedures of the companion study presented a significant risk in addition to the risks of the primary study; in that case, it would seem to be reasonable to say that even if the subject were willing to participate in the primary study, they might reasonably choose not to participate in the companion study. However, if the risks of the companion study were negligible, and the value of the companion study depended on having also participated in the primary study, then perhaps it might not be reasonable to refuse to participate in the companion study too, if there's nothing wrong with agreeing to it. In that instance, the influence exerted by combining participation into the package deal might not be *undue* influence.

There are three factors, then, that contribute to whether the level of influence created by the terms of the package deal are substantial enough to be undue:

1. the potential benefit offered by the primary study to a research subject
2. the strength or weakness of the researcher's rationale for mandating participation in the companion study
3. the kind of objection prospective subjects have to participating in the companion study, including consideration of the risks posed by the companion study.

Since the quality of the influence created by the package deal is a function of the combination of these three factors, some package deals create undue influence for prospective subjects and some package deals do not. The greater the potential benefit of the primary study, the weaker the rationale for combining the studies into a package, and the stronger the objections to participating in the companion study, the greater the likelihood that the package deal will create *undue* influence.

Though there is no clear line for determining when the degree of influence crosses over into being undue, the following type of package deal seems to clearly cross this line: a primary study that offers subjects a potential direct benefit which is likely to occur and could have a significant positive bearing on their health that is coupled with mandatory participation in a biospecimen repository study purely for convenience, and that prospective subjects would choose not to participate in if participation were not mandatory because the research institution has a long history of security breaches and a security breach involving their biospecimen would be troubling in a significant way. In such a case, the package deal is unduly influential because the researcher is creating a situation in which the potential direct benefit of the primary trial is only available if the prospective subject also participates in the companion study. Here the subject knowingly makes a decision about the companion study, but the condition of participation in the primary study is an unwarranted enticement to agree, especially since the potential benefit offered by the primary study is both likely and substantial.

However, it is reasonable to believe that package deals only rarely fit this description. The interest in having the research interventions be in equipoise makes it uncommon that primary studies will offer a likely and substantial potential benefit to subjects, and when such primary studies are conducted they would seldom be paired with a mandatory companion study, merely for convenience, that prospective subjects would find objectionable. In fact, the more typical scenario is probably that the factors are present but only to a lesser degree, or some factors are not present at all. In such circumstances the package deal would not create undue influence.

Consider the first factor, that is, the potential benefit offered to subjects by the primary study. As with threats, some offers are greater than others. The degree to which mandatory banking studies create undue influence will depend in part on the likelihood and nature of the potential benefit offered by the clinical trial. The more likely and substantial the possible benefit of the trial, the more undue the influence could be. The less likely and substantial the potential benefit of the trial, the less forceful the influence, which at some point would be small enough to fail to be undue.

Also, even if the prospect of direct benefit of the trial were significant, a package deal involving such a study might not be undue if an individual could obtain the study intervention that offered subjects the potential direct benefit outside of the research, and obtaining that intervention outside of the research was no more costly or more difficult than obtaining it through the research. In this case, individuals could choose to seek their care outside of the research study and avoid the package deal, and thus a mandatory biospecimen repository companion study would not result in undue influence because an individual's choice about whether to participate in the primary study would not be meaningfully constrained. This circumstance is addressed by the regulatory provision that includes a description of alternatives as an element of informed consent (Code of Federal Regulations, title 45, sec.46.116(a)(4)).

The second factor, the strength or weakness of the rationale for mandating participation in the companion study, is reflected in the first prong of the adapted version of Wertheimer's two-prong theory. As has been noted, whether the researcher's interest in creating the package deal is warranted will vary depending on the relationship of the companion study to the primary study. If there are convincing scientific reasons for using the specific biospecimens obtained from the primary study subjects for secondary research that holds great promise for advancing the understanding and treatment of a medical condition, it may not be undue to compel those subjects to participate in the companion study.

The third factor, the risks or other reasons that prospective subjects may have for objecting to the companion study, is reflected in the second prong of the adapted version of Wertheimer's theory. With regard to the degree of risk posed by a biospecimen repository companion study, it is arguable that the vast majority of biospecimen repository studies pose little risk to subjects, even if identifiable biospecimens and associated data are retained, and that the nature of companion studies is often entirely uncontroversial; in this case there may be no reason to oppose having one's biospecimen used in the companion study, which could contribute to making the package deal's influence not undue.

That said, it must be acknowledged that even if direct identifiers are removed from biospecimens and data before they are shared with researchers, it is not possible to eliminate the risk that individuals will be re-identified. Moreover, the risk of re-identification could grow over time as various sectors of our society collect greater amounts of information about individuals and this information becomes increasingly available to researchers and others. The level of risk would clearly be higher if the companion study involved additional collection of a biospecimen through a risky procedure, say, or if there were significant confidentiality risks involved in having biospecimens or identifiable data preserved in and accessed from a research repository. Or it could be that at some time in the future a novel and controversial research use of the biospecimens could arise that at least some individuals would find objectionable.

This line of thinking about why the prospective subject might want to decline the package deal requires an assessment of the prospective subject's decision making, and the idea that there needs to be a good reason to not want to participate in the companion study. The regulations and OHRP's statements do not venture into this territory. Currently subjects are allowed to decline to participate in research for whatever reason, or for no reason at all. The Belmont Report refers to something short of undue influence called "justifiable persuasion." This would seem to open the door to the possibility that the researcher could present something which might influence the prospective subject to choose to consent without being undue. The modifier of *undue influence* implicitly acknowledges that some influences are not *undue*. These would include reasons or offers which would favor participation in some way, and incline the prospective subject to agree, without overwhelming their capacity to choose. Pointing out that participation in the research might benefit others would be one such reason; offering an appropriate amount of compensation for participation would be another.

Let us now turn to the other interpretation of undue influence: the idea that somehow the prospective subject's will is overtaken by some external pressure. This generally seems to involve contextual factors other than the specific elements of the package deal. For example, the prospective subject might feel compelled to agree to both studies, because the researcher is the prospective subject's mentor or relative, who has a great personal influence over that individual, even though the individual would otherwise have declined. If the prospective subject accepts the package deal on this basis, this would represent undue influence. (This could, of course, even include a package deal in which there is no significant prospect of benefit to the subject in either the primary study or the companion study, no reason for mandating participation in the companion study, and legitimate reasons for wanting to decline participation in the companion study.)

There is a qualification that should be noted regarding both coercion and undue influence under either of the two interpretations discussed here. If the prospective subject is in fact motivated by his or her own personal desire to participate in the companion study, say, because the individual has a heartfelt desire to contribute to science, then that individual is not subject to coercion or undue influence, even if they are presented with a package deal including harm, or are offered a potential direct benefit, or their will would have been overwhelmed by the personal influence of the researcher. If they wanted to participate anyway, there would be no coercion or undue influence, because they were already inclined to consent.

But here again, as with the first interpretation of undue influence, the OHRP statements do not address what considerations might be presented to sway the prospective subject's decision without constituting undue influence. The reasons for avoiding such assessments are likely to include that they would require knowing what the state of mind of each subject is to discern how the considerations would influence them, which is not prospectively feasible. Moreover, since the regulation requires that the possibility of coercion or undue influence be *minimized*, the temptation of the package deal offer or the influence of the researcher should be avoided or reduced as much as possible, *in case* they might improperly affect a prospective subject's motivation. As an ethical matter, however, depending on the circumstances, some package deals involve undue influence, while others do not. We note that our characterization of how undue influence would be relevant to the analysis of a package deal represents a purely ethical perspective, not a regulatory assessment, and is not a statement of OHRP's views.

**Do Package Deals Inhibit Voluntary Informed Consent?**

Even if there is no coercion or undue influence, it may still be true that companion studies presented as package deals of combined consent to the primary study and the companion study serve to inhibit or thwart the prospective subject's ability to make fully voluntary informed decisions about participation, and should therefore be rejected. It may be that there are reasons the subject has for rejecting participation in the companion study which are unrelated to threats, offers, or other kinds of pressure, and are thus not a function of coercion or undue influence, but which might still influence the prospective subject to choose not to participate. For example, people could object to or be suspicious of the possible uses that the researcher's institution might make of their biospecimen or information, and wish to disallow the institution to store the biospecimen for those potential uses. They could have similar concerns about other institutions that might seek their biospecimen or information for other secondary research studies from the original research institution. Or they could have an autonomy interest in their own biospecimen or information, and wish to have the option of providing consent to bank their biospecimens or information for many types of future unspecified research studies, or have the option of providing specific consent for any secondary research study. Or they could simply wish to exercise their privacy in this regard.

Some, including other authors in this volume, have argued that in a learning health-care system people have some obligation to participate in research, and this might include mandatory participation in the kinds of companion studies discussed here (Faden et al. 2013). This would mean that in some cases the individual's personal preferences might be overruled on the basis of the arguments used to justify the obligation to participate in the learning health-care system's research, which appeal to the principles of beneficence and justice. The current HHS human subject protection regulations—in this respect, unchanged in the final rule published in January 2017—do not apply to research using non-identified biospecimens or non-identified data, provided the biospecimens and data were not collected for the specific, present research study. Such research is permissible without consent, regardless of individual preferences. In line with the learning health-care system view, the regulations could be interpreted or modified to permit secondary research use of *identifiable* specimens or data as a condition of participating in a primary study, even if the primary study offered the potential for direct benefit; this would represent a shift toward beneficence achieved through more rapid advances in science, and away from respect for persons insofar as prospective subjects would have to take or leave the package deal.[3] On this view, the tension between beneficence and respect for persons would be resolved in favor of beneficence.

Permitting package deals is not the only way to give beneficence more weight, however. Even if it is assumed that there is an obligation to participate in research, this need not mean eliminating consent; rather this could imply that prospective subjects be urged to consent by pointing out to them their moral obligation (Schaefer et al. 2009). This would be a type of reason that would constitute a good reason for consenting to participate in research, and would represent "justifiable persuasion" rather than undue influence. Other reasons could include appealing to the norm of reciprocity, or to altruistic motives beyond any obligation. Or some measure of compensation could be included, to sweeten the package deal.

This may imply a different model of the investigator/prospective subject interaction in the consent process, in which the investigator acts as a responsible advocate for the prospective subject's participation in the research, rather than as a more neutral presenter of the elements, risks and potential benefits of the research participation. It might lead to greater participation by subjects than under the neutral model, while engendering greater trust in the research enterprise than an arrangement where consent is bypassed. This would be a way of reconciling beneficence and respect for persons, rather than having respect for persons simply give way. Of course this could be more burdensome than eliminating consent, in terms of the time and effort involved in justifiable persuasion, and there would be disadvantages produced by some individuals declining to participate. Still, it is worth considering, both for package deals in particular and for research participation in general.

### Conclusion

The ethical analysis presented in this chapter implies that while some package deals are unethical, some package deals are indeed ethically permissible. If the researcher has good reasons for creating the package deal, and the prospective subjects either would have chosen to participate in the companion study regardless of the offer in the primary study or are not so tempted by the magnitude of the offer that they would ignore that they would have chosen not to participate in the companion study for some legitimate reason, the package deal is ethical. This position is more permissive than an across-the-board prohibition against the use of package deals.

At the same time, this chapter reveals the difficulties inherent in establishing a clear policy to adhere more closely to the line between the ethical and the unethical. To do that would require evaluation of the package deal at two levels: First, package deals would need to be reviewed on a case-by-case basis, to understand and evaluate the likelihood and significance of the potential benefit offered by the primary study, and to understand and evaluate why the researcher wants to create the package deal. Second, the package deal would need to be evaluated at the level of the individuals being offered the package deal, to assess prospective subjects' attitudes about participation in the primary and mandatory companion studies. This would involve far more time and effort than the application of an across-the-board ban, and evaluating the package deal for each individual subject would be practically impossible. Conceptually, such evaluations would also involve more general questions that the regulations have refrained from pursuing: What is the range of acceptable reasons for declining to participate in research? What reasons do not provide a sufficient basis?

Although OHRP does not have published guidance specifically about companion studies and coercion, undue influence and voluntary informed consent, if the existing fact-specific determinations and other communications were extended to infer a general position about package deals, the inference would appear to be that mandatory participation in companion studies involving human subjects research involving the prospect of direct benefit to subjects should seldom if ever be used, because it is likely to run afoul of coercion. In addition, even when there is no coercion or undue influence, our analysis has shown that some might argue that package deals may in other ways inhibit the voluntary informed decision of prospective subjects to participate in companion studies, and that this too should be avoided.

The view that OHRP appears to have embraced in certain case-specific circumstances is that respect for persons entails that any reason for declining to participate in research, or no reason, is reason enough. To take a contrasting view would pose much more far-reaching questions about when and why people should consent to participate in research, questions which unfortunately would involve an analysis far beyond the bounds of this chapter.

## Notes

1. The views expressed are those of the authors and are not necessarily those of the Office for Human Research Protections or the US Department of Health and Human Services.

2. Whether consent for banking biospecimens can be considered broad consent that applies to future secondary studies is beyond the scope of this chapter.

3. There is another way to look at the secondary research use of identified or non-identified biospecimens or data. This would be to say that once the specimens or data have been separated from the subject, that subject is not involved in the secondary research, and there is no decision for them to make. Taking such a position with regard to the research use of identified biospecimens and data would require revising the regulations.

## References

Aristotle. 1962. *Nicomachean Ethics*, revised edition. Bobbs-Merrill.

Department of Health and Human Services (DHHS). 1991. Code of Federal Regulations. Title 45. Basic HHS Policy for the Protection of Human Research Subjects. *Federal Register* 56: 28012–28018.

Department of Homeland Security et al. 2015. Notice of proposed rulemaking. *Federal Register* 80: 53933–54061.

Department of Homeland Security et al. 2017. Final rule. *Federal Register* 82:7149–7274.

Faden, Ruth R., Nancy E. Kass, Steven N. Goodman, Peter Pronovost, Sean Tunis, and Tom L. Beauchamp. 2013. An ethics framework for a learning health care system: A departure from traditional research ethics and clinical ethics. *Hastings Center Report Special Report* 43 (1): 16–26.

National Commission for the Protection of Human Subjects of Biomedical and Behavioral Research. 1979. The Belmont Report: Ethical Principles and Guidelines for Research Involving Human Subjects.

NIH Genomic Data Sharing Policy. 2014. *Federal Register* 79: 51345–51354.

Office for Human Research Protections. 2007/2013. Informed Consent FAQs. http://www.hhs.gov/ohrp/regulations-and-policy/guidance/faq/informed-consent/.

Office for Human Research Protections. 2012. Determination Letter to the University of Michigan. http://wayback.archive-it.org/4657/20150826185104/http://www.hhs.gov/ohrp/detrm_letrs/YR12/aug12a.pdf.

Schaefer, G. Owen, Ezekiel J. Emanuel, and Alan Wertheimer. 2009. The obligation to participate in medical research. *Journal of the American Medical Association* 302 (1): 67–72.

Wertheimer, Alan. 1987. *Coercion*. Princeton University Press.

# IV Special Populations and Contexts

# Introduction

Pamela Gavin

We are entering what promises to be a remarkable period of medical advances. Almost daily there are new discoveries of chemicals or biological products that have the potential to address previously unmet medical needs. The Food and Drug Administration (FDA) is approving new drugs at a record pace. The investment community is committed to companies that seek to develop new therapies, even though the risks of failure are high and it takes a long time for products to be tested and approved.

We who work daily with individuals and families challenged by rare diseases with unmet medical need are extremely encouraged by the remarkable advances we are seeing. Since 1983, when the Orphan Drug Act was passed by Congress, more than 500 new drugs have been approved in the United States, taking advantage of the incentives provided by the legislation.

While we are grateful for the medical progress to date and optimistic about the future, we know that the numbers define an enormous challenge. There are about 7,000 identified rare diseases, according to the National Institutes of Health. Most are genetic and as a consequence affect people starting in childhood. The number of physicians trained to diagnose and treat rare diseases is relatively sparse; often it takes years for a child, or adult, to receive an accurate diagnosis. Many rare diseases are treated with drugs that are used outside of FDA-approved labeling and, therefore, individuals are at increasing risk of being denied access to "off-label" uses as a result of strategies to limit reimbursement.

The topics that are covered in this part are especially relevant to patients with rare diseases. Biospecimens collected during newborn screening programs or during clinical diagnosis and treatment are especially relevant for people with rare diseases, as biospecimens can facilitate diagnosis and help focus treatment in these small patient populations. At the same time, making sure that patient privacy is ensured is of special importance to patients with rare diseases who do not want to be stigmatized, and consent often is difficult because rare diseases are often diagnosed in infancy or childhood.

Beyond that, we see every day how common it is for patients with rare diseases to be left behind as public policies are developed and implemented. Rare diseases are, almost by definition, not widely understood and thus there is a struggle to be sure that the policies that apply to patients with more common diseases are equally applicable—and equally applied— to patients with rare diseases. Inclusion is a very important concept for patients with rare diseases.

What is most important for us to recognize with regard to biospecimens and the consent and related issues is that the changing role of the patient in clinical studies has evolved. Patients are more active participants than ever before in understanding diseases, treatment options and how clinical trials are conducted, including the role of the investigator and how consent is provided. The patients feel empowered as never before. The new role of the patient is reflected very clearly within the rare disease community. In fact, individuals with rare diseases have provided a model for other patients in their engagement in their own diagnosis and treatment, especially with regard to research into targeted and personal therapies. Thus, while the issues of consent and patient involvement affect all patients, regardless of the prevalence of their disease, they affect patients with rare diseases in a special manner, and the chapters in this part reflect this reality.

In chapter 11, Aaron J. Goldenberg and Suzanne M. Rivera address the aforementioned issue head on. They note that studies of precision or personalized medicine will require large numbers of specimens and data from diverse sets of patients, and question whether there will be adequate representation of patients from underserved communities, including "historically underserved racial and ethnic populations." They further address the all-important access question, asking whether these "populations would have less access to the medical advances developed through research using their biospecimens, further exacerbating health disparities and impairing efforts to promote health equity in the United States and globally." As Goldenberg and Rivera point out, researchers and other repository officials must address significant ethical and practical issues as they design, establish, and implement "a large national cohort, or in other biorepositories aimed at achieving the promise of precision medicine." Goldenberg and Rivera identify and discuss these important issues. We who represent individuals with rare diseases recognize that many of the patients we work with are underserved, not only due to racial or ethnic group membership, but also due to the fact that rare diseases often are left behind.

# Introduction to part IV

In chapter 12, Jeffrey R. Botkin, Erin Rothwell, Rebecca A. Anderson, and Aaron J. Goldenberg address the issues associated with biospecimens collected from newborn screening programs. Newborn bloodspot screening is a public health program conducted by all states and territories in the United States to screen infants for 32 or more conditions. As Botkin et al. state, "The purpose of these programs is to identify affected infants before the onset of symptoms in order to reduce morbidity or mortality from the conditions through early intervention. ... After clinical screening is complete, residual blood remains from virtually every newborn screened for varying lengths of time. The residual blood can be analyzed for such things as genetic traits in the child, and for infectious disease or environmental toxin exposures for the mother and the child during pregnancy." The retention of these biospecimens raises ethical issues because of the potential for their use for additional research. Botkin et al. deal with the changing national policies on the storage and research use of newborn screening program biospecimens, with a special emphasis on "whether parental consent can be obtained for secondary uses of bloodspots in a way that does not create major obstacles to the conduct of research."

The rare disease community enthusiastically supports newborn screening programs because of the potential to identify, at birth, many of the genetic diseases that otherwise might go undetected or undiagnosed until extensive, irreversible, physical damage is done. The screening programs have extra value because they are conducted across the entire population, thus enabling us to gain understanding of the prevalence of certain diseases. It is important for all of us not only to support such programs but also to address the issues raised in this chapter with regard to retention of biospecimens and parental consent.

In chapter 13, Sara Chandros Hull addresses similar issues. Her chapter focuses on how best to obtain informed consent for collecting and sharing biospecimens, "especially given that plans for future research with these resources cannot be precisely described at the time that consent is obtained." She asks very directly the question that we all must address: "To what extent should individual research subjects be able to control the use of their samples and data for ongoing genomic research?" Hull points out, quite correctly, that this question is "especially important in the context of research on rare and undiagnosed genetic diseases, for which widespread sharing of elusive samples and data is needed to facilitate research addressing the lack of diagnostics and interventions for these populations."

We who work with individuals and families affected by rare diseases see every day the difficulty of doing research when there are inconsistent standards applying to such basic matters as informed consent. It is important for clinicians and researchers alike to be able to share information readily, without violating patients' privacy.

We, within the rare disease community, agree with Hull's conclusion: "As a starting point, the limited data available at this time suggest that broad consent for the donation of samples and data—that is, a consent process that includes the provision of initial information about broad future research plans, coupled with ethical oversight and provision of ongoing information to participants about future research uses—is consistent with the values of patients with rare diseases and their relatives."

In chapter 14, Geoffrey Lomax and Heide Aungst deal with considerations for the use of biospecimens in induced pluripotent stem (iPS) cell research. Such research has great potential for advancing the development of new therapies. As with all research, there are ethical issues that need to be addressed. Thus far, the researchers have, as Lomax and Aungst point out, "developed a robust donor consent process with the opportunity for donors to withdraw their specimens before cell manipulation." Particular attention is given to "the commercial use of cell lines, the ability to re-contact donors about clinically significant findings, the possibility of donor withdrawal, and the potential for sensitive uses of derived lines." The aim of this process is "to maintain a collection of cell lines that can be used broadly in research for perpetuity. This process fulfills the aspiration of donors to contribute to science and medicine while ensuring the repository meets high ethical and legal standards." Lomax and Aungst provide us with a perspective that carries implications for addressing similar issues in medical research.

Underlying all the chapters in this part are a few fundamental questions:

- How can we see to it that barriers to research are removed to ensure that the promise of medical advancement can be achieved?
- How can we establish rational, uniform, and practical public policies that will serve to advance research while serving the needs and privacy of patients?
- What special considerations must be taken into account to ensure that the needs of the most vulnerable patients are considered?

These chapters provide unique and insightful perspectives on these essential issues, and set the stage for further discussion and hopefully the development of a consensus. We within the rare disease patient community advocate for sound public policies that serve to both advance research as well as protect the interests of the patients we represent.

# 11 Biorepositories and Precision Medicine: Implications for Underserved and Vulnerable Populations

Aaron J. Goldenberg and Suzanne M. Rivera

The goal of precision or personalized medicine is to identify targeted prevention and treatment options for subsets of patients on the basis of unique biological, behavioral, and social determinants of disease. Achieving this goal requires large numbers of specimens and data from diverse sets of patients in order to identify the biological mechanisms for disease across populations. However, there are growing concerns that the cohorts that have been established for purposes of research do not include enough patients from underserved communities, including historically underserved racial and ethnic populations (Haga 2010; Precision Medicine Initiative Working Group 2015). Additionally, as future biorepositories are developed, including the planned establishment of a national research cohort, there are concerns that the exclusion of underserved communities will persist. Even if underserved communities were adequately represented in biospecimen and data repositories, there are further concerns that these populations likely will have less access to the medical advances developed through research using their biospecimens, further exacerbating health disparities and impairing efforts to promote health equity in the United States and globally.

In his 2015 State of the Union address, President Barack Obama introduced a proposal to invest $215 million into a multifaceted research initiative focused on precision medicine (Collins and Varmus 2015). The president's Precision Medicine Initiative (PMI) will include, among other projects, a $130 million effort to build a national research cohort, the *All of Us Research Program*, containing biospecimens and medical data from at least one million Americans (PMI All of Us Research Program 2016). This initiative represents only one research program in which precision medicine may be driven by the collection, storage, and use of large numbers of biospecimens. However, it is a fitting example of a large-scale, population-focused project in which the composition of the cohort itself is a major factor. As study results become generalizable, it will be essential that the biospecimens and data for health research be truly representative of the US population. In September 2015, the National Institutes of Health's Precision Medicine Working Group released a report outlining a guide for the development and initiation of a national cohort (Precision Medicine Initiative Working Group 2015). One of the emphases of this report was the importance of collecting a diverse enough sample to not only accurately represent the population of the United States, but also for providing research opportunities to address the health needs of underserved populations. The Working Group recommended that the "PMI-Cohort Program leverage America's rich diversity, thereby increasing scientific rigor that accounts for individual variation while providing opportunities to advance research that may reduce disparities and move towards health equity" (Precision Medicine Initiative Working Group 2015).

In this chapter, we will address topics associated with biospecimen repositories and their implications for underserved and vulnerable populations. We will identify a number of ethical and practical issues that researchers and other repository officials ought to address in the design, establishment, and implementation of a large national cohort, or in other biorepositories aimed at achieving the promise of precision medicine. These issues are grouped within the chapter to identify and discuss elements of biorepository design and participant recruitment, specimen and data storage and management, and specimen and data research use and translation. We also will discuss potential implications for precision medicine if underserved communities are not adequately included in local and national cohorts and, in turn, the short-term and long-term consequences for the health of underserved communities if they have limited access to benefits of biospecimen research. Additionally, we will discuss how increasing an awareness of both the perspectives of potential donors and the implications of lower participation among underserved and vulnerable populations can promote policies and governance structures for biorepositories that enhance dialogue and foster trust between researchers and communities, increase representativeness of our data and biospecimen repositories, and ensure that *all* patients can access the potential benefits of translational research using biospecimens and data.

## Background: Previous Scholarship on Research and Underserved Populations

Much has been written and said about the degree to which racial and ethnic minorities are underrepresented in clinical trials and other kinds of biomedical research studies (Schmotzer 2012; Wendler et al. 2006; George, Duran, and Norris 2014). The consequences of this underrepresentation are significant, as research participation is associated with better health outcomes for individuals (Wallerstein 2006), and greater heterogeneity of study data yields results that are more broadly representative of the whole population (Ramos and Rotimi 2009; Haga 2010).

Some have attributed the participation disparity to attitudes and beliefs held by minority-group members that resulted from widely publicized incidents of disrespect, exploitation, and even abuse in the name of science (Schmotzer 2012; Wendler et al. 2006; George, Duran, and Norris 2014). Others have reported that subtle and overt biases by investigators are responsible for exclusion of ethnic and racial minorities, citing structural barriers to research enacted by researchers who may not be sensitive to minority group members' needs, a lack of diversity among researchers, and blatantly discriminatory practices (Isler et al. 2013; Bustamante, De la Vega, and Burchard 2011).

A similar pattern of underrepresentation is evident in the literature about minority group participation in biorepositories and other kinds of specimen collection protocols. Numerous studies have found that members of racial and ethnic minority groups are less likely to be included in specimen collection protocols and, consequently, in biorepositories (Partridge 2014; Dang et al. 2014; Ewing et al. 2015; Halverson and Ross 2012; Streicher et al. 2011).

This problem of underrepresentation in research is especially pernicious in light of the fact that minority-group members are disproportionality burdened with chronic diseases such as cancer, diabetes, and heart disease (Ramos and Rotimi 2009; Yancey et al. 2006). For this reason, there are both moral and scientific imperatives to proactively recruit donation of biological specimens (and other health-related data) from members of racial and ethnic minority groups (Cohn et al. 2017; Wallerstein 2006).

Specimens can be obtained for research through prospective study protocols designed to collect samples and associated data, or they can be re-directed to research uses following collection for diagnostic and treatment interventions. Numerous studies have shown that minority-group members are more willing to donate specimens prospectively and to allow their clinically collected samples to be retained and used for research when they have the opportunity to learn about the research to be performed (Partridge 2014; Dang et al. 2014). An even more effective strategy for including minority-group members in specimen-based and other forms of biomedical research is to include them in the conceptualization of research questions and study design processes. (Wallerstein 2006; Cohn et al. 2015;

Partridge 2014; Dang et al. 2014). In the following section we will discuss how biorepositories should address the need for inclusion in the development of their resources.

### Biorepository/Cohort Design and Participant Recruitment

Addressing a biorepository's potential implications for underserved and vulnerable populations requires an examination of its original goals, project design, and recruitment processes. In the subsections that follow, we will discuss a number of issues biorepositories must address in the initial design and implementation stages of their development.

### Defining the Purpose of the Biorepository

If the goal of the repository is to provide specimens and data for general medical research, with the intention of maintaining a *representative sample* of a regional or national population, then the focus concerning underserved groups might be limited to ensuring that traditionally marginalized populations are included within the sample. These actions ultimately would need to include enhancing educational and consent procedures that address the priorities and needs of these populations, expanding opportunities for all willing donors to participate, and increasing the potential for community engagement about the purposes of the repository (Plunkett, Kearns, and Caplan 2015; Johnson et al. 2011; Ewing et al. 2015; Murphy et al. 2009). In this case, the goal would be to ensure that groups that historically have been underrepresented in medical research are included, thus increasing the representativeness of the sample and allowing for research to examine outcomes across different racial or ethnic populations.

Alternatively, if a goal of the biorepository is to increase *health equity* through its work, then inclusiveness in recruitment is necessary to ensure inclusiveness in participation, but may not be sufficient to understanding and ultimately reducing the health disparities experienced by historically underserved or vulnerable populations. Meeting this goal would require the promotion of research questions that focus on identifying or addressing health inequalities within underserved populations. This would include framing "calls for proposals" to use samples and data from the repository and orienting scientific review processes in ways that would promote these types of projects. To do so effectively, it also may be necessary to oversample underserved populations to ensure adequate statistical power within studies utilizing the biorepository's resources (Cohn et al. 2017). Doing so would not only ensure increased inclusion of participants from traditionally underrepresented populations, but could help to better elucidate the fundamental genetic, social and or environmental causes of disparities by increasing the inclusion of patients actually experiencing those disparities.

Determining the goals of a biorepository may be influenced by a number of factors, including the proposed size of the resource, the kinds of researchers involved in the endeavor, whether the repository is aimed at a particular disease or set of diseases, and, of course, where donors are being recruited. Smaller repositories may not be able take on both population representativeness and equity-focused aims. However, larger resources like the proposed PMI national cohort should strive to achieve both types of research goals. By combining both existing cohorts and adding new participants, a national biorepository should be able to design studies that reflect a generalizable and representative sample of the US population and create opportunities for more focused research on health disparities within and between particular populations, including underserved communities. Such an approach would promote the ethical principle of justice.

### Defining Diversity and Underserved Populations

As biorepositories and large cohort studies begin to develop their recruitment plans, they must recognize the importance of defining how their projects are conceptualizing diversity. Generally, most efforts to address diversity focus on the inclusion of racial and ethnic minority populations when designing research plans. This definition appropriately reflects the need to account for the inclusion of populations who historically have been under-represented in research, and who also experience health disparities at exceedingly higher rates. However, if the goal of precision medicine is not only to collect a representative sample of the American public, but also to ensure that all citizens have the right and the opportunity to participate in research, then a broader definition of diversity may be needed to ensure the inclusion of other underserved populations, including the disabled, low-income white populations, the chronically uninsured, immigrants, and non-English speakers. Doing so would not only ensure a broader diversity of participants within the biorepository itself, but might also help encourage broader sets of possible research questions that may be important to these communities. For example, it also may be valuable to include different conceptualizations of diversity beyond demographic differences. As precision medicine focuses on major illnesses with large available cohorts such as heart disease and cancers, there is the potential it may be more difficult to promote research on rare diseases with few available patients. It will be crucial that the US cohorts and other biorepositories include patients with rare conditions to ensure that researchers interested in those diseases also will have an opportunity to use the power of a larger data set for the benefit of disease communities that have traditionally struggled to maintain a robust research program (Lochmüller et al. 2009).

**Designing Recruitment and Consent Processes**
The design of proactively inclusive recruitment strategies and consent processes will be necessary for ensuring that underserved populations and other vulnerable groups are able to participate and that their concerns about having their samples and data in a biorepositories are adequately addressed. When developing consent processes, educational materials, and the consent forms, it is critical that institutions establishing a repository include language that addresses any concerns that underserved or vulnerable populations may have about sample and data donation. However, before collecting these materials, researchers must engage local communities to assess their concerns, questions, and expectations about research and the biorepository. This may be done through a number of engagement strategies including town hall meetings and community dialogue sessions, and more traditional research approaches like focus groups and in-depth interviews (Haldeman et al. 2014; Lemke et al. 2010). There also have been a number of recent studies that have developed tools for assessing underserved communities' attitudes regarding biospecimen repositories and participation in research (Johnson et al. 2011; Streicher et al. 2011; Cohn et al. 2015).

In addition to the recruitment and consent processes, the venues in which participants are recruited can play an important role in promoting the participation of underserved populations in a biorepository. This means recruiting not only at major medical centers, but also at community hospitals, places of worship, local businesses, and community health centers to ensure that all populations have the opportunity to participate. However, many of these venues may not have the time, personnel, or other resources necessary to implement a consent process for a biorepository on their own given financial and other constraints that are already challenges to providing care in underserved settings. It will be important that biorepositories, like the PMI national cohort, build resources into their design and implementation for addressing the resource needs of underserved settings, for example providing multilingual staff to help facilitate an education and consent process for potential donors.

## Biorepository Maintenance and Management

Once participants have been recruited, the ethical and social obligations of the biorepository do not end. The next section of this chapter will discuss aspects of the ongoing storage and management of specimens and data within a biorepository that may relate to participants from underserved and or vulnerable populations.

### Protecting Participants' Rights

Ensuring the security of the samples and data, maintaining confidentiality of participant identifiers and personal health information, and preserving the rights of donors throughout the life of the bank are important steps in sustaining a trusting donor-researcher relationship. Individuals from underserved communities who may already feel vulnerable in health-care settings—historically, underserved and minority populations—have had lower levels of trust in both the health-care system and researchers, and these trends have continued within the context of biorepository research (Dang et al. 2014; Murphy et al. 2009). When making decisions about whether to participate in a bank, underserved populations' willingness to donate samples and data may be mediated by their own trust in researchers and the relationships these institutions have with local communities. The PMI will be further challenged given that the collections of samples and data will be maintained by a government agency. This may have further implications for gaining the trust and buy-in of underserved populations who may be wary of government agencies having access to their samples and data. It will be important that the consent process and education about a repository explicitly lay out the protections in place to ensure participant privacy and when possible, their rights to withdraw their samples and data from the repository at any time. It also will be valuable to describe state and federal legal protections against genetic discrimination, such as the Genetic Information Nondiscrimination Act (GINA) (US Department of Health and Human Services 2009).

### Managing Access to Samples and Data

Once a biorepository establishes a diverse set of samples and data, they also will need to consider how the use of those resources may impact underserved communities. As was discussed above, the repository will need to decide whether research questions that pertain directly to the health of underserved communities will be a priority for research uses of specimens or data. Alternatively, banks will need to at least consider how health disparities or the health needs of underserved communities are approached within the larger scope of research projects associated with the repository. In addition, the bank should work with researchers to address the potential negative implications of approved research protocols on underserved populations, including the potential stigmatization or discrimination of populations based on the results of studies using the bank's resources.

To address these challenges, it will be important for biorepositories to explore how local donors or community leaders may be involved in the development of research questions being proposed using samples and/or data. Alternatively, repositories may wish to consider how to include the values of underserved communities when making decisions about what studies can use bank resources. This may include a formal mechanism to involve representatives of underserved communities on boards or other official roles within the scientific review process, or could be achieved through engagement with local communities to assess local values that may help to drive criteria for study proposals. Either way, it will be important to ensure that the attention paid to the needs of underserved populations are not simply addressed in the donor recruitment process, but continue throughout the collection, storage and management, and research processes of the biorepository.

**Issues Related to Combining Cohorts**

To adequately power population-based research aimed at promoting precision medicine, researchers may need to access large networks of biorepositories that combine new collections of specimens and data with multiple existing cohorts across the country. By linking new and old collections of specimens and data, large networked cohorts will be able to expand existing resources, while at the same time developing new mechanisms to involve wider populations in research studies. While this has never been done on a national level, there are a number of examples where existing biorepositories have linked together to share specimens and data. Though doing so can create a much larger database for research purposes, there may be a number of unanticipated implications for underserved communities. First, many biorepositories have been developed within the context of local communities and with local community engagement and buy-in (Simon, Newbury, and L'Heureux 2011; McCarty et al. 2011). Further, these local repositories may have goals or research protocols aimed at addressing local health issues. While linking smaller banks into a larger network of specimens and data may expand the types of possible research, there is a danger that these local research needs may no longer be a focal point for the bank and thus receive less attention and fewer resources. Additionally, as banks join larger networks of biospecimen repositories it may be more difficult to include local expertise, community leaders, and engaged populations in governance decisions or setting research priorities.

## Research Use and Translation of Results

Given that the purpose of developing and maintaining a biorepository is to support ongoing research, the potential impacts of that research on underserved and vulnerable population must be addressed. In this section we will discuss the implications of policies regarding human subject protections, issues related to return results to biorepository participants, and whether study findings ultimately will benefit underserved communities.

## Research Oversight and Control

The regulatory landscape for collection and use of biospecimens is evolving. In 2015, changes to the Federal policy for protection of human research subjects were proposed that would have altered the circumstances under which biospecimens could be re-used for research when collected for other purposes. These proposed changes promoted the adoption of a "broad" consent model form for specimen collection that would allow donors to agree to future unspecified research uses of their biological material. One of the critiques of this approach had to do with concerns about equitable access to participation in biorepositories and the principle of justice. Because a "broad" consent form for specimen collection probably would be used only in established academic medical centers, people who might otherwise be willing to allow their biospecimens to be shared and used for future unspecified purposes may not have the opportunity. For example, people who seek care in community settings and those who are medically underserved due to a lack of health insurance or geographical distance from large medical facilities may not be approached to give "broad" consent for tissue collection because the federally mandated process may be too unwieldy for local clinics and other places where the underserved seek care. Thus, underserved populations may be left out of biorepository participation because only large, academic medical centers will have the resources and expertise to execute this approach to specimen collection. This could result in an exacerbation of the already socially unequal participation of underserved groups in important biorepositories and may reduce our ability to learn important things about the health of members of minority groups.

The promise of precision medicine will be only a dream without a robust and diverse representation of human biospecimens, available to answer questions about all people who need health care—not just the ones with access to university teaching hospitals. However, addressing this criticism of a "broad" consent approach should not be simply to either leave populations that seek care outside of academic health centers out of efforts like a national cohort, nor should it be to promote a system in which all choice is taken from potential participants and de-identified samples and data are used without the knowledge of underserved patients.

Although the Final Rule published in January, 2017 did not include the most controversial aspects of the 2015 proposal, a perception persists that there is conflict between honoring underserved donors' values and optimizing the number of specimens that may be collected for research purposes. This perception is based on the idea that underserved patients' values are at odds with the promotion of research, an assumption that has not been evidenced by empirical research. Rather, many underserved communities have expressed interest in participating in research and worry that "their" community may be left behind (Buseh et al. 2013; Streicher et al. 2011). More research is needed to better understand the concerns and hopes of underserved patients regarding research participation in a national cohort or some other large research repository. Once these values are better understood, we can design recruitment processes that maximize respect of donor values, attention to underserved community needs, and the promotion of representative research cohorts. After meaningful engagement with interested parties, we may discover that we in fact need a fuller consent process, or a "broad" consent model, or an opt-out approach in which samples and data may be used for research unless patients specify that they do not want their information added to a repository.

In our ongoing discussions about polices that impact specimen and data use, it will be crucial that policy makers consider the unique time and resource needs of community clinics and other institutions providing health care to underserved populations. Further, whether consent is required or not ultimately is less important than promoting trust in researchers, ensuring that patients are educated about the potential uses of their specimens and the choices they have about participation, and empowering interested patients to actively engage in research efforts.

**Return of Results**
One of the most contentious issues facing biorepositories today is the decision about when and how to return results of genomic research using stored

biospecimens and data (Wolf et al. 2012; Forsberg, Hansson, and Eriksson 2009; Bledsoe et al. 2012). Historically, results from research studies using stored samples were rarely returned because samples and data were generally de-identified or anonymized before they were distributed to researchers. However, given the potential importance of certain types of genomic research results for the health of research participants and/or their families, there recently have been calls to encourage researchers to give actionable results back and to develop governance structures within a biorepository to facilitate that process (Murphy et al. 2008). Doing so, however, will require biorepositories to consider a number of issues around communication of complex results, participant values and their desire for results, and the potential need for genetic counseling and follow-up services. All of these concerns present unique challenges or needs with regard to returning results to underserved communities.

First, many participants who belong to medically underserved communities may have a lower level of health literacy, and, therefore, require additional time and resources for understanding complex genetic testing results. Additionally, some participants may have lower levels of access to regular medical services, and therefore would be at a further disadvantage with regard to actionable results that may require increased screening or other medical interventions. Biorepositories will need to be prepared to address how genetic information might be incorporated into the health care of participants who are already dealing with poorer health and health disparities within their communities. And biorepositories must be aware of concerns about the potential negative social implications that returning results may have, such as the stigmatization of, or discrimination against underserved individuals or communities based on genetic results.

Nevertheless, there may be an assumption among researchers or biorepositories that because of these preexisting social and health conditions, that underserved participants may not want results back or that returning results would only serve to exacerbate the disparities already being experienced within their communities. However, recent data show that even when experiencing health disparities, underserved communities want to be included in research and in turn to learn about and benefit from study results (Partridge 2014; Wendler et al. 2006; Halverson and Ross 2012; Ewing et al. 2015). Biorepositories must engage with underserved participants and their communities to better understand individual interest and needs with regard to returning results. They must also work with local health services, genetic counselors, and community leaders to ensure that when results are returned, it is done so with regard to social implications and accompanied by proper access to follow-up services and care.

### Translation of Benefit Back to Communities

In addition to the return of individual results, there are questions about whether the findings from studies using samples and data from biorepositories will be used to address the health concerns of underserved populations or if any medical innovations developed out of this research would be available in underserved settings. In other words, will underserved communities ultimately benefit from the research using their donated samples and data? This is not an entirely new concern. There have been a number of studies showing that underserved individuals have felt less willing to participate in research because of fears that the results of the studies would never benefit their communities, or that they would have less access to new medical technologies and resources developed as a result of the research (Dang et al. 2014; Isler et al. 2013).

While these questions raise fundamental issues about equitable access to new medical discoveries, they also raise an important question about the responsibility of a biorepository to direct and oversee the translation of results from research using samples and data from their collection. Traditionally, biorepositories have served primarily as a research resource with little influence on how research results may get translated from bench to bedside. However, as universities, Clinical and Translational Science awardees, and other local, regional, or national research networks increasingly are building biorepositories that are meant to help achieve the larger goals of their institutions research programs, there may be greater impetus for repositories to take a more active role in directing the ways in which research findings are translated into clinical medicine and public health. This will be particularly true for the PMI, as the development of the national cohort is meant to be an integral part of the program's overall goals of increasing research to develop new approaches to personalized health care for all patients. It is essential that the goals of recruiting diverse populations in a biorepository should not end merely with inclusion itself and a representative sample. Rather, including underserved populations in cohorts should also be accompanied by a commitment to include research questions related to the health of those communities when making decisions regarding how resources from the biorepository are used. This should also help facilitate research to address health disparities and promote access to precision medicine across all populations.

## Conclusions

Solutions to the problems described above ultimately will require building trust and ongoing relationships with members of underserved populations and their communities. As we have discussed, it also will require addressing the implications for underserved and vulnerable populations at all stages of the establishment and implementation of a biorepository. A recent NIH Request for Information on how to best "Address Community Engagement and Health Disparities" within the PMI resulted in suggestions from a number of biorepository stakeholders on how to engage communities, including "involving them in discussions about research questions for the cohort, and ensuring that logistical details are addressed (e.g., covering transportation costs, providing childcare during research visits) to encourage participation" (National Institutes of Health 2015).

Repositories do not have to tackle these challenges alone. Many institutions, including existing biorepositories and health data repositories, already have established active Community Advisory Boards (CABs). The roles of these boards vary. Some serve simply as conduits for engaging a wider set of community members and organizations and educating them about the work of a repository. However, many CABs are significantly more involved in the overall governance of the repository. Some CABs may play a role in helping to establish trust within the local community, may provide advice on developing inclusive recruitment and consent processes, and may have an influence on the kinds of research for which samples and/or data may be used. These CABs represent an invaluable opportunity to better connect with underserved communities, hear concerns and expectations from potential participants, and engage in a dialogue about the future of precision medicine. While new regional or national CABs may be needed for the growth of the PMI, it would be beneficial to learn from existing local CABs about communities' previous experiences with research, and to connect quickly with community members and groups who already have thought deeply about the implications of biorepositories and who already may have established trusting relationships with local researchers and institutions involved in the larger PMI efforts.

Clearly, dialogue alone is not sufficient to facilitate truly meaningful and just inclusion of underserved and vulnerable populations into a biorepository. Engagement can help institutions build governance structures that acknowledge and address the importance of tailoring a repository's standard operating procedures to better meet the needs of underserved populations and to build a resource that is both representative and equitable.

## References

Bledsoe, M. J., W. E. Grizzle, B. J. Clark, and N. Zeps. 2012. Practical implementation issues and challenges for biobanks in the return of individual research results. *Genetics in Medicine* 14 (4): 478–483.

Buseh, Aaron G., Sandra M. Underwood, Patricia E. Stevens, Leolia Townsend, and Sheryl T. Kelber. 2013. Black African immigrant community leaders' views on participation in genomics research and DNA biobanking. *Nursing Outlook* 61 (4): 196–204.

Bustamante, Carols D., Francisco M. De La Vega, and Esteban Gonzalez Burchard. 2011. Genomics for the world. *Nature Reviews Genetics* 475: 163–165.

Cohn, E. G., G. E. Henderson, and P. S. Appelbaum. 2017. Distributive justice, diversity, and inclusion in precision medicine: What will success look like? *Genetics in Medicine* 19 (2): 157–159.

Cohn, Elizabeth Gross, Maryann Husamudeen, Elaine L. Larson, and Janet K. Williams. 2015. Increasing participation in genomic research and biobanking through community-based capacity building. *Journal of Genetic Counseling* 24 (3): 491–502.

Collins, Francis S., and Harold Varmus. 2015. A new initiative on precision medicine. *New England Journal of Medicine* 372 (9): 793–795.

Dang, Julie H. T., Elisa M. Rodriguez, John S. Luque, Deborah O. Erwin, Cathy D. Meade, and Moon S. Chen. 2014. Engaging diverse populations about biospecimen donation for cancer research. *Journal of Community Genetics* 5 (4): 313–327.

Ewing, Altovise T., Lori A. Erby, Juli Bollinger, Eva Tetteyfio, Luisel J. Ricks-Santi, and David Kaufman. 2015. Demographic differences in willingness to provide broad and narrow consent for biobank research. *Biopreservation and Biobanking* 13 (2): 98–106.

Forsberg, J. S., M. G. Hansson, and S. Eriksson. 2009. Changing perspectives in biobank research: From individual rights to concerns about public health regarding the return of results. *European Journal of Human Genetics* 17 (12): 1544–1549.

George, Sheba, Nelida Duran, and Keith Norris. 2014. A systematic review of barriers and facilitators to minority research participation among African Americans, Latinos, Asian Americans, and Pacific Islanders. *American Journal of Public Health* 104 (2): e16–e31.

Haga, S. B., and L. M. Beskow. 2008. Ethical, legal, and social implications of biobanks for genetics research. *Advances in Genetics* 60: 505–544.

Haga, S. B. 2010. Impact of limited population diversity of genome-wide association studies. *Genetics in Medicine* 12 (2): 81–84.

Haldeman, Kaaren M., R. Jean Cadigan, Arlene Davis, Aaron Goldenberg, Gail E. Henderson, Dragana Lassiter, and Erik Reavely. 2014. Community engagement in US biobanking: Multiplicity of meaning and method. *Public Health Genomics* 17 (2): 84–94.

Halverson, C., and L. Ross. 2012. Engaging African-Americans about biobanks and the return of research results. *Journal of Community Genetics* 3 (4): 275–283.

Isler, Malika Roman, Karey Sutton, R. Jean Cadigan, and Giselle Corbie-Smith. 2013. Community perceptions of genomic research: Implications for addressing health disparities. *North Carolina Medical Journal* 74 (6): 470–476.

Johnson, Vanessa A., Yolanda M. Powell-Young, Elisa R. Torres, and Ida J. Spruill. 2011. A systematic review of strategies that increase the recruitment and retention of African American Adults in genetic and genomic studies. *Journal of the Association of Black Nursing Faculty* 22 (4).

Lemke, Amy A., Joel T. Wu, Carol Waudby, Jill Pulley, Carol P. Somkin, and Susan Brown Trinidad. 2010. Community engagement in biobanking: Experiences from the eMERGE Network. *Genomics, Society, and Policy* 6 (3): 35.

Lochmüller, H., S. Aymé, F. Pampinella, B. Melegh, K. A. Kuhn, S. E. Antonarakis, and T. Meitinger. 2009. The role of biobanking in rare diseases: European Consensus Expert Group report. *Biopreservation and Biobanking* 7 (3): 155–156.

McCarty, C. A., A. Garber, J. C. Reeser, and N. C. Fost. 2011. Study newsletters, community and ethics advisory boards, and focus group discussions provide ongoing feedback for a large biobank. *American Journal of Medical Genetics* 155A (4): 737–741.

Murphy, E., and A. Thompson. 2009. An exploration of attitudes among black Americans towards psychiatric genetic research. *Psychiatry* 72 (2): 177–194.

Murphy, J., J. Scott, D. Kaufman, G. Geller, L. LeRoy, and K. Hudson. 2008. Public expectations for return of results from large-cohort genetic research. *American Journal of Bioethics* 8 (11): 36–43.

Murphy, Juli, Joan Scott, David Kaufman, Gail Geller, Lisa Leroy, and Kathy Hudson. 2009. Public perspectives on informed consent for biobanking. *American Journal of Public Health* 99 (12): 2128–2134.

National Institutes of Health. 2015. Request for Information: NIH Precision Medicine Cohort—Strategies to Address Community Engagement and Health Disparities. https://grants.nih.gov/grants/guide/notice-files/NOT-OD-15-107.html.

Partridge, E. E. 2014. Yes, minority and underserved populations will participate in biospecimen collection. *Cancer Epidemiology, Biomarkers & Prevention* 23 (6): 895–897.

Plunkett, C., L. Kearns, and A. Caplan. 2015. Worth the Money? Paying to Ensure a Representative Cohort in the Precision Medicine Initiative. *Health Affairs* Blog, July 30, 2015. http://healthaffairs.org/blog/2015/07/30/worth-the-money-paying-to-ensure-a-representative-cohort-in-the-precision-medicine-initiative/.

Precision Medicine Initiative (PMI)-All of Us Research Program. 2016. https://www.nih.gov/research-training/allofus-research-program.

Precision Medicine Initiative Working Group. 2015. The Precision Medicine Initiative Cohort Program—Building a Research Foundation for 21st Century Medicine. https://www.nih.gov/sites/default/files/research-training/initiatives/pmi/pmi-working-group-report-20150917-2.pdf.

Ramos, E., and C. Rotimi. 2009. The A's, G's, C's, and T's of health disparities. *BMC Medical Genomics* 2 (1): 29.

Schmotzer, G. L. 2012. Barriers and facilitators to participation of minorities in clinical trials. *Ethnicity & Disease* 22 (2): 226–230.

Simon, C. M., E. Newbury, and J. L'Heureux. 2011. Protecting participants, promoting progress: Public perspectives on Community Advisory Boards (CABs) in biobanking. *Journal of Empirical Research on Human Research Ethics* 6 (3): 19–30.

Streicher, Samantha A., Saskia C. Sanderson, Ethylin Wang Jabs, Michael Diefenbach, Meg Smirnoff, Inga Peter, Carol R. Horowitz, Barbara Brenner, and Lynne D. Richardson. 2011. Reasons for participating and genetic information needs among racially and ethnically diverse biobank participants: A focus group study. *Journal of Community Genetics* 2 (3): 153–163.

US Department of Health and Human Services. 2009. "GINA": The Genetic Information Nondiscrimination Act of 2008. https://www.genome.gov/Pages/PolicyEthics/GeneticDiscrimination/GINAInfoDoc.pdf.

Wallerstein, N. B. 2006. Using community-based participatory research to address health disparities. *Health Promotion Practice* 7 (3): 312–323.

Wendler, D., R. Kington, J. Madans, G. Van Wye, H. Christ-Schmidt, L. A. Pratt, O. W. Brawley, C. P. Gross, and E. Emanuel. 2006. Are racial and ethnic minorities less willing to participate in health research? *PLoS Medicine* 3 (2): e19.

White House Press Office. 2015. Fact Sheet: President Obama's Precision Medicine Initiative. January 30. https://www.whitehouse.gov/the-press-office/2015/01/30/fact-sheet-president-obama-s-precision-medicine-initiative.

Williams, Brett A., and Leslie E. Wolf. 2014. Biobanking, consent, and certificates of confidentiality: Does the ANPRM muddy the water? In *Human Subjects Research*

*Regulation Perspectives on the Future*, ed. I. Glenn Cohen and Holly Fernandez Lynch. MIT Press.

Wolf, S. M., B. N. Crock, B. Van Ness, F. Lawrenz, J. P. Kahn, L. M. Beskow, and W. A. Wolf. 2012. Managing incidental findings and research results in genomic research involving biobanks and archived data sets. *Genetics in Medicine* 14 (4): 361–384.

Yancey, A. K., A. N. Ortega, and S. K. Kumanyika. 2006. Effective recruitment and retention of minority research participants. *Annual Review of Public Health* 27:1–28.

# 12 The Ethical Management of Residual Newborn Screening Bloodspots

Jeffrey R. Botkin, Erin Rothwell, Rebecca A. Anderson, and Aaron J. Goldenberg

Biospecimens and associated data are enormously valuable resources for biomedical research.[1] Biospecimens used in research come from two primary sources: specimens collected for research purposes (and often used for secondary research) and specimens collected in the course of clinical care. Residual dried bloodspots (DBS) from newborn screening programs are valuable but controversial sets of biospecimens. These biospecimens are retained after clinical screening tests of newborns have been conducted.

In this chapter, we will address the ethical issues and current controversies with respect to these bloodspots. We will review the history and purpose of state newborn screening programs and the secondary uses of DBS, discuss a number of empirical studies that assess parental attitudes toward the use of DBS, and review professional statements and the potential impact of changing national policies on the storage and research use of biospecimens on newborn screening programs. Throughout the chapter we will focus on policy implications of whether parental consent can be obtained for secondary uses of bloodspots in a way that does not create major obstacles to the conduct of research.

Newborn bloodspot screening is a public health program conducted by all states and territories in the United States to screen infants for 32 or more conditions. The purpose of these programs is to identify affected infants before the onset of symptoms in order to reduce morbidity or mortality from the conditions through early intervention. Since 1963, these programs have been run by state health departments to benefit the 4 million children born per year in the US and their families. After clinical screening is complete, residual blood remains from virtually every newborn screened for varying lengths of time. The residual blood can be analyzed for such things as genetic traits in the child, and for infectious disease or environmental toxin exposures for the mother and the child during pregnancy. A broad range of research has been conducted using these resources over the years to better understand serious health conditions affecting mothers and infants (Barbi et al. 1998; Burse et al. 1997; Chaudhuri, Butala, Ball, and Braniff 2009; Olney, Moore, Ojodu, Lindegren, and Hannon 2006).

Residual bloodspots from newborn screening programs are particularly valuable because they represent virtually the entire population of newborns in a state and, as such, are free from selection bias. This collection enables investigators to test thousands of bloodspots when attempting to determine the prevalence of a rare condition in a state's population. Before a new test is included in the public health program, it is important to know how many positive tests there will be in a population using a specific test platform. Further, it may be important to measure exposures to infectious diseases or toxins of women during pregnancy. Other research use include potential public health concerns. For example, one of the first research uses of residual bloodspots was conducted in New England in the 1980s to determine how many pregnant women were infected with HIV (Checko, Abshire, Roach, Markowski, and Backman 1991; Hoff et al. 1988). In addition to overall infection rates, bloodspots from different state regions were used to determine where higher and lower rates of infection occurred within the state. A more recent example of research using DBS evaluated the extent of prenatal mercury exposure in the Lake Superior Basin of Minnesota; such exposure may be related to pregnant women's consumption of fish from the surrounding lakes (Minnesota Department of Health 2016). The value of this bloodspot resource declines substantially if a large percentage of the overall population is not represented. It is particularly concerning if large segments of specific communities are not represented, such as underserved communities that may already be less represented in health-related research—an issue discussed in greater depth in chapter 11 of this volume.

Newborn bloodspot screening is conducted by state-based public health programs, so decisions about the retention and potential research use of these bloodspots are made at the state level (see figure 12.1). Although many states store the specimens only long enough to ensure that accurate clinical testing is complete (about 6 months or less), a number of states store the bloodspots for various lengths of time for several purposes, including biomedical research, quality assurance, and, rarely, forensic uses. (Forensic uses are usually to help identify the remains of a child or to investigate the cause of death for a child who has been deceased for some time.)

An important policy aspect of newborn screening programs is that almost all states conduct screening under the authority of the state and without parental permission (i.e., parens patriae). The "mandatory" approach to

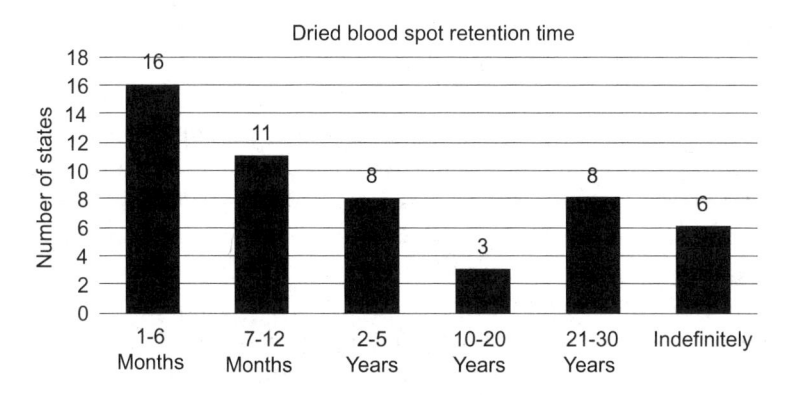

## Figure 12.1
Retention Time for Dried Bloodspots. Data from NewSTEPs, August 2015.

[Bar chart: Dried blood spot retention time. Number of states by retention period: 1-6 Months: 16; 7-12 Months: 11; 2-5 Years: 8; 10-20 Years: 3; 21-30 Years: 8; Indefinitely: 6.]

screening has been a hallmark of these programs since their inception in the 1960s, the justification being that the value of the screening is so great and the burden to the infant and family is so low that state authority is justified. Most states permit parents to decline screening of their newborns for religious or philosophical reasons. All states provide parents with informational material about newborn screening, usually in the form of a brochure distributed in the post-partum period. In states that retain dried bloodspots, the brochures may contain a few sentences about state policy and practice in this regard. However, the brochures are typically mingled with other items and informational materials given at the hospital during birth so this approach to education has been largely ineffective (Arnold et al. 2006; Rothwell, Anderson, and Botkin 2010). As such, most new parents have little or no understanding of newborn screening or of states' practices regarding DBS.

The mandatory aspect of newborn screening programs have been controversial for decades, although this approach to screening is broadly supported by health departments because it facilitates screening and maximizes the number of infants screened (Faden, Holtzman, and Chwalow 1982; Ross 2010). However, because newborn screening itself is conducted without parental permission, the secondary uses of DBS also are conducted without the knowledge or permission of the vast majority parents in most states. Some have argued that the secondary uses of DBS are outside the authority of the state even when used for to develop new screens or testing platforms (Lewis, Goldenberg, Anderson, Rothwell, and Botkin 2011).

Historically, residual bloodspots have been used for research under federal regulatory standards without informed consent under two sets of circumstances. If the DBS are de-identified, the research with the bloodspots (before 2014—see below) was not considered human subjects research and was not under the regulations or under oversight by an institutional review board (IRB). The vast majority of research conducted with DBS has used de-identified bloodspots, although the spots may be linked to the information on the collection form by a sample number. In this context, the investigator using the DBS would not have access to the key linking the sample number to the child's or family's identifying information. Alternatively, DBS have been used under a waiver of consent if an IRB determines that the research meets the criteria in the regulations for a waiver, basically that the research is minimal risk and re-contact of parents for permission would not be practicable. These regulatory provisions have worked very well, at least to the extent that there have been no published reports of adverse events or unethical research conducted using DBS.

The American Academy of Pediatrics (AAP) published a broad analysis and set of policy recommendations regarding newborn screening in 2000 (American Academy of Pediatrics 2000). The AAP recommended that states develop policies and procedures to address research with DBS and that parents be informed that DBS might be used for research purposes that support public health programs. Further, the AAP supported the ability of investigators to conduct research with DBS without consent if the spots are de-identified but recommended that consent be obtained if the DBS were identifiable. There were no broad-scale changes in state policies or procedures after the AAP statement.

Public controversy recently arose over the retention and use of residual dried bloodspots (DBS) when three states—Indiana, Minnesota, and Texas—were sued by groups of parents for this practice (Lewis 2015). Their concerns arose specifically because the bloodspots were retained and used without parental knowledge or permission. The parents claimed that the state governments were engaged in developing a comprehensive DNA bank on the state population, and that doing so was a violation of their privacy and individual rights. While the details of the legal claims and court proceedings are beyond the scope of this chapter, suffice it to note that, owing to a settlement in Texas and a court decision in Minnesota, millions of retained DBS were destroyed (Lewis 2015). Texas has subsequently instituted an informed consent process for the retention and use of DBS, and it is still unclear what the impact will be in Indiana. These suits had an enormous impact on field of newborn screening. These lawsuits put programs on notice nationally that they could not simply maintain traditional DBS policies and practices without public awareness and support (Lewis 2015).

Public Attitudes

Our research group undertook a large national survey of public attitudes regarding the retention and use of DBS in 2012 (Botkin et al. 2012). Recognizing that most parents and the general public have poor awareness of both newborn screening and DBS use, we developed a 22-minute educational movie about these policies and practices. One important question was whether increased education increased or decreased support for DBS retention and research use. That is, might increased knowledge create a higher level of concern and disapproval, or would additional information lead to decreased concern and a higher level of approval?

We engaged 3,855 members of the public through three mechanisms: focus groups ($n = 157$), paper or telephone survey ($n = 1,418$), and a Knowledge Networks panel ($n = 2,280$). Knowledge Networks uses a large, nationally representative pool of participants for survey-based research. Forty-six percent of the participants viewed our educational movie about DBS, and this group constituted our enhanced knowledge group. Our results showed that a large majority of the respondents were supportive of newborn screening programs in general. The respondents were almost evenly divided on the question of whether should be conducted with parental permission. With respect to state policies regarding the retention and use of DBS, a large majority of respondents were supportive of research uses of DBS. Nevertheless, respondents fell across a spectrum regarding their level of concern about state retention and use. While 25 percent stated that they were not concerned at all, 30 percent stated that they were very concerned (see figure 12.2).

We asked our respondents whether state policies should require parents to sign a form to enable research with DBS (an opt-in approach) or whether states should notify parents of the practice and not use DBS only if parents express an objection (an opt-out approach). Responses indicated a clear preference for the opt-in approach; 62 percent selected that option.

A key finding for this project was that individuals who watched the educational movie were substantially more supportive of newborn screening and use of DBS than those who had limited education on the topics. This suggests that public health programs can raise public awareness and knowledge of these programs without jeopardizing public support.

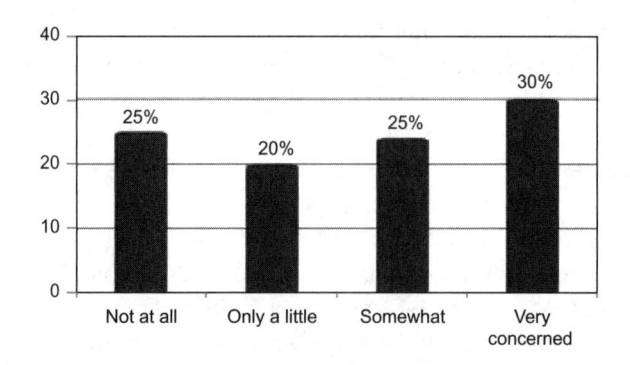

**Figure 12.2**
Responses to the question "How concerned would you be if the Health Department saved the leftover blood samples from babies after the tests are done?"

Our results were similar to those reported by a group of Canadian investigators who addressed these issues with eight focus groups ($n = 60$) (Bombard et al. 2012). They found that the Canadian public also was highly supportive of secondary uses of anonymous DBS for quality control and public health research. However, there were mixed opinions over whether parents should be asked to consent to such uses.

In other contexts involving the use of biospecimens in research, public attitudes are consistent on two elements: people want to know about these practices and they want a choice (Chen et al. 2005; Hull et al. 2008; Wendler and Emanuel 2002). Surveys tend to suggest that people prefer an informed consent process (an opt-in process) rather than an opt-out process, although these opinions are not uniform (Botkin et al. 2015a).

**Professional Statements**

A number of professional organizations have taken stands regarding appropriate policies and practices for DBS retention and research use. The DHHS Secretary's Advisory Committee for Heritable Diseases in Newborns and Children (SACHDNC) published a statement in 2011 (Therrell et al. 2011) highlighting the value of these specimens and recommending that all states develop policies addressing the disposition of DBS, whether or not the DBS are retained. The Advisory Committee did not take a stand on parental permission but stated that parents should be aware of these activities and that states should consider whether parental permission should be required for research activities. Enhancing provider education and developing model consent and dissent processes were recommended.

In 2012 the National Society of Genetic Counselors issued a statement supporting the retention and use of DBS (National Society of Genetic Counselors 2012). According to that statement, "parents should be fully informed of their options through comprehensive education during the prenatal and immediate postnatal period regarding blood spot storage and use policies." In 2015 the American Society of Human Genetics published a broad statement governing a variety of ethical, legal, and psychosocial issues relevant to genetic testing in children and adolescents (Botkin et al. 2015b).[2] Among the conclusions were the following:

- The ASHG encourages states to retain DBS and to make specimens available to investigators and to public health programs under carefully developed guidelines.
- Parents should be informed of state policy and practices regarding the retention and use of DBSs.
- Parents should be offered a choice regarding the retention and use of their child's DBSs for purposes beyond the clinical newborn screening program and QA uses. This choice ought to be clearly separated from the decision to participate in Newborn Screening.

These professional statements indicate that there was no clear professional standard regarding parental permission for DBS retention and use as of 2014. All of the statements support the importance of parental knowledge about the storage and uses of DBS, and that parents should have the ability to control whether their newborns bloodspots are used in future research. However, none specifically advocated for a formal informed consent process in which parents must "opt in"—that is, must make an active choice to allow, or not allow, their newborn's bloodspots to be used. Notably, however, the American Academy of Pediatrics, through its Committee on Bioethics, advocated for parental permission for newborn screening in a 2013 statement on genetic testing for children and adolescents (Committee on Bioethics 2013). This statement did not address DBS retention and use.

## Changes in Federal Policy

The landscape for the retention and use of DBS changed suddenly and dramatically in December 2014. A federal bill, the Newborn Screening Saves Lives Act, was originally passed in 2008 with strong bipartisan support. The bill provides support for various aspects of newborn screening including resources to evaluate newborn screening effectiveness, quality-assurance activities, professional and public education, and support for the SACHDNC. Upon renewal of the bill in 2014, a new section was added addressing DBS and requiring that all federally funded research using DBS be conducted with parental permission(Roybal-Allard 2014). The bill stated that the research must be considered human subjects research even if the bloodspots are de-identified and that waivers of consent are not permitted even if the research meets the traditional criteria under the federal regulations governing human subjects research, known as the Common Rule. This new policy language also stipulated that these requirements would remain until such a time that the common rule is amended with regard to the secondary use of biospecimens. Additionally, these stipulations are relevant to all DBS collected after March 2015, meaning that DBS collected before that date can still be used without parental permission. This section of the bill was included without public comment and without previous knowledge of the newborn screening community.

This dramatic change in federal policy creates and illustrates a challenging dilemma for this particular set of research resources. As we have seen, the general public clearly wants to know about these practices and expects a choice. Yet newborn screening programs are conducted without parental permission and with limited parental and public education. This means that an informed consent process for DBS storage and research use would need to be created from scratch in almost all states. For the DBS to be valuable, the consent rates of the population must be high enough to ensure an accurate representative sample of the states newborn population. It will be crucial to ensure that there are not segments of the population consenting at significantly lower rates to avoid bias within the overall sample, which may lead to the under-representation of those populations in research using DBS. Further, education and consent for DBS retention and use must be obtained in a manner that does not lead a substantial number of parents to refuse newborn screening altogether, adversely impacting the public health benefits of these programs for newborns and families.

Informed consent forms and processes are familiar in medicine so this expectation might not seem like much of a challenge. But there are several aspects of this context that are different. First, the hours and days after the birth of a baby are hectic and exhausting for new parents and for nursery staff. It is common for babies to be discharged home within 24 to 48 hours after delivery, and thus much has to be done in a short period of time for both the mother and the baby. The topic of newborn screening must compete with other more immediate health-care demands of a newborn, and the management of residual clinical samples will be an even lower priority for clinicians and parents. Downstream uses of DBS are not relevant to the immediate care of the baby or family so an informed consent process is extra work for the staff with no direct clinical benefit. This means that this extra task may not get done when the nursery is busy or when other priorities fill the baby's brief time in the hospital. Second, for these same reasons, many staff will have only a limited understanding of the topic, making it challenging for them to answer questions or address concerns with confidence.

Other challenges are more technical but are nonetheless substantial. In order for the parents' choice about retention to be effectively communicated to the health department, it is ideal to have the consent form physically attached to the filter paper card with the bloodspot. All of the bloodspots are sent to the laboratory for clinical testing, but then the residual DBS must be clearly labeled with a parental choice to guide subsequent storage and use. However, the federal regulations governing human subjects research require that signatures for consent must be part of the consent form itself (C.F.R., 46.116–17). That is, it is not acceptable to provide, say, a brochure with the consent information and have the signature on a separate form. In the context of state newborn bloodspot screening, this means that the whole consent form must be physically attached to, or incorporated in, the filter paper cards used to collect the DBS. Because state newborn screening programs print the cards for DBS collection, states must develop entirely new forms and processes to support a consent process in birthing facilities. Further, the consent information must be streamlined in such a fashion to fit in the space yet remains readable, understandable, and sufficient to foster informed decision making.

Certainly none of these challenges are impossible to overcome, and it can be argued that public expectations and our respect for patient autonomy justify the effort. On the other hand, such an effort will clearly require substantial resources to accomplish these tasks effectively. Because of the limited budgets of many health departments, resources may be needed more for other public health goals. Further, there is a substantial risk that the collection of specimens will fall dramatically, not necessarily because many parents are saying "No," but because birthing facility staff cannot find the time or motivation to complete the consent process—other chapters in this volume raise similar concerns about requirements for broad informed consent as to biospecimens more generally. Also recall that there have been no reported instances of abuse or adverse events from the use of DBS for research so the argument cannot be made that a consent process is necessary to protect the welfare of infants and families. If the retention rates fall dramatically, the ability to conduct research on rare conditions affecting infants and mothers will be jeopardized. Thus we may have a significant ethical dilemma: respect for persons (offering choice) versus beneficence (conducting valid research).

Public sensitivities to the secondary research uses of residual clinical specimens are also reflected in the publication of the Notice of Proposed Rulemaking (Notice of Proposed Rulemaking 2016) that proposes changes in the Common Rule governing human subjects research. The NPRM included sweeping changes in the oversight of biospecimen-related research, including requirements for informed consent for all secondary uses of specimens. These proposed changes were highly controversial in the biomedical field and the Office of Human Research Protection received a large volume of negative comments on these elements. The final rules published in January 2017 do not contain many of the NPRM's most controversial regarding stored biospecimens, leaving the status quo intact for research with non-identified biospecimens (OHRP 2017). These new Common Rule regulations are scheduled to become effective in January 2018 but there is, at the time of this writing, some uncertainty about the fate of many new federal regulations in the transition between the Obama and Trump administrations. Nevertheless, the proposed elements in the NPRM relevant to biospecimen repositories and the secondary use of biospecimens illustrate how ethical standards may be evolving. The final rule includes language that would supersede the requirements in the 2014 Newborn Screening Saves Lives Act relevant to research with residual bloodspots when the new Common Rule regulations go into effect. Given that the new rule maintains the status quo, whether this suspension will be acceptable to legislators who proposed this provision in the 2014 legislation remains to be seen. If this issue is still on the national radar for Congress, new legislation might well re-institute the consent requirements for DBS use that have been superceded by the new Common Rule.

**Conclusions**

As one of us has argued elsewhere (Botkin 2015), there is a substantial probability that consent process for DBS uses, when required by states, will be perfunctory for the reasons noted above. Clinicians are very busy and work hard to achieve clinical goals with the patient. Discussion of secondary uses of clinical specimens is unlikely to be a priority, unless the clinician happens to be an investigator, yet the topic is sufficiently complex that it should not be assigned to a clerk or student. In the context of health care and in other contexts, we are all familiar with signing (or clicking approval for) long, complex agreements that we do not read or understand (for example, HIPPA forms). In the context of newborn screening, an additional concern is that a "broad" consent form, used in the absence of the time and resources for adequate education, may only confuse parents about the distinctions between newborn screening and the use of residual DBS. Confused parents may decline newborn screening when their primary concern is the secondary uses of DBS. This would be a serious problem for these highly successful public health programs and the families they serve.

The controversies and policy initiatives over residual DBS from newborn screening and the changes in the federal regulations governing NBS biospecimens are challenging because they reflect respect for competing, deeply held values. The failure of the biomedical research enterprise to be transparent over the years has led to problems with trust, and to be remedied in these contexts by informed consent. But trust will not be rebuilt with perfunctory consent processes. The larger challenge is to develop more sophisticated approaches to information and education that foster a higher level of true understanding about how research is conducted and, potentially, the need to collaborate with communities to support research. We have been successful in making biospecimen-related research safe but we have not been successful in fostering transparency or public trust that adequate systems are in place to promote and protect the interests of patients.

## Notes

1. Biospecimens have little value for research if they are not linked with data about the specimen and the individual from which it came. For brevity, we will use the term "biospecimens" in this chapter when we mean the specimens plus linked data.

2. The first author of this chapter was also the first author of the ASHG statement.

## References

American Academy of Pediatrics. 2000. Serving the family from birth to the medical home. Newborn screening: A blueprint for the future—a call for a national agenda on state newborn screening programs. *Pediatrics* 106 (2 Pt. 2): 389–422. doi:10.1542/peds.106.2.S1.389.

Arnold, C. L., T. C. Davis, J. O. Frempong, S. G. Humiston, A. Bocchini, E. M. Kennen, and M. Lloyd-Puryear. 2006. Assessment of newborn screening parent education materials. *Pediatrics* 117 (5 Pt. 2): S320–S325. doi:10.1542/peds.2005-2633L.

Barbi, M., S. Binda, V. Primache, A. Tettamanti, C. Negri, and C. Brambilla. 1998. Use of Guthrie cards for the early diagnosis of neonatal herpes simplex virus disease. *Pediatric Infectious Disease Journal* 17 (3): 251–252.

Bombard, Y., F. A. Miller, R. Z. Hayeems, J. C. Carroll, D. Avard, B. J. Wilson, J. Little, et al. 2012. Citizens' values regarding research with stored samples from newborn screening in Canada. *Pediatrics* 129:239–247.

Botkin, J. R. 2015. Crushing consent under the weight of expectations. *American Journal of Bioethics* 15 (9): 47–49. doi:10.1080/15265161.2015.1062181.

Botkin, J. R., E. Rothwell, R. Anderson, L. Stark, A. Goldenberg, M. Lewis, et al. 2012. Public attitudes regarding the use of residual newborn screening specimens for research. *Pediatrics* 129 (2): 231–238. doi:10.1542/peds.2011-0970.

Botkin, J. R., J. W. Belmont, J. S. Berg, B. E. Berkman, Y. Bombard, I. A. Holm, et al. 2015a. Points to consider: Ethical, legal, and psychosocial implications of genetic testing in children and adolescents. *American Journal of Human Genetics* 57 (1): 1233–1241. doi:10.1016/j.ajhg.2015.05.022.

Botkin, J. R., E. Rothwell, R. Anderson, L. A. Stark, and J. Mitchell. 2015b. Public attitudes regarding the use of electronic health information and residual clinical tissues for research. *Journal of Community Genetics* 6 (2): 183. doi:10.1007/s12687-015-0216-6.

Burse, V. W., M. R. DeGuzman, M. P. Korver, A. R. Najam, C. C. Williams, W. H. Hannon, and B. L. Therrell. 1997. Preliminary investigation of the use of dried-blood spots for the assessment of *in utero* exposure to environmental pollutants. *Biochemical and Molecular Medicine* 61 (2): 236–239.

Chaudhuri, S. N., S. J. M. Butala, R. W. Ball, and C. T. Braniff. 2009. Pilot study for utilization of dried blood spots for screening of lead, mercury and cadmium in newborns. *Journal of Exposure Science & Environmental Epidemiology* 19 (3): 298–316. doi:10.1038/jesee.2008.19.

Checko, P. J., J. P. Abshire, A. I. Roach, M. Markowski, and L. Backman. 1991. HIV seroprevalence among childbearing women in Connecticut. *Connecticut Medicine* 55 (1): 9–14.

Chen, D. T., F. G. Miller, E. J. Emanuel, D. Wendler, D. L. Osenstein, P. Muthappan, and S. G. Hilsenbeck. 2005. Research with stored biological samples. *October* 165: 652–655.

Committee on Bioethics, Committee on Genetics, American College of Medical Genetics & Genomics Social Ethical and Legal Issues Committee. 2013. Ethical and policy issues in genetic testing and screening of children. *Pediatrics* 131 (3): 620–622. doi:10.1542/peds.2012-3680.

Faden, R. R., N. A. Holtzman, and A. J. Chwalow. 1982. Parental rights, child welfare, and public health: The case of PKU screening. *Am J Public Health* 72 (12): 1396–400. http://www.ncbi.nlm.nih.gov/entrez/query.fcgi?cmd=Retrieve&db=PubMed&dopt=Citation&list_uids=7137438.

Hayeems, R. Z., F. A. Miller, Y. Bombard, D. Avard, J. Carroll, B. Wilson, et al. 2015. Expectations and values about expanded newborn screening: A public engagement study. *Health Expectations* 18 (3): 419–429. doi:10.1111/hex.12047.

Hoff, R., V. P. Berardi, B. J. Weiblen, L. Mahoney, M. L. Mitchell, and G. F. Grady. 1988. Seroprevalence of human immunodeficiency virus among childbearing women. *New England Journal of Medicine* 318: 525–530.

Hull, S. C., R. R. Sharp, J. R. Botkin, M. Brown, M. Hughes, J. Sugarman, et al. 2008. Patients' views on identifiability of samples and informed consent for genetic research. *American Journal of Bioethics* 8 (10): 62–70. doi:10.1080/15265160802478404.

Lewis, M. H. 2015. Lessons from the residual newborn screening dried blood sample litigation. *Journal of Law, Medicine & Ethics* 43 (Suppl. 1): 32–35.

Lewis, M. H., A. Goldenberg, R. Anderson, E. Rothwell, and J. Botkin. 2011. State laws regarding the retention and use of residual newborn screening blood samples. *Pediatrics* 127 (4): 703–712. doi:10.1542/peds.2010-1468.

Minnesota Department of Health. 2016. Mercury in Newborns in the Lake Superior Basin. http://www.health.state.mn.us/divs/eh/hazardous/topics/studies/newbornhglsp.html.

National Society of Genetic Counselors. 2012. Blood Spot Storage and Use blog. http://nsgc.org/p/bl/et/blogaid=21.

Notice of Proposed Rulemaking. 2016. https://www.gpo.gov/fdsys/pkg/FR-2015-09-08/pdf/2015-21756.pdf.

Office for Human Research Protections (OHRP), Department of Health and Human Services. 2017. https://www.federalregister.gov/documents/2017/01/19/2017-01058/federal-policy-for-protection-of-human-subjects/.

Olney, R. S., C. A. Moore, J. A. Ojodu, M. Lou Lindegren, and W. H. Hannon. 2006. Storage and use of residual dried blood spots from state newborn screening programs. *Journal of Pediatrics* 148 (5): 618–622. doi:10.1016/j.jpeds.2005.12.053.

Protection of Human Subjects. 45 C.F.R. §46.116–46.117. 2009.

Ross, L. F. 2010. Mandatory versus voluntary consent for newborn screening? *Kennedy Institute of Ethics Journal.* doi:10.1353/ken.2010.0010.

Rothwell, E., R. Anderson, and J. Botkin. 2010. Policy issues and stakeholder concerns regarding the storage and use of residual newborn dried blood samples for research. *Policy, Politics & Nursing Practice* 11 (1): 5–12.

Roybal-Allard, L. 2014. Newborn Screening Saves Lives Reauthorization Act of 2014. H.R.1281–113th Congress. https://www.congress.gov/bill/113th-congress/house-bill/1281/text.

Therrell, B. L., W. H. Hannon, D. B. Bailey, E. B. Goldman, J. Monaco, B. Norgaard-Pedersen, et al. 2011. Committee report: Considerations and recommendations for national guidance regarding the retention and use of residual dried blood spot specimens after newborn screening. *Genetics in Medicine* 13 (7): 621–624. doi:10.1097/GIM.0b013e3182147639.

Wendler, D., and E. Emanuel. 2002. The debate over research on stored biological samples: What do sources think? *Archives of Internal Medicine* 162 (13): 1457–1462. doi:10.1001/archinte.162.13.1457.

# 13 Informed Consent for Genetic Research on Rare Diseases: Insights from Empirical Research

Sara Chandros Hull[1]

In the past decade, genetic research has expanded exponentially in its scope, with both the adoption of next-generation sequencing technologies and concerted efforts to create large-scale repositories of human biospecimens and data for research (Green, Guyer, and NHGRI 2011). This expansion has raised challenging questions about how best to get informed consent for the collection and broad sharing of biological samples and associated data, especially given that plans for future research with these resources cannot be precisely described at the time that consent is obtained. To what extent should individual research subjects be able to control the use of their samples and data for ongoing genomic research?

These questions are at the heart of this volume, but they are especially important in the context of research on rare and undiagnosed genetic diseases, for which widespread sharing of elusive samples and data is needed to facilitate research addressing the lack of diagnostics and interventions for these populations. Up to 50 percent of persons with rare diseases, defined as disorders or conditions with a prevalence of fewer than 200,000 people in the United States (Groft 2013), never receive a formal diagnosis, and many have faced a long and costly "diagnostic odyssey" in pursuit of a correct molecular diagnosis and possible treatments (Sawyer et al. 2015). Rare disease research is difficult to implement because of the relatively small pool of patients from which to recruit research participants, the need to share biospecimens and data collected from this limited pool as broadly as possible to maximize the ability to conduct valid research, and the increased privacy risks that persons with rare diseases may face in the research context because of uniquely identifiable aspects of their diseases (Grady, Rubinstein, and Groft 2011).

Various governance frameworks and consent models have been implemented in response to these challenges; however, there is growing recognition that inconsistent approaches to informed consent can be a substantial barrier to effective transnational sharing of valuable research resources contributed by rare disease populations (Rubinstein et al. 2012; Colledge, Elger, and Howard 2013). The success of initiatives to facilitate rare and undiagnosed research has depended, in part, on the ability to implement consistent informed consent processes across multiple institutions to allow research data to be shared broadly between participating sites and investigators (Brownstein et al. 2015).

There is controversy about the optimal consent model to achieve these goals and whether various proposals are adequately tailored for genetic research on rare or highly stigmatized disorders (Grady et al. 2015). Yet there has been little discussion of how different models might play out in the context of rare disease research, nor of the attitudes, preferences, and decisions of persons affected with rare diseases regarding informed consent for research with their samples and data. This chapter attempts to address this gap by reviewing the available empirical data on informed consent for genetic research on rare and undiagnosed diseases, with a focus on whether these data support recent consent proposals and biorepository practices, or if instead the data support the idea of treating rare disease research as a special case.

I begin this chapter by framing the debate about broad consent and how it relates to the context of rare disease research. I then provide a synopsis of empirical research on public attitudes about broad consent to identify areas of consensus and disagreement, and to establish a general backdrop against which the views of rare disease populations can be understood. I then review five qualitative and mixed-methods studies that explore the attitudes and preferences of persons with rare diseases and their relatives regarding different dimensions of informed consent for broad sharing of their samples and data for genetic research. Finally, I offer tentative conclusions about the appropriateness of broad consent for genetic research on rare diseases in light of these findings, and identify knowledge gaps that would benefit from further research.

## Broad vs. Specific Consent

The tension between broader and more specific approaches to informed consent for research on rare genetic disease has been described as follows:

Except in exceptional circumstances, research participants should be included in studies only if they have given their valid consent. What ought to count as valid consent in research into rare conditions? To be valid, consent must be voluntary, informed, and competently given. Research ethic [sic] committees have tended to interpret the requirement for informed consent to imply that such consent must be specific and closed. Consent is informed only if participants know in detail what is going to happen to them, what is going to happen to any sample taken from them, and when the research will be completed. In research into rare conditions, it is not always possible to provide the participant with a detailed account of this kind. This need not imply, however, that informed consent is not possible. Given adequate support, research participants are able to understand the nature of rare disease research sufficiently well to enable them to give valid consent. The true test of validity, we suggest, is whether the participants have sufficient understanding of the research and of their part in it to enable them to make a reasoned and balanced assessment about whether to participate. (Parker et al. 2004)

Recent proposals have also promoted a standard of "broad consent" for the collection of biospecimens and data in genetic research with the intent to use and share these resources as broadly as possible,[2] as was discussed in part III of this volume. Broad-consent proposals, which have been applied to biospecimens collected in either a clinical or research context, have varied in their requirements for governance of secondary research, ongoing communications with participants, and scope of potential future uses. One detailed broad-consent proposal, described in chapter 8 of this volume by Grady et al., includes three components: initial broad consent for an unspecified range of future research, subject to a few content restrictions; a process of oversight and approval for future research activities; and, wherever feasible, an ongoing process of providing information to or communicating with donors (Grady et al. 2015).

Emerging research suggests a trend toward adopting variations of broad consent for the collection of human samples and data (Allen and Foulkes 2011). However, controversy remains about whether broad consent is an optimal model for facilitating autonomous decision making about future research and preserving trust between participants and researchers, with some advocating for an alternative model of "dynamic consent" that allows participants to make specific choices about future research projects (Kaye et al. 2015). Ultimately, the success of genetic research on rare and undiagnosed diseases will depend on identifying a consistent approach to informed consent that both supports the autonomous interests of potential participants and maximizes the ability to share contributed samples and data as widely as possible. Whether this is better achieved through broad or specific approaches to informed consent is a question that can be informed by empirical data on the willingness of persons with rare diseases and their family members to consent to genetic research, their preferences among different consent approaches that have been proposed, and how well different consent processes facilitate an informed decision about participation in genetic research. Understanding whether rare disease populations have heightened concerns about the risks to their privacy, for example, can provide insights into the likelihood that they will be willing to allow their samples and data to be shared with a broad range of secondary users. Empirical research is also an important way to respond to calls to engage stakeholders in discussions about the process of obtaining informed consent for biospecimen repositories to facilitate genetic research on rare diseases (Grady, Rubinstein, and Groft 2011).

## Systematic Reviews of Public Attitudes about Broad Consent

Empirical research has been conducted with thousands of stakeholders, including the general public, patients, family members, and research participants, about their views on informed consent for the future research uses of their specimens (Grady et al. 2015). Although these studies have employed different methodologies and have framed their questions in vastly different ways, several systematic reviews and commentaries that have been published within the past decade provide useful synopses and comparisons of the empirical literature on the acceptability of broad consent, attitudes toward genomic data sharing, and willingness to participate in biospecimen repository-based research (Wendler 2006; Hoeyer 2010; Master et al. 2012; Husedzinovic et al. 2015; Shabani, Bezuidenhout, and Borry 2014; Garrison et al. 2015).

A review by Wendler (2006) concluded that respondents are generally willing both to donate specimens to research and to provide broad consent, defined as "one-time general consent, on the understanding that an ethics committee will review and approve future projects." The review, which encompassed thirty studies that were conducted in different populations around the world using different methodologies, found that at least 80 percent of respondents would donate a sample for research in seventeen of the twenty studies that assessed willingness to donate. In the subset of studies that assessed preferences regarding different consent options, between 79 percent and 95 percent of respondents expressed a willingness to provide one-time general consent and rely on ethics committees to review subsequent research uses of their samples (Wendler 2006).

More recent reviews suggest that although there is general support for broad consent, variously defined, there are differences in the attitudes of patients vs. the general public, hypothetical vs. actual research scenarios, and majority vs. minority populations. One review found positive attitudes toward broad consent among patients, with 80–90 percent agreeing to broad consent; while up to 44 percent of respondents from the general public would request to be re-contacted to give consent for new studies with their samples (Husedzinovic et al. 2015). Another review concluded that while the majority of the public as well as research participants appreciate the benefits of broad data sharing, concerns about the risks varied according to a range of contextual factors (Shabani, Bezuidenhout, and Borry 2014). There are significant gaps in what is known about factors such as race, ethnicity, gender, and other sociodemographic factors that affect people's attitudes and decisions regarding broad consent (Garrison et al. 2015). In chapter 11 of the present volume, Aaron Goldenberg and Suzanne Rivera discuss the views of underserved medical communities in greater depth.

Such variations have led others to question the existence of consensus on the acceptability of broad consent in the literature. One such review identified ten empirical studies published between 2006 and 2012 that reported a range of views on broad consent, with 34–79 percent of respondents preferring broad consent to other forms of consent (Master et al. 2012). Another commentary suggests that instead of searching for an underlying consensus in the available data, important insights can be gained by looking at differences across multiple studies, such as methodological differences that are likely to be responsible for the different trends that have been observed (Hoeyer 2010). For example, although some studies have found that reconsent is preferred by many respondents, a study that took a deliberative engagement approach found that informants were less likely to require reconsent after negotiating their views in a group debate, suggesting that there is a difference between asking people about their personal preferences among multiple options and asking them to choose the best policy option on behalf of a group.

These reviews suggest that broad consent is viewed as an acceptable option to many potential donors of samples and data, even if it is not always the most preferred approach among multiple options. The reasons for differences in attitudes and preferences regarding informed consent have not been well explored, however, and some populations have not been well-represented in these studies. None of the studies included in these systematic reviews explored the attitudes or experiences of persons with rare diseases. Although these reviews provide a helpful overview of the perspectives of the general public and persons with more common medical conditions about the acceptability of broad consent, it remains unclear whether these perspectives are shared by rare disease populations.

## Attitudes of Persons with Rare Diseases and Their Family Members

Only a small number of empirical studies have focused on the perspectives and experiences of persons with rare diseases related to consent for broad sharing and ongoing uses of samples and data for genomic research. The few studies that have been conducted with rare disease populations have employed qualitative methods, which are appropriate for exploring a phenomenon (i.e., informed consent for genomic research) in populations about whom little is already known, especially for purposes of exploring experiences, beliefs, and attitudes (Creswell 1998). Although qualitative research does not allow for broad generalizations to be drawn from the reported findings, these studies offer nuanced insights about the unique experiences and perspectives of this population that can be considered against the backdrop of quantitative research that has been conducted in larger populations. They can also form the basis for larger studies downstream.

A total of five studies were identified via a comprehensive search of PubMed Central.[3] No additional studies were identified by a comparable search in GenETHX and a search of PubMed Central articles that have cited those articles initially identified. The five studies included in this analysis report findings from qualitative interviews or mixed-method surveys that were conducted with persons affected with a rare genetic disorder ("probands"), their parents, and other relatives (table 13.1) All five studies occurred in the context an actual (i.e., non-hypothetical) informed consent process that was either being developed or that the respondents had already undergone for genetic research on rare diseases, focusing on topics related to motivations and willingness to participate in genetic research, attitudes and preferences about sharing and future use of samples and/or data, and associated attitudes about privacy and confidentiality. Respondent quotes that correspond to these themes are included in table 13.2.

### Willingness and Motivations to Participate

Most of the studies in this analysis included probands and family members who had already agreed to participate in a rare disease research protocol, so it is a given that they were willing to participate in such research at least at one point in time. However, one study that was conducted as part of an effort to optimize the consent process for an emerging leukodystrophy (LD) research database assessed respondents' willingness to participate in rare disease research before enrolling (Darquy et al. 2015). The majority of

# Informed Consent for Genetic Research on Rare Diseases

**Table 13.1**
Empirical studies on broad consent in rare disease populations

| | Population | Methodology | Focus of analysis |
|---|---|---|---|
| Ponder et al. 2008 | 78 individuals from families with two or more males with intellectual disability of unknown cause who enrolled in Genetics of Learning Disabilities (GOLD) study | Semi-structured interviews | Experiences, memory, understanding of consent process for GOLD study |
| Tabor et al. 2012 | Two families, including parents and children with Miller Syndrome | Semi-structured open-ended interviews | Perceptions of informed consent process, including purpose of research, motivations for participation, benefits and risks, confidentiality, data sharing |
| Bergner et al. 2014 | 11 adults and 4 parents of affected children enrolled in Mendel Project | Semi-structured interviews | Understanding of research, perception of risks/benefits, opinions about confidentiality, overall impressions of consent process, and interest in re-consent |
| Jamal et al. 2014 | Participants in ClinSeq and Whole Genome Medical Sequencing (WGMS, $n = 13$) protocols (*Note: this chapter selectively focuses on the WGMS cohort, which includes persons with rare disorders*) | In-depth semi-structured phone interviews | Attitudes towards confidentiality and motivations for sharing data with researchers |
| Darquy et al. 2015 | Patients with leukodystrophies ($n = 46$) and relatives ($n = 149$) between ages 40–64 | Mixed-method questionnaire with open ended responses | Patient/family motivations and reluctances to share health data at European level |

**Table 13.2**
Selected quotations from Qualitative Studies on Consent for Rare Genetic Disease Research

| Theme | |
|---|---|
| Motivations to enroll | ***To find cause/risk factors related to disease*** |
| | I mean for me … it's really important to get an answer as to why, you know, because we have this huge void. … I just kind of want that 'why' answer. (proband, B) |
| | I don't think they'll find "the" cause of the condition. I think they might find some contributing risk factors. But I think it's complicated. I think it's genetic, and I think there's probably something really that's environmental, or like virus, or something else that triggers it, as well as the genetic component. That's probably going to be much harder to nail down than just by looking at the genome. (proband, C) |
| | ***To Facilitate Research*** |
| | Patients are key to advancing research by providing data to researchers. (E) |
| | The more information collected, the more it will promote advancement of research. (E) |
| | [I]n a rare disease like this, maximum participation is required for effective research. (E) |
| | ***Altruism*** |
| | … of course, I mean it it's going to benefit people in years to come. Like there's no benefit for us apart from the knowledge that … if they can find something out. (parent, A) |
| | We really feel strongly that if some other human being can benefit from us participating in this study and found out the missing link, that satisfies us at a very deep level. (parent, B) |
| | The benefits outweigh. You can help people. I'm probably considered a mild case to some people out there who have all these horrible diseases. I just think if I could help someone in the future, it has to be done, really. … Yes for me, it's because I've had this condition since I was born. I'm very familiar with medical stuff, how important research and genetics is. (proband, D) |
| Access to data by secondary researchers | Sharing data with a lot of researchers in different countries is a plus to improve research (E) |
| | It is necessary to multiply, federate and pool research (E) |
| | The rights of individuals should be respected (E) |

Informed Consent for Genetic Research on Rare Diseases 265

Table 13.2 (continued)

| Theme | |
|---|---|
| Privacy and confidentiality | **Lack of Concern**<br>"[T]here's more things to be worried about in our life right now than whether we might be known or not, in a study." (parent, B)<br>We've never kept our kids hidden. ... Our kids have been published. Our life has been published. ... It's there for people to see, and it's there for other medical professions to be involved with. (Parent, B)<br>[Researchers] don't understand what little privacy we have, every day we step out the door. ... Our perception of privacy is a lot different than, than somebody who has a normal body. (Parent, B)<br>No. It's not gonna do anything to me. (proband, C)<br>**Balancing of Risk and Benefits**<br>I understand the amount [of privacy protection]. ... And beyond that, nobody can control anything. So it's worth the risk. (proband, B)<br>**Potential Concerns**<br>I think it is the way that I am about myself. I think I'm more comfortable about myself and what people know about me. And some people could be embarrassed if it fell into the wrong hands and they could be discriminated against in some way if they're on file somewhere, maybe have a disability. Those are the only things I could think of from someone else's point of view, that the information could be used against them. For me, I don't have those sorts of issues. (D)<br>I guess I would think 'What would happen in 50 or 100 years, and you're all over a medical journal? I actually don't think I would mind that necessarily, but it just made me think twice about it, you know? I guess I would want to be asked first. (D)<br>**Use of Research Findings in Prenatal Diagnosis**<br>I don't want to contribute to the fact that if they know what causes [my child's condition] now when they test an embryo that has [my child's condition], they're throwing it away because that's just—I feel like that's basically saying [my child] is insignificant. That was one of the things I had to really come to terms with. (parent, D)<br>**Pharmaceutical Industry Involvement in Research**<br>If there is such a partnership, I refuse to participate in the database. The pharma industry orients research in their own interests, not in the interests of patients." (E) |
| Non-welfare interests in controlling future uses of data | |
| Length and burden of consent | ... taking a long time about things that maybe I don't necessarily worry about. (Family, B)<br>I'm just surprised that so much effort has to go into the consent process. ... It just sounds like the whole process is more burdensome than it should be. (proband, C) |

*Note:* Quotes are identified by type of respondent (e.g., proband, parent, etc.), if known, and the study from which the quote was extracted (A-E, as listed in table 13.1) respondents in this study indicated that they would agree to participate in any research that collected data for the study of LD, with affected patients expressing somewhat higher rates of willingness (89 percent) compared to their relatives (84 percent). In their open-ended comments on this survey, respondents highlighted the importance of data sharing for the conduct of research on this rare disease.

Respondents in studies by Tabor et al. (2012), Bergner et al. (2014), Ponder et al. (2008), and Jamal et al. (2014) were motivated by multiple reasons to enroll in research. Many hoped to learn information about the rare disease with which they or their family members are affected, such as identifying the gene that causes a particular syndrome or a diagnosis for an unexplained condition, while some respondents anticipated uncertainty in the ability to find the cause of the disease under study. Respondents understood that research was unlikely to be beneficial for their family members and those currently affected with rare diseases, however, and expressed altruistic motivations for their participation with an eye toward other families who were at risk for similar medical conditions.

**Access to Data by Secondary Researchers**
Participants in the Darquy et al. LD survey (2015) expressed a high degree of willingness to permit researchers outside of the primary genetic disease network to have access to their research data, with 90 percent of relatives and 76 percent of probands in the survey supporting access to further study of LDs; while a somewhat lower proportion (75 percent of relatives and 64 percent of probands) supported access for the study of other diseases. The reasons offered in support of sharing with secondary researchers mirrored their reasons for participating in research in general, i.e., to promote the advancement of research and broaden the scope of research being conducted globally, with caution expressed regarding the confidentiality of international research and protecting the interests of individual participants.

**Privacy and Confidentiality**
Views about possible risks to the confidentiality and privacy of research participation varied across these studies, often depending on the unique disease-related experiences of respondents. On the one hand, some respondents indicated that they were not concerned about such research risks relative to other disease-related issues with which they were dealing. For example, they were already quite open about the diseases that affect them or their family, their view of privacy is shaped by having a disease with

visible manifestations that cannot be kept private, or they viewed the benefits of participation as outweighing the risks (Tabor et al. 2012; Bergner et al. 2014).

However, respondents also viewed adequate confidentiality measures as important features of research and expressed some concerns about the risks of a breach of confidentiality. Three fourths of participants in the LD survey indicated that data security and confidentiality would be essential prerequisites for participation in research (Darquy et al. 2015). Respondents in a study by Jamal et al. (2014) were aware of concerns that others may have about embarrassment or potential discrimination that could occur if research findings are released to unauthorized third parties, although this seemed to be less of a concern for the particular respondents themselves.

**Non-Welfare Interests**

A small number of respondents in these studies identified concerns about possible future uses of their samples and data that can be classified as related to their non-welfare interests. Non-welfare interests have been defined in the biospecimen context by some as donors' interest in "preserving the moral significance" of donated samples and data, i.e., beyond concerns about personal harms and benefits (Tomlinson 2013). One such concern that was raised by a parent of an affected child related to the possibility that findings from research would lead to the ability to diagnose prenatally the rare condition that affected the respondent's child, potentially leading someone to discard an affected embryo (Jamal et al. 2014). This possibility complicated the decision to participate in open-ended research for this parent.

Several respondents in another study expressed concern about pharmaceutical companies having access to their research data. These respondents questioned profit-oriented motives of pharmaceutical companies and were concerned about ownership of the data. In at least one case, the involvement of partnerships with the pharmaceutical industry would motivate a potential participant to decline research participation (Darquy et al. 2015).

**Length of Consent Process**

Participants in several studies expressed surprise or frustration about the length and burden of the consent process, even while acknowledging the importance of the process. These findings highlight the challenge in balancing the provision of complex information to participants about genomic research and its implications with time constraints and informational needs of potential research participants.

## Discussion

The limited data available about the attitudes and preferences of potential participants in research on rare genetic diseases are generally consistent with robust frameworks of broad consent that have been described in recent policy proposals and are being used with increasing frequency. Participants are generally willing to engage in genetic research and have altruistic motivations for participation that are closely aligned with the structural goals of rare disease research, i.e., to maximize the ability to aggregate precious samples and data for purposes of generating novel insights about genetic disorders. They desire information about confidentiality and security protections, understand the risks and tradeoffs associated with participation, and are willing to proceed with research that involves the sharing of their samples and data broadly.

Consultation with rare disease patients and relatives about their values and preferences using empirical methods can help inform the appropriate framework to implement for informed consent and the associated governance of biospecimen repositories. For example, findings from the survey of potential participants in an LD research database directly contributed to the development of the database. Based on the survey results, organizers of the database chose to combine broad initial consent language with a commitment to providing additional information over time as well as oversight by an ethics steering committee (Darquy et al. 2015, 5). The attitudes expressed by potential enrollees in the database informed the content domains to be included in the initial consent form, including the following:

- nature of data collected and purposes of the database
- data security and confidentiality
- length of storage
- database ownership and governance
- conditions governing academic and pharma-industry partnerships
- commitment to give ongoing information
- existence of an ethics steering committee

Furthermore, the findings from Darquy et al. (2015) suggested that the database governance structure should include the provision of updates to enrollees about relevant topics, including any changes in the nature of collaborations, governance, or ownership of the data, and the availability of research results.

Jamal et al. (2014, 967), who observed that patients' preferences about informed consent may evolve over time as the conduct of research with the samples and data, write: "Our results suggest that beliefs about information use are informed by factors related to situational security and uncertainty, altruism, personality traits, illness histories, and other attributes of context. Because these factors are dynamic and interrelated, we hypothesize that they may change over time."

Although some broad-consent proposals suggest that ongoing communication with participants is an optional, resource-dependent component of broad consent (see chapter 8 in this volume), the dynamic nature of participant preferences revealed in these studies suggests that the provision of updated information may be important in the context of rare disease research.

The data also point to the need for efficient models that do not unduly burden respondents with more information than is required to make autonomous decisions about research participation. Respondents in more than one study found the consent process to be lengthier and more detailed than they had expected—a finding that is consistent with proposals to streamline the consent process for next-generation sequencing research (Levenseller et al. 2014; Bergner et al. 2014) and that lends support to broad-consent models to the extent that they are more efficient and less burdensome than other models.

This analysis is necessarily tentative, given the small number of qualitative studies that have been conducted in rare disease populations. There is a need for further research on whether these views are consistent across a variety of rare disease populations in different national and cultural contexts, such as emerging focus group research with rare disease populations in Europe (McCormack and Cole 2014). In addition, data are needed on whether various consent models that are being implemented to facilitate rare disease research actually result in adequate enrollment while supporting the autonomous decision making of participants. It will be important both to evaluate whether emerging broad-consent policies are adequately responsive to the values and concerns of rare disease populations, and to ensure that emerging programs of research on rare and undiagnosed diseases support the autonomous interests of potential participants while maximizing the ability to conduct valuable research with their contributed samples and data. As a starting point, the limited data available at this time suggest that broad consent for the donation of samples and data—that is, a consent process that includes the provision of initial information about broad future research plans, coupled with ethical oversight and provision of ongoing information to participants about future research uses—is consistent with the values of patients with rare diseases and their relatives.

## Notes

1. The author gratefully acknowledges the expert analysis of Paul Lee, MD, in determining which studies to include in this chapter based on whether the conditions involved qualified as rare disorders, the critical review of an earlier version of the chapter by Benjamin E. Berkman, JD, and the assistance of Carlisle Runge in formatting the manuscript. The preparation of this chapter was funded in part by the intramural research programs of the National Human Genome Research Institute and by the Department of Bioethics at the Clinical Center, National Institutes of Health. The views expressed herein are those of the author and not necessarily a reflection of the policies of the National Institutes of Health or the US Department of Health and Human Services.

2. https://gds.nih.gov/PDF/NIH_GDS_Policy.pdf.

3. Search terms that were used in various combinations included *rare diseases, genetic research, informed consent, broad consent, data sharing, attitudes, health care surveys, research subjects, research participants, genome sequencing, genomic sequencing, exome sequencing,* and *confidentiality.*

## References

Allen, Clarissa, and William D. Foulkes. 2011. Qualitative thematic analysis of consent forms used in cancer genome sequencing. *BMC Medical Ethics* 12: 14. doi:10.1186/1472-6939-12-14.

Bergner, Amanda L., Juli Bollinger, Karen S. Raraigh, Crystal Tichnell, Brittney Murray, Carrie Lynn Blout, Aida Bytyci Telegrafi, and Cynthia A. James. 2014. Informed consent for exome sequencing research in families with genetic disease: The emerging issue of incidental findings. *American Journal of Medical Genetics* 164A (11): 2745–2752. doi:10.1002/ajmg.a.36706 .

Brownstein, Catherine A., Ingrid A. Holm, Rachel Ramoni, and David B. Goldstein, and the Members of the UDN. 2015. Data sharing in the Undiagnosed Diseases Network. *Human Mutation* 36 (10): 985–988. doi:10.1002/humu.22667.

Colledge, Flora, Bernice Elger, and Heidi C. Howard. 2013. A review of the barriers to sharing in biobanking. *Biopreservation and Biobanking* 11 (6): 339–346. doi:10.1089/bio.2013.0039.

Creswell, John W. 1998. *Qualitative Inquiry and Research Design: Choosing Among Five Traditions.* SAGE.

Darquy, Sylviane, Grégoire Moutel, Anne-Sophie Lapointe, Diane D'Audiffret, Julie Champagnat, Samia Guerroui, Marie-Louise Vendeville, Odile Boespflug-Tanguy, and Nathalie Duchange. 2015. Patient/family views on data sharing in rare diseases: Study in the European LeukoTreat project. *European Journal of Human Genetics*. doi:10.1038/ejhg.2015.115 .

Garrison, Nanibaa', Nila A. Sathe, Armand H. Matheny Antommaria, Ingrid A. Holm, Saskia C. Sanderson, Maureen E. Smith, Melissa L. McPheeters, and Ellen W. Clayton. 2016. A systematic literature review of individuals' perspectives on broad consent and data sharing in the United States. *Genetics in Medicine* (November): 18 (7): 663–671. doi:10.1038/gim.2015.138.

Grady, Christine, Yaffa Rubinstein, and Stephen Groft. 2011. Informed consent and patient registry for the rare disease community. *Contemporary Clinical Trials* 33 (1): 3–4. doi:10.1016/j.cct.2011.10.005.

Grady, Christine, Lisa Eckstein, Dan Brock Ben Berkman, Robert Cook-Deegan, Stephanie M. Fullerton, Hank Greely, Mats G. Hansson, et al. 2015. Broad consent for research with biological samples: Workshop conclusions. *American Journal of Bioethics* 15 (9): 34–42. doi:10.1080/15265161.2015.1062162.

Green, Eric, and Mark Guyer, and the National Human Genome Research Institute. 2011. Charting a course for genomic medicine from basepairs to bedside. *Nature* 470 (7333): 204–213. doi:10.1038/nature09764.

Groft, Stephen. 2013. Rare disease research: Expanding collaborative translational research opportunities. *Chest* 144 (1): 16–23. doi:10.1378/chest.13-0606.

Hoeyer, K. 2010. Donors perceptions of consent and feedback from biobank research: Time to acknowledge diversity? *Public Health Genomics* 13 (6): 345–352. doi:10.1159/000262329.

Husedzinovic, Alma, Dominik Ose, Christoph Schickhardt, Stefan Fröhling, and Eva C. Winkler. 2015. Stakeholders' perspectives on biobank-based genomic research: Systematic review of the literature. *European Journal of Human Genetics* 23 (12): 1607–1614. doi:10.1038/ejhg.27.

Jamal, Leila, Julie C. Sapp, Katie Lewis, Tatiane Yanes, Flavia M. Facio, Leslie G. Biesecker, and Barbara B. Biesecker. 2014. Research participants' attitudes towards the confidentiality of genomic sequence information. *European Journal of Human Genetics* 22: 964–968. doi:10.1038/ejhg.2013.276.

Kaye, Jane, Edgar A. Whitley, David Lund, Michael Morrison, Harriet Teare, and Karen Melham. 2015. Dynamic consent: A patient interface for twenty-first century research networks. *European Journal of Human Genetics* 23 (2): 141–146. doi:10.1038/ejhg.2014.71.

Levenseller, Brooke L., Danielle J. Soucier, Victoria A. Miller, Diana Harris, Laura Conway, Barbara A. Bernhardt. 2014. Stakeholders' opinions on the implementation of pediatric whole exome sequencing: Implications for informed consent. *Journal of Genetic Counseling* 23 (4): 552–565. doi:10.1007/s10897-013-9626-y.

Master, Zubin, Eric Nelson, Blake Murdoch, and Timothy Caulfield. 2012. Biobanks, consent and claims of consensus. *Nature Methods* 9 (9): 885–888. doi:10.1038/nmeth.2142.

McCormack, Pauline, and Anna Kole. 2014. Setting up strategies: Patient inclusion in biobank and genomics research in Europe. *Orphanet Journal of Rare Diseases* 9 (Suppl. 1): 2. doi:10.1186/1750-1172-9-S1-P2.

Parker, M., R. Ashcroft, A. O. M. Wilkie, and A. Kent. 2004. Ethical review of research into rare genetic disorders. *BMJ (Clinical Research Ed.)* 329 (7460): 288–289. doi:10.1136/bmj.329.7460.288.

Ponder, M., H. Statham, N. Hallowell, J. A. Moon, M. Richards, and F. L. Raymond. 2008. Genetic research on rare familial disorders: Consent and the blurred boundaries between clinical service and research. *Journal of Medical Ethics* 34 (9): 690–694. doi:10.1136/jme.2006.018564.

Rubinstein, Yaffa, Stephen C. Groft, Sara Chandros Hull, Julie Kaneshiro, Barbara Karp, Nicole Lockhart, Patricia Marshall, et al. 2012. Informed consent process for patient participation in rare disease registries linked to biorepositories. *Contemporary Clinical Trials* 33 (1): 5–11. doi:10.1016/j.cct.2011.10.004.

Sawyer, S. L., et al. 2016. Utility of whole-exome sequencing for those near the end of the diagnostic odyssey: Time to address gaps in care. *Clinical Genetics* 89 (3): 275–284. doi:10.1111/cge.12654.

Shabani, Mahsa, Louise Bezuidenhout, and Pascal Borry. 2014. Attitudes of research participants and the general public towards genomic data sharing: A systematic literature review. *Expert Review of Molecular Diagnostics* 14 (8): 1053–1065. doi:10.1586/14737159.2014.961917.

Tabor, Holly K., Jacquie Stock, Tracy Brazg, Margaret J. McMillin, Karin M. Dent, Joon-Ho Yu, Jay Shendure, and Michael J. Bamshad. 2012. Informed consent for whole genome sequencing: A qualitative analysis of participant expectations and perceptions of risks, benefits, and harms. *American Journal of Medical Genetics* 158A (6): 1310–1319. doi:10.1002/ajmg.a.35328.

Tomlinson, Tom. 2013. Respecting donors to biobank research. *Hastings Center Report* 43 (1): 41–47. doi:10.1002/hast.115.

Wendler, David. 2006. One-time general consent for research on biological samples. *BMJ (Clinical Research Ed.)* 332 (7540): 544–547. doi:10.1136/bmj.332.7540.544.

# 14 Considerations for the Use of Biospecimens in Induced Pluripotent Stem (iPS) Cell Research

Geoffrey Lomax and Heide Aungst

In 2006, the Japanese stem cell researcher Shinya Yamanaka became the first to generate induced pluripotent stem cells, or iPS cells, from mouse fibroblast (skin) cells (Takahashi and Yamanaka 2006). The potential advantage of these cells is that they can be manipulated to behave exactly like human embryonic stem cells to become any type of cell in the body. Yamanaka and other scientists from around the world recognized the promise this discovery had for disease and drug research, as well as potential therapies. Three papers followed Yamanaka's, published in 2007, confirming that, in mice, these skin cells could be reprogrammed to behave like embryonic stem cells (Okita 2007; Wernig 2007; Maherali 2007). Soon after, Yamanaka proved this same pluripotent state could be achieved with human skin cells. Since the pluripotent state represents one of the earliest stages of development where cells can multiply and become any cell in the body, this discovery meant that iPS cells could be used for multiple research purposes, including transforming disease research and drug development, and providing new types of regenerative therapies.

Because skin cells are readily available from a human donor with a known phenotype, the research community recognized that iPS cells can inform research intended to understand the basis for genetically complex diseases. For example, the iPS cells from donors with a similar diagnosis now can be studied to consider how subtle genetic differences may influence disease phenotype. In 2012, when Dr. Yamanaka received the Nobel Prize for this discovery, he acknowledged this possibility, stating, "I believe that the biggest potential of [iPS cell] technology resides in disease modeling and drug screening. Hundreds of diseases can be studied this way. Progress has been made in modeling intractable diseases while searching for new drugs with patient-derived iPS cells by many groups all over the world" (Yamanaka 2012).

As the potential of iPS cells for the study of genetically complex diseases became apparent and the science behind reprogramming cells became routine, the need emerged to curate and distribute collections of derived lines. Biospecimen repositories provide logistical, economic, and scientific advantages (as compared to individual laboratories) for storage and distribution. By 2011, recognizing these advantages, numerous iPS cell biospecimen repositories opened internationally with the expectation that tens of thousands of lines would be available for basic research, disease modeling, and drug screening (Grens 2014).

With the promise of this growing demand, the California Institute for Regenerative Medicine (CIRM) set the goal of sponsoring one of the world's largest repositories of iPS cells (Grens 2014). Since its beginnings, CIRM has had a history of being on the leading edge of stem cell research. CIRM started in response to President George W. Bush's August 2001 ban on using federal funds to pay for research that would develop new embryonic stem cell lines. Concerned about the quality of the existing lines that could be used for federally funded research, scientists believed that the United States would fall behind other countries. In 2004, California voters passed Proposition 71, a referendum authorizing $3 billion to fund research in the state to develop stem cell therapies. This authorization meant that California could fund embryonic stem cell research, attract preeminent researchers, and keep the important scientific momentum going in the field of regenerative medicine.

From its beginning, CIRM's mission has been to accelerate stem cell treatments to patients with unmet medical needs. Substantial and sustained funding has placed California at the center of an international effort to advance stem cell science and regenerative medicine—beginning with embryonic stem cells and today, evolving, along with the science, to include iPS cells.

In 2012, with the potential for iPS cells to inform disease modeling and drug discovery evident, CIRM recognized the advantages of enhancing standardized cell collections. After consulting with stem cell scientists, CIRM determined that existing collections lacked cell numbers and disease diversity (CIRM 2010). In addition, these collections had been created using a variety of cell line derivation procedures. Heterogeneity in methods may contribute to differences in (1) the degree of reprogramming and (2) the behavior of cells in downstream transformation experiments (CIRM 2010; NIH 2014). CIRM realized that this lack of standardization ultimately could impact the ability to replicate studies and compare findings from cell lines derived from different donors.

In an effort to address these operational issues, CIRM committed $32 million to create an iPS cell repository to be populated with up to 3,000 lines of high and consistent quality iPS cells. Through a competitive peer review process, CIRM asked organizations interested in deriving and banking iPS cell lines to propose a set of rigorous methods of quality control that could be applied through supply chain management—across biospecimen procurement, derivation, expansion and storage, and distribution. In addition, CIRM determined the verification of cell identity, purity, viability, and sterility (CIRM 2015a) was equally important.

In 2013, CIRM awarded funds to a consortium of clinical sites, for the collection of primary biospecimens, a derivation center, and a biorepository. Seven clinician scientists from four California institutions led efforts to recruit tissue donors who suffered from specific diseases, along with healthy controls. The target diseases included neurodevelopmental disorders (epilepsy, autism, cerebral palsy), pulmonary fibrosis, viral hepatitis, heart disease, Alzheimer's disease, and blinding eye disease.

The process of creating an iPS cell line for the CIRM repository begins with the collection of blood or a small piece of skin from a donor. The collection site sends the sample to the company Cellular Dynamics International (CDI). CDI generates iPS cells from the samples using a standardized process and then transfers the iPS cells to the Coriell Institute for Medical Research. Coriell operates the CIRM repository and makes the iPS cells available, for a fee, to researchers at academic and other nonprofit institutions, as well as to pharmaceutical companies that may want to use them to find new drugs for the diseases represented in the samples. In order to comply with the requirements that CIRM funds be expended in California, Coriell, headquartered in Camden, NJ, and CDI, headquartered in Madison, Wisconsin, set up facilities in the state to generate and bank the iPS cells for this initiative.

**Unique Ethics Policy Considerations**

CIRM created the iPS cell repository to address technical limitations with existing lines. During the formative phase of development, however, CIRM discovered unique ethics policy considerations associated with the creation and distribution of iPS cell collections. These considerations extend beyond those concerns associated with biorepositories in general. These additional issues emanate from the fact that (1) donor cells are transformed, (2) the resulting iPS cells can renew themselves with (3) the potential to differentiate into any cell in the body (Lo et al. 2010; Sugarman 2008; Zarzeczny 2009;

Dasgupta 2014). These properties—transformation, self-renewal (immortalization), and plasticity with differentiation potential—allow researchers to conduct novel experiments. For example, to better understand the natural history of disease, a researcher can engraft stem cells into animals to simulate human disease. Some experiments involve implanting human cells into animals prenatally during gestation. Transplantation during early development allows the cells to integrate into the mature animal tissue. As a result, substantial human animal chimerism may occur. Such integration is qualitatively different from the simple injection of human cells into an animal—a technique commonly used in research that some may be familiar with. This procedure, while necessary to advance some types of research, raised ethical concerns to the point that a select committee of the International Society for Stem Cell Research issued "Ethical Standards for Human-to-Animal Chimera Experiments in Stem Cell Research" in 2007 (ISSCR 2007). The unique properties of pluripotent stem cells, combined with the potential for novel scientific application, requires special consideration in the procurement of donor specimens and the management and distribution of resulting iPS lines.

Before the initial collection of biospecimens from research participants, CIRM consulted a standing body of advisors—the Medical and Ethical Standards Working Group (CIRM 2015b) to address ethical and policy issues associated with the development and operation of the iPS cell repository. Scientists, ethicists, legal scholars, and patient advocates participated in the working group. As part of its ongoing tasks, the working group frequently consults outside experts and includes the public in its deliberations. Over a three-year period, CIRM conducted a series of public meetings and workshops to address a number of ethics and policy considerations associated with the iPS cell biospecimen repository program. CIRM designed these meetings specifically to include perspectives from researchers, patients, potential donors, and the public.

Through this deliberative and inclusive policy development process, CIRM has addressed the scientific and social concerns associated with the research it funds. As a policy baseline, CIRM requires all funded research to comply with the federal Common Rule and California law designed to protect research subjects (CIRM 2005). Since its beginnings in 2004, CIRM has strived for consistency with federal regulations governing human subjects research. However, with the restrictions placed on embryonic stem cell research using federal funds during CIRM's development, CIRM had to be at the forefront of creating many of its own ethical policies. In the years that followed CIRM's 2004 formation, other professional and governmental organizations also established guidelines and policies for stem cell research. The National Academy of Sciences set Guidelines for Human Embryonic Research in 2005 (followed by amendments in 2007, 2008, and 2010). The National Institutes of Health published its Guidelines for Human Stem Cell Research in 2009, coinciding with President Barack Obama's executive order reversing President Bush's imposed limitation on using federal funds for the development of new stem cell lines. In his order, President Obama wrote, "These Guidelines apply to the expenditure of NIH funds for research using human embryonic stem cells and certain uses of human induced pluripotent stem cells."

The mention of induced pluripotent stem cells in the executive order proved that federal policy finally recognized both the scientific potential of iPS cells and the possibility of ethical issues in their usage. Understanding that iPS cell research continues to push the boundaries, CIRM aims for a uniform approach to protecting research participants. For example, there are unique ethics policy considerations associated with iPS cell derivation and banking, including the potential to create germline cells. In December 2014, researchers from the UK and Israel announced that they had made human sperm and egg precursor cells in a laboratory dish from iPS cells (Irie et al. 2015). CIRM had identified this potential years earlier, integrating it into a comprehensive donor consent.

**Donor Consent**

The CIRM iPS cell repository is a complex undertaking operationally. There are seven different tissue collection programs with the aim of prospectively collecting primary specimens (blood or skin) from 3,000 donors. Each of the seven programs represents a different disease condition. A major priority of the program is to ensure that all iPS cells may be used for broad research purposes, including development of commercial products. Therefore, before initiating collection of specimens, CIRM decided it was essential to develop a model donor consent process to describe the nature and purpose of the iPS cell repository. CIRM held a series of public meetings to develop, in consultation with the Standards Working Group, a model consent form (CIRM 2011a; CIRM 2012).

Research teams from the seven different tissue collection protocols adapted the model consent form to their clinical sites. Before initiating collection of specimens, CIRM, in consultation with the consortium, re-reviewed the consent forms. The collection sites typically had adapted consent protocols used for biospecimen collections to be retained at the host institution. In some instances, there were statements in these forms that were inconsistent with the iPS cell banking protocol. The most common inaccuracies were statements that the principal investigator or attending physician would retain control of the specimens or derived cells or "approve" future uses of the materials. In fact, the iPS cell bank (at the daily management level operated by Coriell) has that responsibility. In addition, some consent forms indicated that CIRM (as the guiding body) would approve cell usage. We alerted collection teams to these inaccuracies, and they modified the forms accordingly. This consent "calibration" procedure is an essential step in large-scale, multisite specimen collection programs.

The final CIRM template addressed basic issues related to the collection, processing, and distribution of biospecimens, including research aims, risks to the donor, and the potential uses of derived cell lines. The consent form also emphasized multiple other important considerations in stem cell research, including the commercial use of cell lines, return of results, withdrawal from research, and sensitive use of cells in research.

**Commercial Use**

One aim of the iPS cell repository was to spur commercial use, particularly, for drug development. CIRM selected academic, nonprofit, and commercial organizations to develop the iPS cell repository. In large part, CIRM relied on CDI and the Coriell Institute since both have considerable experience with customers using cells to develop commercial products. Given the substantial investment required to market new research and medical products, these customers are particularly attentive to consent language pertaining to the use of specimens for commercial applications. CIRM deemed these two issues as critical: (1) the ability to use cells for a broad range of commercial applications and (2) clear disclosure that the donor would not retain a financial interest in any resulting products. Based on these considerations, CIRM added language to the model consent form disclosing the following:

The iPS cells may be useful for developing drugs, medical products or other commercial products. The products may be patentable or have commercial value and you will not own or have a financial interest in them.

Although drug development is an immediate aim of the repository, thus specifically stated, research involving the cell lines also could spur the development of a wide array of products. For example, banked lines have been used to develop imaging technologies that enable injected cells to be tracked within the body (Wang et al. 2013). There are numerous ways in which the cells may be instrumental in creating products. To attract commercial users, robust consent should reflect the range of potential uses. In addition to the consent form, CIRM emphasized these potential uses in an educational brochure that the collection sites provided to donors as part of the informed consent protocol.

Robust consent for commercial use should not be viewed as solely benefitting private interests. CIRM involved patients and patient advocates in policy discussions relating to the use of donated specimens. While perhaps not representative of the population as a whole, participants in these discussions expressed a strong desire for their donated materials to be applied toward the advancement of science and medicine. Disease had touched these donors' lives, and they expressed trust and confidence in the research community (CIRM 2011b). A principle aim of participation is to spur the development of new treatments for disease. For example, a Parkinson's patient speaking at a CIRM seminar (unpublished) indicated that trust in his doctor, who recruited him for an iPS study, was his primary motivator, and he hoped his cells could contribute to science and medicine.

The aspirations of certain donors to contribute to science and medicine, coupled with a robust consent process and the opportunity to withdraw, can accelerate the development of therapies. CIRM believes the iPS cell repository protocol effectively advances both donors' and societal aspirations for new therapies in an ethically responsible and legally compliant manner. The issue of biospecimen use in research is contested in the contemporary policy arena. Alternative proposals have called for re-review and/or consent for secondary studies (OHRP 2015), and the final revision to the regulations published in 2017 set forth a range of options for research with specimens, from specific consent and IRB review to broad consent and limited IRB review to no consent at all, but improved explanation of how specimens might later be used for research at the time they are collected (Final Rule 2017).

**Return of Results**

The CIRM iPSC initiative employs a repository research system model where we use coded donor tissue samples to generate iPS cell lines that we then deposit in the repository for curation and subsequent distribution to secondary researchers (Wolf 2012). Coding enables the potential for re-contact because the cell line can be linked back to one of the seven collection sites that provide ongoing patient care. As CIRM distributes cell lines, new information may become available. For example, CIRM sponsors new projects with the aim of performing genetic characterization of the iPS cell collection. Also, genetic sequencing or expression analysis may be routinely performed in independent studies where researchers use banked lines. Besides genetic characterizations, banked lines may be used to evaluate the therapeutic potential of drugs or molecular agents. The genetic research, evaluation of drug efficacy, or a range of unforeseen studies may reveal information that has potential therapeutic benefit to the donor.

Therefore, CIRM questioned under what conditions, if any, such findings would warrant re-contacting the original donor and/or donors with the same condition or a related one. First, CIRM considered whether previously published ethical criteria for the return of research results also applied to this type of genomic research (Wolf 2012) using iPS cell lines. In the case of genomic research, the literature has focused on the predictive value of genetic markers for future risk of disease. Applying the standards for genomic research, we considered whether iPS cell lines might reveal risks to the original cell donors that would warrant re-contact. To address this question, we reviewed the literature to ascertain whether markers in derived iPS cell lines would be sufficiently predictive. This review revealed the following (Lomax and Shepard 2013):

- *Genetic Instability*: iPS cells have shown significant genetic variability when reprogrammed and cultured (Young 2012). Therefore, the resulting genotype does not always accurately reflect the donor's genotype.
- *Validation Criteria*: Protocols do not exist to harmonize results from research laboratories using iPS cell lines with those of Clinical Laboratory Improvement Amendments (CLIA)-approved clinical laboratories that are necessary to validate findings.
- *Etiologic Complexity*: Genetically complex conditions, such as neurological and heart disease, result from a poorly understood confluence of pleiotropic gene effects. Methodologies for the quantification of relative risk and penetrance for such disorders are limited and generally have not been sufficiently validated.

Based on these factors, CIRM determined that iPS cell research would be unlikely to satisfy criteria applied in genomic research for clinically relevant research findings, and therefore, would not return results to the specimen donors based only on genetic sequencing information.

There may be non-genetic findings, however, that may have clinical significance. For example, researchers routinely use iPS cells as a drug screening tool. Screening may reveal a positive or adverse drug reaction in a subpopulation. If we had recruited cell donors within this population, this knowledge would inform the patients' care. There is consensus that research results should be disseminated if participants and others potentially stand to benefit (Johnson et al. 2012). Further, Ulrich (2013) suggests that if a researcher possesses knowledge that could mitigate harm or advance the health of a research participant, there may be a duty to return information. The imperative to return information may be amplified when there is little risk involved in providing such information. Finally, studies suggest there is a desire among participants and the public to receive research results (Bollinger et al. 2012; Mester 2015).

In addition, patient advocates and members of the public who participated in public meetings and workshops expressed a strong desire for CIRM to disseminate and share research findings. Patient advocates actually expressed frustration over privacy laws that prohibit the sharing of biospecimens and research findings (CIRM 2011b). CIRM's Standards Working Group concurred with the conclusion that if one found information through subsequent biospecimen analysis that may be clinically actionable, a duty to notify may exist. In view of this potential benefit, there was consensus that the informed consent template should leave open the possibility of re-contacting the principal investigator at the collection site. Such an approach embodies the principles for general return of results articulated by the Secretary's Advisory Committee for Human Research Protection (SACHRP 2014).

Thus, in the model consent template, donors have the option of being re-contacted. The statement reads:

In the future, we may want to contact you to (1) obtain additional samples or updates on your health or (2) inform you about significant new findings that may impact you, or (3) to get your permission for research not covered in this consent form.

In addition, the form states that donors may opt out of re-contact or they may "partially withdraw" their participation so re-contact would not be possible. Partial withdrawal occurs when the participant requests that the code linking the iPS cell to their donated skin or blood be removed. If that linkage is removed, there is no way to link the iPS cell to the donor and re-contact is not possible.

CIRM advised the IRBs for each of the collection sites of the rationale for this language, and each approved the provision for allowing re-contact. One noteworthy consideration is whether participants should "opt in" or "opt out" of re-contact. Under the "opt in" model, donors must express the desire to be re-contacted. Under the "opt out" model, there is a presumption that re-contact would occur unless donors indicate they do not wish to be re-contacted. A number of institutions indicated that they had an "opt in" policy for all research. Regardless of the model used, all donors have the opportunity to be re-contacted if they choose.

During the early planning phase of the iPS Initiative, the consortium developed a protocol to be followed in the event that a biospecimen met the conditions for re-contacting the donor. If there were findings of sufficient strength where donor re-contact may be warranted, an investigator could contact the repository. The repository has the ability to link the iPS cell line back to one of the seven collection sites. The link is a code that associates the specimen to the individual donor. The collection site's principal investigator then mediates re-contact. However, typically, an IRB would review the request for donor re-contact first. If the IRB approved it, then donor re-contact could proceed.

**Withdrawal from Research**

CIRM's Standards Working Group also advised on policies and procedures for participant withdrawal. The CIRM iPS cell biospecimen repository maintains (1) personal/medical information associated with the donor, (2) primary biospecimens (blood or skin) and (3) derived iPS cell lines. CIRM's primary concern was balancing respect for participants while maintaining an important scientific resource representing a public investment of approximately $32 million. Beyond the financial investment, ethically, this balance between donor autonomy and societal beneficence is a critical consideration for all repositories, not just those with iPS cells (see chapter 6).

CIRM policies incorporate the federal Common Rule standards for withdrawal from research. Under this framework, identifiers may be removed as a means of withdrawing the participant from research while the biospecimen can continue to be used in research. The Standards Working Group did not recommend removal of identifiers as a policy for the CIRM bank because they viewed it as "not respecting the participant." But the working group did agree that if a patient chose to withdraw from research, then any remaining donated materials should be destroyed and there would be no further contact.

From the standpoint of donor withdrawal, CIRM views derivative iPS cell lines as distinct from the donated skin and blood biospecimens. First, iPS cells result from the genetic transformation of donated materials, since manipulation of the cells to a pluripotent state requires the use of additional source DNA usually injected into the donor cells. Thus, they are a derivative product of the primary specimen. These iPS cell lines would not exist without the agency of the investigator, and the resulting lines are not genetically identical to the primary specimens. Perhaps more importantly, iPS cell derivation requires a substantial commitment of resources. The commitment of public resources for sample collection and line derivation to CIRM's technical specifications is approximately $7,000 per cell line. Based on this, CIRM informs donors during the consent process that derived iPS cell lines cannot be withdrawn from the bank. Once derived, the de-identified iPS cell line will continue to be distributed. If donors express a desire withdraw, they only have the choice of having their original blood or skin specimens destroyed, having identifiers stripped from their donated specimens and iPS cell lines, or both.

To implement this policy, CIRM requires the biorepository to determine whether the participant has withdrawn consent before or after the iPS cell derivation. In the context of the current protocol, an iPS cell line is considered derived when at least one clone at passage five (of dividing and growing the cells) has been transferred to the repository.

## Sensitive Use of Specimens

The differentiation potential of iPS cells creates concerns over possible sensitive downstream use. There is anecdotal and empirical evidence suggesting that donors for iPS cell derivation have concerns over how materials may be used. For example, findings from a US-based focus-group study (Sugarman 2008) that elicited patients' attitudes toward the donation of cells for iPS cell derivation and banking identified a number of specific concerns, including the following:

• re-identification of the donor, privacy infringements, and the potential for this information to be used in an unfair or discriminatory manner;
• inability to control the downstream use of cells and prevent their inappropriate commercial use;
• reprogramming cells to create gametes or to perform human cloning;
• somatic cell nuclear transfer studies without explicit donor consent.

As is discussed in other chapters in this volume, some of these issues are not unique to iPS cells and various mechanisms exist to address donor concerns, including a robust consent process and data use transfer agreements designed to control downstream uses of cell lines.

Gamete and embryo creation is, however, one sensitive use that is unique to stem cell research. For example, scientists have observed that stem cells can undergo self-organized development in vitro into structures that mimic the body plan of the post-implantation embryo (Pera 2015). Given the potential for this area of research and the stated concerns of donors, CIRM decided to explicitly ask for consent to gamete and embryo research as an option to tissue donors in CIRM's model consent protocol despite a general desire to limit the number of donor options in a consent form to reduce operational complexity. Given the important, yet sensitive, nature of this research, particularly of gamete research, there was a strong consensus that donors should be offered a choice as to whether to consent to this specific direction of research. Parents and guardians viewed this choice as particularly important when consenting minors.

**Previously Banked Research Specimens**

CIRM also has worked in collaboration with other research organizations through a research project called DISCUSS (standing for Deriving Induced Stem Cells Using Stored Specimens) that had the goal of developing consensus for the use of banked biospecimens that don't have explicit consent for iPSC derivation and banking (Lomax et al. 2013). CIRM formed these partnerships because we believe that stem cell research may be consistent with the underlying consents for many established biospecimen collections. CIRM found that many donors had consented for broader biomedical research purposes for their specimens, in accordance with accepted principles for ethics and oversight (NIGMS 2015; NIMH 2015). For example, donors may consent to have specimens archived for research aimed at understanding disease processes or informing therapy development. In this context, cellular reprogramming represents one method, among many, of studying disease etiology or potential therapies.

CIRM adopted and applied what we learned through the DISCUSS project: a consistency standard for banked specimens, by evaluating the previously obtained consent against the research aim, rather than to require consent tailored to a specific experimental procedure. Some experts have suggested that the unique qualities of stem cells necessitates the need for control of cell lines by donors or dynamic consent when researchers propose new experimental procedures (Holm 2006; Mascalzoni et al. 2008). In effect, the act of deriving an iPS cell line necessitates specific consent. We should exercise caution in "expanding obligations" to comparatively narrow areas of research (e.g., derivation of pluripotent human cell lines) that may create unintended consequences (Gunsalus 2006). Well-intended, stem cell-specific regulatory actions may actually inhibit research, thus undermining beneficence, without providing clear social benefit (Kawakami 2010).

Determining the consistency of a particular scientific application with the donor consent serves to protect research participants, while supporting the beneficent use of specimens. Every biorepository should have a mechanism in place to systematically review proposed research uses to ensure appropriate utilization by qualified researchers. Material transfer agreements should be used to document and enforce conditions for use. Repositories routinely apply such mechanisms to curate existing collections.

As Dr. Yamanaka suggested in 2012, disease modeling and drug discovery remain primary applications of iPS cell research. Many biorepositories established their collections (typically blood) before 2007. These collections include more than 100,000 well-characterized specimens representing a spectrum of disease genotypes (NIMH 2015). Many of these specimens have consents for broader biomedical research, in accordance with accepted principles for ethics and oversight (NIGMS 2015). However, consent-related considerations may impact the ability to create or otherwise use iPS cell lines.

**Conclusion**

Induced pluripotent stem cells are a powerful tool for advancing biomedical research and therapy development. CIRM is committed to expanding its iPS cell repository to accelerate the development of therapies for unmet medical needs. Recognizing the potential for iPS cells to advance research and medicine, donors are willing to participate in CIRM's biorepository initiative. To ensure informed donors, the consent process addresses a number of critical topics that arise in the context of cell line immortalization and differentiation. The CIRM repository initiative has developed a robust donor consent process with the opportunity for donors to withdraw their specimens before cell manipulation. CIRM pays close attention to several critical issues, including the commercial use of cell lines, the ability to re-contact donors about clinically significant findings, the possibility of donor withdrawal, and the potential for sensitive uses of derived lines. CIRM's aim is to maintain a collection of cell lines that can be used broadly in research for perpetuity. This process fulfills the aspiration of donors to contribute to science and medicine while ensuring the repository meets high ethical and legal standards.

## References

Bollinger, Juli, Joan Scott, Rachel Dvoskin, and David Kaufman. 2012. Public preferences regarding the return of individual genetic research results: Findings from a qualitative focus group study. *Genetics in Medicine* 14 (4): 451–457.

CIRM. 2005. Scientific and Medically Accountability Standards. https://www.cirm.ca.gov/our-funding/chapter-2-scientific-and-medical-accountability-standards.

CIRM. 2010. Summary and Recommendations of the CIRM Human iPS Cell Banking Workshop San Francisco. November 17–18. https://www.cirm.ca.gov/sites/default/files/files/about_cirm/iPSC_Banking_Report.pdf.

CIRM. 2011a. CIRM Medical and Ethical Standards Working Group. April 29. https://www.cirm.ca.gov/sites/default/files/files/agenda/transcripts/042911_SWG.pdf.

CIRM. 2011b. Stem cell banking and the making of a patient "advocist." https://blog.cirm.ca.gov/2011/06/02/stem-cell-banking-and-making-of-patient-2/.

CIRM. 2012. CIRM Model Consent Form. https://www.cirm.ca.gov/sites/default/files/files/Appendix_A.pdf.

CIRM. 2015a. Induced Pluripotent Stem Cell Initiative. https://www.cirm.ca.gov/researchers/ipsc-initiative.

CIRM. 2015b. Scientific & Medical Accountability Standards Working Group. https://www.cirm.ca.gov/node/3425.

Dasgupta, Ishan, Juli Bollinger, Debra J.H. Mathews, Neil M. Neumann, Abbas Rattani, and Jeremy Sugarman. 2014. Patients' attitudes toward the donation of biological materials for the derivation of induced pluripotent stem cells. *Cell Stem Cell* 14 (9): 9–12.

Department of Homeland Security et al. 2017. Final rule. Federal policy for the protection of human subjects. *Federal Register* 82 (12): 7149–7274.

Grens, Kerry. 2014. Banking on iPSCs. *The Scientist*. http://www.the-scientist.com/?articles.view/articleNo/40376/title/Banking-on-iPSCs/.

Gunsalus, C. K., Edward Bruner et al. 2006. Mission creep and the IRB world. *Science* 9: 1441.

Holm, Soren. 2006. Who should control the use of human embryonic stem cell lines: A defense of the donors' ability to control. *Bioethical Inquiry*. (3): 55–68.

Hyun, Insoo, Konrad Hochedlinger, Rudolf Jaenisch, and Shinya Yamanaka. 2007. New advances in IPS cell research do not obviate the need for human embryonic stem cells. *Cell Stem Cell* 4 (1): 367–368.

Hyun, Insoo, Patrick Taylor, Giuseppe Testa, Bernard Dickens, Kyu Won Jung, Angela Mcnab, John Robertson, Loane Skene, and Laurie Zoloth. 2007. Ethical standards for human-to-animal chimera experiments in stem cell research. *Cell Stem Cell* 2 (1): 159–163.

Irie, Naoko, Leehee Weinberger, Walfred W. C. Tang, Toshihiro Kobayashi, Sergey Viukov, Yair S. Manor, Sabine Dietmann, Jacob H. Hanna, and M. Azim Surani. 2015. SOX17 is a critical specifier of human primordial germ cell fate. *Cell* 1–2 (160): 253–268.

Johnson, Gina, Frances Lawrenz, and Mao Tao. 2012. An empirical examination of the management of return of individual research results and incidental findings in genomic biobanks. *Genetics in Medicine* 14 (4): 444–450.

Kawakami, Mashairo, Douglas Sipp, and Kazuto Kato. 2010. Regulatory impacts on stem cell research in Japan. *Cell Stem Cell* 6(5): 415–418.

Lo, Bernard, et al. 2010. Cloning mice and men: Prohibiting the use of iPS cells for human reproductive cloning. *Cell Stem Cell* 6 (1): 16–20.

Loh, Yuin-Han, Odelya Hartung, Hu Li, Chunguang Guo, Julie M. Sahalie, Philip D. Manos, Achia Urbach, et al. 2010. Reprogramming of T cells from human peripheral blood. *Cell Stem Cell* 1 (7): 15–19.

Lomax, Geoffrey, and Kelly Shepard. 2013. Return of results in translational iPS cell research: Considerations for donor informed consent. *Stem Cell Research & Therapy* 6 (4): 1–5.

Lomax, Geoffrey, Sara Chandros Hull, Justin Lowenthal, Mahendra Rao, and Rosario Isasi. 2013. The DISCUSS Project: Induced pluripotent stem cell lines from previously collected research biospecimens and informed consent: Points to consider. *Stem Cells Translational Medicine* 2 (10): 727–730.

Maherali, Nimet, Rupa Sridharan, Wei Xie, Jochen Utikal, Sarah Eminli, Katrin Arnold, Matthias Stadtfeld, et al. 2007. Directly reprogrammed fibroblasts show global epigenetic remodeling and widespread tissue contribution. *Cell Stem Cell* 1 (1): 55–70.

Mascalzoni, Deborah, Andrew Hicks, Peter Pramstaller, and Matthias Wjst. 2008. Informed consent in the genomics era. *PLoS Medicine* 5 (9): 1302–1305.

Mester, Jessica, MaryBeth Mercer, Aaron Goldenberg, Rebekah Moore, Charis Eng, and Richard Sharp. 2015. Communicating with biobank participants: Preferences for receiving and providing updates to researchers. *Cancer Epidemiology, Biomarkers & Prevention* 24 (2): 708–712.

NIGMS. 2015. National Institute of General Medical Sciences consent protocol. https://catalog.coriell.org/0/Sections/Support/NIGMS/InfConsent.aspx?PgId=216.

NIH. 2014. NIH Common Fund Center for Regenerative Medicine (CRM) NIH CRM Virtual Workshop Report—May 6, 2014. https://commonfund.nih.gov/sites/default/files/CRM_May_6_2014_Summary_final.pdf.

NIMH. 2015. The NIMH Repository and Genomics Resource. https://www.nimhgenetics.org/about/faqs.php.

OHRP. 2015. NPRM 2015 Summary. http://www.hhs.gov/ohrp/regulations-and-policy/regulations/nprm-2015-summary/.

Okita, Keisuke, Tomoko Ichisaka, and Shinya Yamanaka. 2007. Generation of germline-competent induced pluripotent stem cells. *Nature* 448 (7151): 313–317.

Pera, Martin F., Guido de Wert, Wybo Dondorp, Robin Lovell-Badge, Christine L Mummery, Megan Munsie, and Patrick P. Tam. 2015. What if stem cells turn into embryos in a dish? *Nature Methods* 12 (10): 917–919.

Robinton, Daisy A., and George Q. Daley. 2012. The promise of induced pluripotent stem cells in research and therapy. *Nature* 7381 (481): 295–305.

SACHRP. 2014. Recommendations Regarding Return of General Research Results. http://www.hhs.gov/ohrp/sachrp/commsec/sharing_study_data_and_results.html.

Sugarman, Jeremy. 2008. Human stem cell ethics: Beyond the embryo. *Cell Stem Cell* 2 (6): 529–533.

Takahashi, Kazutoshi, Koji Tanabe, Mari Ohnuki, Megumi Narita, Tomoko Ichisaka, Kiichiro Tomoda, and Shinya Yamanaka. 2007. Induction of pluripotent stem cells from adult human fibroblasts by defined factors. *Cell* 131 (5): 861–872.

Takahashi, Kazutoshi, and Shinya Yamanaka. 2006. Induction of pluripotent stem cells from mouse embryonic and adult fibroblast cultures by defined factors. *Cell* 126 (4): 663–676.

Ulrich, Michael. 2013. The duty to rescue in genomic research. *American Journal of Bioethics* 13 (2): 50–51.

Wang, Yaqi, Chenjie Xu, and Hooisweng Ow. 2013. Commercial nanoparticles for stem cell labeling and tracking. *Theranostics* 3 (8): 544–560.

Wernig, Marius, Alexander Meissner, Ruth Foreman, Tobias Brambrink, Manching Ku, Konrad Hochedlinger, Bradley E. Bernstein, and Rudolf Jaenisch. 2007. In vitro reprogramming of fibroblasts into a pluripotent ES-cell-like state. *Nature* 448 (7151): 318–324.

Wolf, Susan, Brittney Crock, Brian Van Ness, Frances Lawrenz, Jeffrey P. Kahn, Laura M. Beskow, Mildred K. Cho, et al. 2012. Managing incidental findings and research results in genomic research involving biobanks and archived data sets. *Genetics in Medicine* 14 (4): 361–384.

Yamanaka, Shinya. 2012. Nobel Lecture. http://www.nobelprize.org/nobel_prizes/medicine/laureates/2012/yamanaka-lecture.html.

Young, Margaret, David Larson, Chiao-Wang Sun, Daniel R. George, Li Ding, Christopher A. Miller, Ling Lin, et al. 2012. Background mutations in parental cells account for most of the genetic heterogeneity of induced pluripotent stem cells. *Cell Stem Cell* 10 (5): 570–582.

Zarzeczny, Amy, Chris Scott, Insoo Hyun, Jami Bennett, Jennifer Chandler, Sophie Chargé, Heather Heine, et al. 2009. iPS cells: Mapping the policy issues. *Cell* 139 (6): 1032–1037.

# V Governance, Accountability, and Operational Considerations

# Introduction

Barbara E. Bierer

Since the issuance of the Belmont Report, in 1979, discovery and understanding of pathobiology, of mechanisms of disease origin and progression, of correlative biomarkers, of therapeutic targets, and of the genomic foundations of risk, health, and disease are progressing at an unprecedented pace. Notably, much of the research relies upon the availability of well-curated biospecimens and associated phenotypic data. Biospecimens donated from individuals enrolled in clinical research protocols and those that remain after their appropriate clinical use (excess clinical specimens) are collected and stored in biospecimen repositories for future use (specified or unspecified). The value and the utility of those biospecimens are enhanced substantially when the specimens are associated with personal health information, often identifiable. The speed of research is accelerated by the availability of such biospecimen repositories, and the growth of and investment in biorepositories over the last twenty years testify to their value.

In the United States, biospecimen repositories are governed by ethical principles and are subject to review and oversight by institutional review boards. Any biorepository must conform to these principles and comply with human research regulations. But a biorepository has many stakeholders, including the institution housing the repository, the sponsor funding the repository, the IRB with oversight responsibilities, the investigators responsible for initial collection of biospecimens, the investigators who utilize the banked biospecimens for secondary research, and the participants who donate their biospecimens and data.

Responsibility for a biorepository, once solely in the hands of investigators and their institutions, is shifting. With increasing impact and expectation, participants, their families, and patient advocates are challenging the status quo. The chapters in this part review current requirements for the creation and operations of a biospecimen repository, then go on to argue that the current requirements are largely inadequate, and that the role of the participant and of the patient advocate should be respected, not simply as a passive voice in consent discussions but as an active and important partner throughout the process. The boundaries of this partnership are yet to be defined.

The creation of any biospecimen repository begins with careful consideration of what specimens will be collected and how they will be transferred to the biospecimen repository and stored. These considerations include how and where potential participants will be approached, how informed consent will be obtained, and how each repository will be managed and governed. Therein lies the first set of issues to be considered, addressed in the chapter by Heffernan and colleagues and in that by Ostrom and Barnholtz-Sloan. Beyond good planning and execution, review and approval of the biorepository protocol, and standard operating procedures for each process-driven procedure and transaction (including ensuring quality specimens are delivered and maintained in the biorepository), the architects of the biospecimen repository must consider—from the outset—who will have access to the collected specimens for secondary research and the process for gaining such access.

The nature of these considerations changes if the specimens are specifically collected for research (and thus with an initial informed consent, to be discussed below), if they remain after clinical use and must then, without research consent, be de-identified before distribution and secondary use, or if they will be provided after IRB waiver of informed consent. Any collection of excess clinical biospecimens ("discarded" specimens) represents an invaluable resource for future research and yet must conform to applicable internal hospital or institutional policies and procedures. Large-scale transcriptional profiling and, later, whole genome sequencing of rare diseases (e.g., malignant pleural mesothelioma) were first performed using tissue specimens collected over many years in the course of clinical care. This research, which yielded information about genetic mutations and potential avenues for therapies, would not have been possible without pathologists' careful stewardship of biospecimens collected and used in the course of standard clinical care. Under current regulatory approaches, these specimens are routinely stripped of identifiers, and thus informational risk is minimized.

On the other hand, when biospecimens are collected for research purposes *ab initio*, and when biospecimens will be linked to potentially identifiable personal health information, informed consent from the subject/donor (or IRB waiver of informed consent) is typically needed. Many of the specific issues in informed consent, including whether "broad" consent can be truly informed at all, are discussed elsewhere in this volume, but in this part a resonant theme is that of participants' trust in research and science, and in investigators' specific use of their specimens and data. Rothwell and Johnson explore that issue in detail and present an interesting and novel approach. They suggest, and present empirical data for, the value of an "Investigator's Oath," modeled on FDA Form 1572 ("Statement of Investigator") and reminiscent of the Hippocratic Oath, that would concretize the responsibilities of investigators in research. This oath would serve multiple purposes: reminding investigators of their commitments, securing written confirmation of responsibilities that the institution could use as documentation in the event of misconduct, and clarifying for participants the roles and responsibilities of investigators, hopefully to increase trust in the process and the actors.

Additional consent considerations are reviewed by Ostrom and Barnholtz-Sloan, including whether the subject/donor chooses to permit re-contact and under what circumstances. Notably, as Wolf and Kohane discuss, re-contact may be particularly appropriate in circumstances where return of research results or notice of incidental findings may have tangible, significant effects on health decision making or on health planning. Two derivative issues follow from the notion of re-contact for reporting results and must be considered. First, enabling re-contact means that identifiers must be maintained with each biospecimen throughout its use, from storage to release from the biobank to the user and through finalizing the results. Yet the very maintenance of identifiers with the biospecimen increases the risk of inadvertent re-identification and breach of privacy. The second of these issues is the determination of what, when, how, by whom, and to whom results should be returned. Though some situations may seem obvious (e.g., a BRCA1 mutation indicates a 65 percent cumulative risk of breast cancer), the implications may not be. For the woman with a BRCA1 mutation, should the finding simply increase her diligence for periodic mammography? Should she seek counseling or undergo prophylactic surgery to remove her breasts and perhaps her ovaries? How certain must the investigator be of the accuracy of the test and of the clinical significance before making such a disclosure? What follow-up (e.g., confirmatory testing, genetic counseling, referral for treatment) should be offered, and who will bear the cost? And must one then also inform family members who might carry the same genetic risk factor? Complexity increases as the clinical significance and personal utility of many genetic and other findings are unknown. What is the appropriate threshold for informing the donor of certain results derived from his or her specimens? And how certain must the researcher be regarding the accuracy of the result, no less its significance, for any action to be taken? If the results are incorrect or the interpretation inaccurate, who bears responsibility? Should a research result, at a minimum, be confirmed in a clinical laboratory, and who bears the costs of confirmatory testing?

The answers to many of these questions must balance different ethical principles—the right of the individual to his or her information; the potential harm done by communicating inaccurate, wrong, or uncertain information; the fiduciary responsibility of the clinician to the patient; and the challenge of expending scarce resources on pursuing and delivering potentially insignificant findings. For too long, clinicians and researchers have made these and other decisions for their patients. However, the "paternalistic" attitude, as Wolf and Kohane state, should be replaced by an active partnership between researcher and patient. Central to the discussion is the involvement of participants themselves.

Perlmutter and Aungst argue that the voices of the patient and the patient advocate have been inappropriately absent, or have been trivialized, in all aspects of the governance of specimen science. Only by attentive listening to and engagement with a patient or a patient advocate will the researcher understand the needs of the individual, his or her family, and the community. Importantly, Perlmutter and Aungst distinguish the role of the patient from that of an informed patient advocate, arguing further that the involvement of the patient advocate is necessary and helpful in designing the educational process that should precede any request for donation, the informed consent document and process, and the research question itself, to ensure that the answer will be of relevance to the participant. Beyond simple consultation, patient advocates should be represented on the IRB, in the biospecimen repository, and in the body making decisions about the repository's policies and processes for the release of biospecimens—indeed, in all aspects of governance.

The partnership of the participant with the researcher is a central feature of the UK Biobank and of the Precision Medicine Initiative (PMI) Cohort Program (PMI-CP)[1] in the United States; respect for that partnership has evolved into a model of participatory governance in both jurisdictions, as discussed by Maschke and further illuminated by Wolf and Kohane. In both the UK and US national biospecimen repositories, there is the appearance of meaningful engagement of participants in all facets of governance. However, the UK Biobank has been criticized for the involvement of patient/patient advocates in an advisory role only, and not on the most senior Ethics and Governance Council or the Board of Directors. The complaints of tokenism resonate with Perlmutter and Aungst's admonition that true partnership must be reflected at all levels of control and accountability.

The rapid diffusion of social media coupled with the availability of information and empowerment of the public has resulted not only in patient empowerment in governance but also in the aggregation of patient-led and patient-driven research agendas. Wolf and Kohane review trends, over the last decade, of patient-based organizations not only creating biospecimen repositories and sharing personal health data but also in nucleating patient-led cohorts (e.g., in the area of rare diseases) and in exercising decisional control over the use of specimens and data. Patients and families increasingly control the terms of collaborations with researchers, and in some circumstances they actually lead research efforts. It is not clear whether or how the safeguards that have been in place for decades—study protocols, IRB review and approval, institutional oversight—will apply to these new models of patient-led research and how accountability for complications (e.g., breaches of privacy) as they arise will be established. Patient engagement in both biospecimen repositories specifically and in research generally is rapidly changing, driven in part by advancing technology that enables patients and their family members to connect with and share with one another, and one in which historical relationships of health-care institutions, investigators, patients, and families are challenged and disrupted. An era of a new partnership between participant and researcher, including communication, dynamic consent, shared governance, and transparency, has begun.

### Note

1. The National Institutes of Health renamed the PMI-CP in October 2016. The new name is the *All of Us* Research Program.

# 15 Governance Issues for Biorepositories and Biospecimen Research

Karen J. Maschke[1]

On January 30, 2015, the White House Office of the Press Secretary issued a fact sheet about the Precision Medicine Initiative (PMI) that President Obama had announced more than a week earlier in his State of the Union Address. The fact sheet explained that the president's FY 2016 budget request to congress earmarked $130 million to embark on "a bold new research effort to revolutionize how we improve health and treat disease" (White House Press Office 2015). The president's proposal calls for the National Institutes of Health (NIH) to receive some of the earmarked funds to develop the PMI Cohort Program (PMI-CP),[2] an endeavor that involves creating a biospecimen repository and collecting biospecimens, genetic research data, lifestyle information, and electronic health records data from at least a million Americans for longitudinal research that will generate "knowledge applicable to the whole range of health and disease" (Collins and Varmus 2015, 793). The expectation of the PMI-CP is that research conducted with individuals' biospecimens and associated data will lead to disease prevention and treatment interventions that can be tailored to a specific patient's genetic makeup, including cancer interventions tailored to the genomic profile of specific cancer tumors (White House Press Office 2015).

Of interest for this chapter is the proposed "participatory governance" model a Working Group for the PMI-CP recommended barely eight months after the president formally announced the PMI. In September 2015, the PMI Working Group released a report that described how the PMI-CP's research model involves a "true partnership between participants and researchers" (PMI Working Group 2015, 39). The governance model the Working Group endorsed includes direct participation of research participants (the "PMI cohort") in the PMI-CP's governance framework. That framework includes oversight of the initiative's biospecimen repository and the use of the biospecimens and associated data collected and stored in the repository.

This chapter will identify themes from the biobanking literature about biospecimen repository governance, summarize the current biospecimen repository landscape in the United States, and examine the PMI Working Group's vision of participatory governance for the PMI-CP. The goal of the chapter is to promote discussion about various forms of participatory biospecimen repository governance, as well as discussion about whether the PMI Working Group's version of participatory biospecimen repository governance is a governance model that all repositories should consider adopting.

## The UK Biobank and the Road to Governance

There is a large and growing literature on the concept and practice of governance in medicine, science, and technology (Boeckhout and Douglas 2015; Braun et al. 2010; Guston 2014; Isasi and Knoppers 2011; Kaye et al. 2012; Landerweerd et al. 2015; Metzler and Webster 2011; Prainsack 2014; Prainsack and Naue 2006). Governance is not the same as "a government" that consists of elected or appointed officials who are leaders of a nation state or other geographic area and who enact and enforce the laws and rules for those jurisdictions (Graham et al. 2003). Rather, governance is "partly about how governments and other social organizations interact, how they relate to citizens, and how decisions are taken in a complex world" (ibid.). The literature on governance reflects the view that "governance is a process whereby societies or organizations make their important decisions, determine whom they involve in the process and how they render account" (ibid.). Graham and colleagues point out that, because a process is hard to observe, emphasis is placed on "the governance system or framework upon which the process rests—that is, the agreements, procedures, conventions or policies that define who gets power, how decisions are taken and how accountability is rendered" (ibid.).

The discussion about governance for biospecimen repositories and biospecimen research emerged full force during the development and implementation of the UK Biobank, a population-based biospecimen repository designed to collect and store biospecimens from 500,000 people in the United Kingdom between the ages of 40 and 69. The UK Biobank's funders undertook an ambitious effort to integrate into the planning and development of a large-scale research enterprise a number of ethical considerations about research with biospecimens, including attention to issues of repository governance.

Established officially in 2001, the UK Biobank began recruiting research donors in 2006 and reached its goal of 500,000 donors by mid 2010 (Huzair and Papaioannou 2012). The UK Biobank is a not-for-profit organization whose two principal funders are the Wellcome Trust (a UK charitable organization) and the UK government's Medical Research Council. The central objective of the UK Biobank is to be a major research resource of biospecimens and lifestyle, physical, and genetic information "to support *health-related research*, nationally and internationally" (Hunter and Laurie 2009, 152).

The funders of the UK Biobank were aware that collecting and storing people's biospecimens for research, including genetic research, raised ethical challenges for which there was little precedence or guidance. The challenges included meeting the ethical standard of informed consent for research when individuals are asked to give broad consent for future unspecified research with their biospecimens and associated data, providing a wide range of researchers with access to repository donors' materials, and safeguarding the confidentiality of donors' genetic research information. The UK Biobank's funders also were aware of growing public distrust in the UK of the scientific enterprise—distrust stemming partly from controversy about the safety of genetically modified foods. There was also evidence that many people in the UK were concerned about their biospecimens' being used for certain types of genetic research, particularly human cloning research (Levitt 2005).

Given these concerns, during the planning for the UK Biobank its funders held several types of public engagement activities to learn about the public's perspectives and concerns regarding the use of biospecimens and associated data for genetic and other medical research. Surveys, focus groups, and a public workshop were conducted over a three-year period before the UK Biobank began recruiting donors (Levitt 2005). In response to input from these public engagement activities, the UK Biobank's funders created an "ethics+" governance structure that goes beyond the standard governance approach to human subjects research involving prior review and approval of research by a research ethics committee (Laurie, Bruce, and Lyall 2009). The linchpin of this structure is the Ethics and Governance Framework that outlines the UK Biobank's societal commitments, its commitments to donors and researchers, and the ethical principles for conducting research with biospecimens and associated data (Hunter and Laurie 2009, 153). The UK Biobank's funders also created an independent Ethics and Governance Council, whose members are selected through a formal process that publicly advertises open positions (Laurie 2009; UK Biobank

Annual Review 2014). The Ethics and Governance Framework charges the Ethics and Governance Council with the task of monitoring the UK Biobank's compliance with the framework and providing "independent, ongoing ethical oversight and advice" to the UK Biobank (ibid.).

The public consultations during the UK Biobank's planning stages revealed public support for an "ethics+" approach given the differences between biospecimen repository research and traditional clinical trials research. Individuals recruited to enroll in clinical trials are told about the potential risks of a study drug or device, what will happen during the study and when their participation will end. Donors in the UK Biobank would give broad consent for future, unspecified research with their biospecimens without knowing what specific risks they may be assuming by participating or what studies researchers would be conducting. Moreover, their biospecimens and associated data would be available to many researchers for a long period of time, with no specified end to the research endeavor.

Many would agree with the assessment of the former head of the Ethics and Governance Council that the UK Biobank established a groundbreaking governance approach for a large-scale, publicly funded repository (Hunter and Laurie 2009, 151). However, some commentators have raised concerns that the UK Biobank's public engagement activities were limited to getting input on a narrow range of issues and the "ethics+" governance structure did not include a participatory role for repository donors and the public beyond an advisory one (Godard et al. 2004; Levitt 2005; Petersen 2007; Tutton et al. 2004; Wallace 2005; Winickoff 2007). For instance, Levitt (2005, 79) notes that the public was not consulted about how the UK Biobank should assess and resolve the tension between commercial entities profiting from research that meets their goals versus the public's interest in wanting research conducted that meets their needs as patients. Others have suggested that engagement activities were more about "managing perceived *mistrust* and engineering consent than creating the conditions for trust" (Petersen 2007) and about shifting attention away from substantive ethical issues about consent, biospecimen and data sharing, and commercial partners accruing financial benefits from research with the UK Biobank's specimens and data (Petersen 2007; Wallace 2005). Others have pointed out that the formal governance structure for the UK Biobank did not include direct representation of repository donors on the Board of Directors or the Ethics and Governance Council (Tutton et al. 2004; Winickoff and Winickoff 2003).

## What Should Biospecimen Repository Governance Look Like?

Since the UK Biobank was developed, much has been written about the governance issue for repositories and biospecimen research, though no consensus exists about what repository governance should look like, especially about the role for biospecimen contributors and the general public in a repository's governance structure (Fullerton et al. 2010; Gottweis and Petersen 2008; Hawkins and O'Doherty 2010; Hunter and Laurie 2009; Kaye and Stranger 2012; Laurie 2011; Mascalzoni 2015; Solberg and Steinsbekk 2015; Winickoff and Winickoff 2003). What a "participatory" governance approach should look like is open to debate. On the one hand, a participatory governance approach could be satisfied by engaging the general public and actual repository donors in consultation activities such as focus groups and workshops and/or including them as members of an external advisory committee (Goldenberg and Rivera 2016; Haldeman et al. 2014; Hunter and Laurie 2009). On the other hand, participatory governance could mean that biospecimen contributors are engaged directly in controlling the use of their biospecimens (e.g., dynamic consent) and in developing repository policies and practices, including decisions regarding who gets access to a repository's biospecimens and associated data and for what purposes.

The following list outlines several proposals for participatory repository governance approaches that give biospecimen contributors and/or the general public a decision-making role in a repository's policies and practices.

Winickoff and Winickoff (2003)

- Charitable trust model: a flexible legal model in which the general public acts as the beneficiary
- Governance by a board of trustees with biospecimen contributors participating in governance of the trust
- Biospecimen contributors could participate on the trust's ethics review committee, on a donor committee with veto power over research projects, and as elected member on the board of trustees

Winickoff and Neumann (2005)

- Clarification of charitable trust model toward a Biotrust model that opens pathways for democratic governance
- Biotrust model a legal structure to handle property rights and manage genetic and informational resources from biospecimen contributors

- Biotrust model also a social structure to bolster community participation, representation, and trust in biospecimen repository governance
- Ethical Review Committee (with biospecimen contributors as members) and Donor Advisory Committee would review and approve the use of the "trust property"

Winickoff (2009; in Kaye and Stranger 2012)

- Partnership governance—a shared governance structure over a collective resource
- Procedural justice and pragmatism require that repository donors have "real rights not of benefits, but of partial control of repositories as common-pool resources"; involved in making distributive decisions over "value collections of informational, genetic, and social capital."
- Establishment of a Participant Association bound to represent the donor collection on relevant governing body or bodies

Hawkins and O'Doherty (2010)

- Public deliberative event in Canada revealed support for repositories and view that a governing body should be independent from researchers and funders
- Donors said governance about a mechanism for: monitoring research and ensure harmful and undesirable consequences could be prevented; controlling vested interests; to manage potential public benefits of biospecimen research; protecting against the misuse of data, research results and technology; addressing potential problems that cross national jurisdictions and boundaries.
- Expectation that governance mechanism will ensure accountability, provide oversight of repository and biospecimen research, enforce relevant rules and regulations, ensure that public values, opinions and viewpoints taken into account
- Repository governance also a means to develop and maintain public trust in repositories and biospecimen research, especially through accountability, transparency and control
- Authors' conclusion: "Trustworthy governance" should be tailored to fit the context of a given repository

O'Doherty et al. (2011)

- Existing ethical conventions for protecting important interests (e.g., consent and privacy interests) of research donors cannot adequately guide repositories' collection, storage and use of biospecimens and associated data

- Governance may be a solution to the ethical, legal, and social challenges posed by repositories because they cannot offer privacy or meaningful consent to donors
- Adaptive governance approach recommended for a proposed repository as part of a longitudinal cohort study: 300,000 donors
- Four principles of repository governance: recognize research donors and publics as a collective body; trustworthiness; adaptivity; governance approach that fits the particular repository.
- Adaptive governance approach: includes direct donor representation on a Board of Directors, a Participant Association with an elected Participant Board, and Deliberative Public Engagement

Dove, Joly, and Knoppers (2012)

- Building and sustaining public trust requires meaningful engagement with the public, especially for large-scale, population-based repository projects
- Wiki-governance: non-hierarchical engagement and oversight for taxpayer-funded large-scale population-based repositories
- Citizens would "collaborate in proposing, drafting and amending repository digital governance structures, protocols, strategies and policies."

A decision-making role could mean that biospecimen contributors are members of a repository's primary governing body, members of a separate repository donor committee, and/or are members of multiple repository governing bodies (O'Doherty et al. 2011; Winickoff and Winickoff 2003; Winickoff 2007). Winickoff's proposals would give biospecimen contributors a decision-making role in one or more entities that make up a repository's governance structure (Winickoff and Winickoff 2003; Winickoff and Neumann 2005; Winickoff 2007; Winickoff 2012). His governance proposals are more legalistic than the others in that they are grounded in charitable trust law and assert that biospecimen contributors have property-like interests in their biological materials, though to date courts in the United States have not recognized property rights in the human body.

The participatory governance structure Dove and colleagues propose takes the concept of "engagement" to the level of the general public, not just for biospecimen contributors. They propose that large publicly funded repositories should use an online wiki forum that gives all citizens an opportunity to "collaborate in proposing, drafting and amending repository digital governance structures, protocols, strategies and policies" (Dove et al. 2012, 2). However, because their participatory model lodges final decision-making authority in the repository's management committee, citizens who participate in the wiki-governance activities would not have a direct vote in the repository's policy outcomes. Dove and colleagues contend that if a repository fails to uphold the policies formulated by the wiki-citizens, there probably would be a loss of public support and trust for the repository, which could jeopardize its viability and the viability of future repositories (ibid., 4).

An important question is whether all repositories should give biospecimen contributors (if not the general public) some decision-making role in the repository's governance structure. Or should certain characteristics of a repository determine what type of governance structure it adopts? Before turning to these and other questions about the role of biospecimen contributors and the public in repository governance, I will summarize recent empirical findings about the repository landscape in the United States.

## The Biospecimen Repository Landscape in the United States

The most recent attempt to map the landscape of biospecimen repositories in the United States comes from Henderson et al. (2013), who defined them as "repositories that assemble, store, and manage collections of human specimens and related data." Of the 456 repositories that responded to the survey by Henderson et al., 75 percent said that the biospecimens were from direct contributions by individuals, 57 percent that the biospecimens were from clinical settings as residual samples after routine clinical uses, and 41 percent that the biospecimens were from both sources. Thirty-six percent of repositories said that they also stored human cell lines. Henderson et al. also found that two-thirds of the repositories obtained most of their biospecimens from individuals with a particular disease or a particular type of disease. Less than half of the repositories said that they obtained biospecimens from individuals enrolled in cohort studies or clinical trials.

Eighty-eight percent of the repository respondents reported being affiliated with a larger organization, and about a quarter said they had multiple affiliations, most frequently with academic institutions. With regard to the source of funding, for 36 percent of the repositories surveyed, the federal government provides the largest source of their funding, whereas 30 percent said their largest funding source was the larger organization in which they were embedded (Henderson et al. 2013). Other reports about repositories in the United States reveal that several were created by large academic medical centers, health systems, disease-based organizations, state agencies, and countries' science funding bodies (Holm et al. 2014; Lemke and Harris-Wai 2015; McCarthy et al. 2008; Olson et al. 2013; Pulley et al. 2008).

The repository for the PMI-CP will enter the institution-based landscape of repositories in the United States as a large-scale federally funded "national biobank." The PMI Working Group recommended that all individuals regardless of age who live in the US and meet several requirements should be eligible to enroll in the PMI-CP. Individuals who enroll in the PMI-CP will give broad consent for future unspecified research with their biospecimens, genetic data, and associated medical records and lifestyle data. The PMI Working Group also recommended that the PMI-CP establish a central facility to process, store, retrieve, and track donors' biospecimens and that individual donors should be recruited from healthcare provider organizations (HPOs) throughout the country. Although the Working Group considered recommending a recruitment approach that would draw donors from existing cohorts (e.g., individuals who are enrolled in existing repositories and longitudinal studies), it concluded that recruitment would be faster and more cost efficient if the PMI-CP invited large numbers of individuals from HPOs that held their dense health-care data. The Working Group pointed out that, because HPOs have primary access to patients' comprehensive health data and the ability to share those data, they "have exceedingly strong potential to be highly effective partners with the PMI-CP, functioning as sites or 'nodes' within the PMI cohort for recruitment, communication, biospecimen collection, and healthcare data collection (through their clinical care relationship)" (PMI Working Group 2015, 31).

### The PMI-CP Governance Model: Raising the Bar for Biospecimen Repository Governance?

Between February and July of 2015, the PMI-CP Working Group sponsored several public workshops to obtain input from various stakeholders about their concerns and perspectives regarding a large-scale publicly funded initiative involving research with biospecimens and associated data. The workshops covered issues related to building a large research cohort, the scientific opportunities such a cohort presented, the collection and use of digital health data, the use of mobile and personal technologies to support recruitment, research, and donor engagement, and donor engagement and health equity.

Drawing from the discussions and points raised at the workshops, the PMI Working Group outlined in its report a vision for the design and utility of the PMI-CP and made recommendations covering six areas "critical to the development, implementation, and oversight" of the project: "cohort assembly, participant engagement, data, biobanking, policy, and governance" (PMI Working Group 2015, 1). As is shown in the list below, the proposed governance structure for the PMI-CP includes biospecimen contributors represented on a PMI-CP Steering Committee, on the Steering Committee's Executive Committee, and on all of the Steering Committee's subcommittees.

**NIH Director**

Has final authority for policy determination, priority setting, and oversight of the implementation of the PMI-CP

**PMI Cohort Program (CP) Director**: be responsible to the NIH Director

PMI-CP Steering Committee

- Chaired by the PMI-CP director
- Reports to an Executive Committee
- Provides coordination of the PMI-CP's activities
- Members should include PMI awardees, research participants and their representatives, academic and private researchers who will use the PMI cohort platform, NIH programmatic staff

Executive Committee of the PMI-CP Steering Committee

- Small group of Steering Committee members
- Chaired by PMI-CP Director
- Should include strong participant representation

Subcommittees of the PMI-CP Steering Committee

- Data Subcommittee
- Resource Access Subcommittee
- Return of Results Subcommittee
- Biobanking Subcommittee
- Security Subcommittee
- All subcommittees should include participant representation

Independent Advisory Board

- Provides external oversight for the PMI-CP
- Should include experts in areas of relevance to the PMI-CP
- Should report to the PMI-CP director and the NIH Director

• NIH should consider using a multicouncil working group structure such as that used by the NIH Brain Research through Advancing Innovative Neurotechnologies (BRAIN) Initiative

The Working Group also recommended that "a substantial number of members of the public and representatives of the participant community" be members of a single institutional review board (IRB) for the PMI-CP (PMI Working Group 2015, 80).

The Working Group's recommendations were motivated by the recognition that many patients, research donors, and advocacy organizations want meaningful engagement with researchers and physicians, not superficial and symbolic "public engagement" activities (PMI Working Group 2015). If the Working Group is correct, then the governance challenge is whether and to what extent biospecimen contributors should be members of internal bodies that have oversight of institutional-based repositories or, at minimum, provide input to these entities in some sort of advisory capacity. Ginsburg et al. (2008, 1361) suggest that institutions that hold and share biospecimens for research should at least include the patients they serve in the governance structure:

Any institutional core resource that will be made available to all researchers requires a governing board with key stakeholder representation including clinicians, researchers, patients, technologists, and health system leadership. This represents a shift from the view that individual investigators or research consortia govern the samples and reflects the reality that such repositories require cooperation among research participants, their health care system, and the many individuals with special scientific or technical expertise.

A normative justification for a participatory approach that gives biospecimen contributors (and possibly the general public ) a decision-making role in a repository's governance structure is that regardless of a repository's funding source or its institutional setting, individuals who will be affected by a repository's decisions should be given an opportunity to directly participate in decision making (Dove et al. 2012; Marris and Rose 2012; Winickoff and Winickoff 2003). Yet other commentators caution against a "one-size-fits-all" governance approach (O'Doherty et al. 2011, 370; Kaye and Stranger 2012). According to O'Doherty et al. (2011, 370), ten factors have important implications for designing repository governance:

1. who created the repository and how
2. the repository's purpose
3. the kind of biospecimens and data the repository holds

4. whether biospecimens are linked prospectively or retrospectively to individuals from whom they were obtained
5. what racial, ethnic, or other types of "populations" are represented in the biospecimen collection
6. who funds the repository and how
7. the size of the repository or the relative biospecimen cohort
8. the extent to which there is social cohesion/political identity among the biospecimen contributors
9. whether the repository is a hospital-based resource, or has a national and/or international reach
10. whether internal and/or external researchers, including researchers from commercial entities, will have access to the biospecimens and associated data.

Since most repositories in the United States are embedded in medical institutions, one approach to governance for institution-based repositories would be to explain during the informed consent process that by providing consent for the collection, storage, and use of their biospecimens for research, individuals also are opting into a governance approach the repository and its institutional partner deems appropriate, even if that approach does not include a direct participatory role for biospecimen contributors.[3] Because a "consent to governance" approach does not include biospecimen contributors or the general public as partners in the biospecimen research enterprise in the way Winickoff and others envision, greater attention to developing transparent and accountable policies and practices may be required. Along with an institutional review board that reviews and approves protocols for establishing a repository and for research with biospecimens, many institutions probably have other internal entities that play an oversight role. For example, an institution's material transfer committee would oversee the distribution of biospecimens to external researchers (Vaught 2015). At minimum, a transparent "consent to governance" approach should include public disclosure about the governance structure in place and about the specific research studies conducted with the repository's biospecimens and associated data, including reference to the publications that describe the findings of the research conducted with those materials. Although the "consent to governance" approach does not accommodate normative expectations of participatory democracy, it gives individuals the option to decline to provide their biospecimens to repositories that do not give some contributors the opportunity to participate on decision-making bodies for that research enterprise.

In the absence of empirical data about what governance means to repository donors, to the general public, and to the institutions in which repositories are embedded—as well as about what governance approaches are currently in use (Henderson et al. 2013)—it is difficult to know whether the ten factors O'Doherty et al. outline are the right factors to take into account when considering how to govern a repository and the biospecimen research enterprise it supports. It is not self-evident that all ten factors together, or in some combination, preclude a participatory approach like the PMI-CP for all types of repositories. On the other hand, some commentators suggest that "engaging too much" with repository donors "can take resources away from the real obligation of researchers, which is to maximize the research effort in order to give back to patients in the future" (Solberg and Steinsbekk 2015, 29). Haldeman and colleagues point out that, for several reasons, there is danger that the term "community engagement" in biospecimen repositories will become a "useless buzzword." In their view, the term "is ambiguous and almost completely devoid of empirical evidence as to what works, in what contexts, and importantly, to what effects. Moreover, there are no agreed-upon ways to evaluate [community engagement]" (Haldeman et al. 2014, 92). The same could be said for "participant engagement" in repositories and "biobank governance." Indeed, more than ten years after the UK Biobank introduced an innovative governance framework for repositories there is little empirical evidence whether a particular model of donor engagement in a repository's governance framework has a direct effect on meeting repository enrollment goals or establishing donor and public trust in repositories and research with biospecimens (Mitchell et al. 2015).

**Conclusion**

The proposed governance approach for the PMI-CP raises the governance bar, at least for publicly funded, population-based repositories. If implemented, the proposed governance approach may also raise public expectations in the United States about the role that biospecimen contributors should play as partners in a repository-related research enterprise, regardless of the size, purpose, and location of the repository. Even if institution-based repositories do not adopt a participatory governance approach like that of the proposed PMI-CP model, they should have a process in place for ongoing evaluation of whether their governance approach adequately meets increasing public expectations for meaningful engagement in the biospecimen research enterprise.

## Notes

1. Funding for this work was supported by grant R01 HG005691 (PI-Rivera) from the National Human Genome Research Institute.

2. The National Institutes of Health renamed the PMI-CP in October 2016. The new name is the *All of Us* Research Program.

3. Thanks to Jeffrey Botkin for pointing out the "consent to governance" option at the symposium, "Specimen Science: Ethics and Policy Implications," Petrie-Flom Center, Harvard Law School, November 16, 2015.

## References

Boeckhout, Martin, and Conor M. W. Douglas. 2015. Governing the research-care divide in clinical biobanking: Dutch perspectives. *Life Sciences, Society and Policy* 11 (1): 1–16.

Braun, Kathrin, Alfred Moore, Svea Luise Herrmann, and Sabine Könninger. 2010. Science governance and the politics of proper talk: Governmental bioethics as a new technology of reflexive government. *Economy and Society* 39 (4): 510–533.

Collins, Francis S., and Harold Varmus. 2015. A new initiative on precision medicine. *New England Journal of Medicine* 372 (9): 793–795.

Dove, E. S., Y. Joly, and B. M. Knoppers. 2012. Power to the people: A wiki-governance model for biobanks. *Genome Biology* 13:158.

Fullerton, Stephanie M., Nicholas R. Anderson, Greg Guzauskas, Dena Freeman, and Kelly Fryer-Edwards. 2010. Meeting the governance challenge of next-generation biorepository research. *Science Translational Medicine* 2 (15): 15cm3–15cm3.

Ginsburg, Geoffrey, Thomas W. Burke, and Febbo Phillip. 2008. Centralized biorepositories for genetic and genomic research. *Journal of the American Medical Association* 299 (110): 1359–1361.

Godard, B., J. Marshall, C. Laberge, and B. M. Knoppers. 2004. Strategies for consulting with the community: The case of four large-scale genetic databases. *Science and Engineering Ethics* 10 (3): 457–477.

Goldenberg, Aaron J., and Suzanne M. Rivera. 2017. Biorepositories and precision medicine: Implications for underserved and vulnerable populations. In this volume.

Gottweis, Herbert, and Georg Lauss. 2010. Biobank governance in the post-genomic age. *Personalized Medicine* 7 (2): 187–195.

Gottweis, Herbert, and Alan Petersen. 2008. *Biobanks: Governance in Comparative Perspective*. Routledge.

Graham, John, Bruce Amos, and Tim Plumptre. 2003. *Principles for Good Governance in the 21st Century*. Ottawa: Institute on Governance.

Guston, D. H. 2014. Understanding "anticipatory governance." *Social Studies of Science* 44 (2): 218–242.

Haldeman, Kaaren M., R. Jean Cadigan, Arlene Davis, Aaron Goldenberg, Gail E. Henderson, Dragana Lassiter, and Erik Reavely. 2014. Community engagement in US biobanking: Multiplicity of meaning and method. *Public Health Genomics* 17 (2): 84–94.

Hawkins, Alice K., and Kiernan O'Doherty. 2010. Biobank governance: A lesson in trust. *New Genetics & Society* 29 (3): 311–327.

Henderson, Gail E., Jean R. Cadigan, Teresa P. Edwards, Ian Conlon, Anders G. Nelson, James P. Evans, Arlene M. Davis, Catherine Zimmer, and Bryan J. Weiner. 2013. Characterizing biobank organizations in the US: Results from a national survey. *Genome Medicine* 5 (1): 3.

Holm, Ingrid A., Sarah K. Savage, Robert C. Green, Eric Juengst, Amy McGuire, Susan Kornetsky, Stephanie J. Brewster, Steven Joffe, and Patrick Taylor. 2014. Guidelines for return of research results from pediatric genomic studies: Deliberations of the Boston Children's Hospital Gene Partnership Informed Cohort Oversight Board. *Genetics in Medicine* 16 (7): 547–552.

Horn, E. J., K. Edwards, and S. F. Terry. 2011. Engaging research participants and building trust. *Science* 312: 370–371.

Hunter, K. G., and G. Laurie. 2009. Involving publics in biobank governance: Moving beyond existing approaches. In *The Governance of Genetic Information: Who Decides?* ed. H. Widdows and C. Muller. Cambridge University Press.

Huzair, Farah, and Theo Papaioannou. 2012. UK Biobank: Consequences for commons and innovation. *Science & Public Policy* 39: 500–512.

Isasi, Rosario, and Bartha M. Knoppers. 2011. From biobanking to international governance: Fostering innovation in stem cell research. *Stem Cells International* 498132. doi:10.4061/2011/498132.

Kaiser, Jocelyn. 2015. NIH opens precision medicine study to nation. *Science* 349 (6255): 1433.

Kaye, Jane, Liam Curren, Nick Anderson, Kelly Edwards, Stephanie M. Fullerton, Nadja Kanellopoulou, David Lund, et al. 2012. From patients to partners: Participant-centric initiatives in biomedical research. *Nature Reviews Genetics* 13 (5): 371–376.

Kaye, Jane, and Mark Stranger, eds. 2012. *Principles and Practice in Biobank Governance*. Ashgate.

Landerweerd, Laurens, David Townend, Jessica Mesman, and Ine Van Hoyweghen. 2015. Reflections on different governance styles in regulating science: A contribution to "Responsible Research and Innovation." *Life Sciences, Society and Policy* 11 (1): 1–22.

Laurie, Graeme. 2009. The role of the UK Biobank Ethics and Governance Council. *Lancet* 374 (9702): 1676.

Laurie, Graeme. 2011. Reflexive governance in biobanking: On the value of policy led approaches and the need to recognise the limits of law. *Human Genetics* 130: 347–356.

Laurie, G., A. Bruce, and C. Lyall. 2009. The role of values and interests in the governance of the life sciences: Learning lessons from the "Ethics+" approach of UK Biobank. In *The Limits to Governance: The Challenge of Policy-making for the New Life Sciences*, ed. C. Lyall, T. Papaioannou, and J. Smith. Ashgate.

Lemke, Amy A., and Julie N. Harris-Wai. 2015. Stakeholder engagement in policy development: Challenges and opportunities for human genomics. *Genetics in Medicine* 17 (12): 949–957.

Levitt, Mairi. 2005. UK Biobank: A model for public engagement? *Genomics, Society, and Policy* 1 (3): 78–81.

Marris, Claire, and Nikolas Rose. 2012. Open engagement: Exploring public participation in the biosciences. *PLoS Biology* 8 (11): 310000549.

Mascalzoni, Deborah, ed. 2015. *Ethics, Law and Governance of Biobanking: National, European and International Approaches.* Springer.

McCarthy, Catherine A., Donna Chapman-Stone, Teresa Derfus, Philip F. Giampietro, and Norman Fost. 2008. Community consultation and communication for a population-based DNA biobank: The Marshfield Clinic Personalized Medicine Research Project. *American Journal of Medical Genetics* 146A: 3026–3033.

Metzler, Ingrid, and Andrew Webster. 2011. Bio-objects and their boundaries: Governing matters at the intersection of society, politics, and science. *Croatian Medical Journal* 52 (5): 648–650.

Mitchell, Derick, Jan Geissler, Alison Parry-Jones, Hans Keulen, Doris C. Schmitt, Rosaria Vavassoir, and Balwir Matharoo-Ball. 2015. Biobanking from the patient perspective. *Research Involvement and Engagement* 1:4.

National Institutes of Health. 2015. Participant Engagement and Health Equity Workshop. https://www.nih.gov/precision-medicine-initiative-cohort-program/events.

O'Doherty, Kieran C., Michael M. Burgess, Kelly Edwards, Richard P. Gallagher, Alice K., Hawkins, Jane Kaye, Veronica McCaffrey, and David E. Winickoff. 2011. From consent to institutions: Designing adaptive governance for genomic biobanks. *Social Science & Medicine* 73 (3): 367–374.

Olson, Janet E., Euijung Ryu, Kiley J. Johnson, Barbara A. Koenig, Karen J. Maschke, Jody A. Morrisette, Mark Liebow, et al. 2013. The Mayo Clinic Biobank: A building block for individualized medicine. *Mayo Clinic Proceedings* 88 (9): 952–962.

Petersen, Alan. 2007. Biobanks' "engagements": Engendering trust or engineering consent? *Genomics, Society, and Policy* 3 (1): 31–43.

PMI Working Group. 2015. Working Group Report to the Advisory Committee to the Director, NIH. The Precision Medicine Initiative Cohort Program—Building a Research Foundation 21$^{st}$ Century Medicine. September 17. https://www.nih.gov/precisionmedicine/09172015-pmi-working-group-report.pdf.

Prainsack, Barbara. 2014. The powers of participatory medicine. *PLoS Biology* 12 (4): e1001987.

Prainsack, B., and U. Naue. 2006. Relocating health governance: Personalized medicine in times of "global genes." *Personalized Medicine* 3 (3): 349–355.

Pulley, J. M., M. M. Brace, G. R. Bernard, and D. R. Masys. 2008. Attitudes and perceptions of patients towards method of establishing a DNA biobank. *Cell and Tissue Banking* 9: 55–65.

Solberg, Berge, and Kristin Solum Steinsbekk. 2015. Biobank consent models–Are we moving toward increased participant engagement in biobanking? *Journal of Biorepository Science for Applied Medicine* 3: 23–33.

Terry, Sharon F. 2013. Don't just invite us to the table: Authentic community engagement. *Genetic Testing and Molecular Biomarkers* 17 (6): 443–445.

Tutton, Richard, Jane Kaye, and Klaus Hoeyer. 2004. Governing UK Biobank: The importance of ensuring public trust. *Trends in Biotechnology* 22 (6): 284–285.

UK Biobank Annual Review. 2014. https://egcukbiobank.org.uk/sites/default/files/UKBEGC%20Annual%20Review%202014.pdf.

Vaught, Jim B. 2015. Building better biorepositories and biobanks. *Annual Review of Pharmacology and Toxicology* 56 (1).

Wallace, H. M. 2005. The development of the UK Biobank: Excluding scientific controversy from ethical debate. *Critical Public Health* 15 (4): 323–333.

White House Press Office. 2015. Fact Sheet: President Obama's Precision Medicine Initiative. January 30. https://www.whitehouse.gov/the-press-office/2015/01/30/fact-sheet-president-obama-s-precision-medicine-initiative.

Winickoff, David E. 2003. Governing population genomics: Law, bioethics, and biopolitics in three case studies. *Jurimetrics* 43: 187–228.

Winickoff, David E. 2007. Partnership in the UK Biobank: A third way for genomic property? *Journal of Law, Medicine & Ethics* 35 (3): 440–456.

Winickoff, David E. 2012. From benefit sharing to power sharing: Partnership governance in population genomic research. In *Principles and Practice in Biobank Governance*, ed. Jane Kaye and Mark Stranger. Ashgate.

Winickoff, David E., and Larissa B. Neumann. 2005. Towards a social contract for genomics: Property and the public in the "biotrust" model. *Genomics, Society, and Policy* 1 (3): 8–21.

Winickoff, David E., and Richard N. Winickoff. 2003. The Charitable Trust as a model for genomic biobanks. *New England Journal of Medicine* 349: 1180–1184.

# 16 The Rise of Patient-Driven Research on Biospecimens and Data: The Second Revolution

Susan M. Wolf and Isaac S. Kohane[1]

Patients and families are increasingly driving genomic research and striving for new control over the biorepositories that power this research. Families with children suffering from puzzling and serious neurodevelopmental abnormalities are seeking genome sequencing to aid diagnosis, research, and treatment. Cancer patients who have failed conventional treatments are seeking analysis of tumor sequence to find new treatment targets and generate research insights into their disease. Patient advocacy groups are assembling needed biorepositories and claiming patent rights.

The genomics revolution is not just a profound reorganization of our understanding of disease and potential interventions. It is a fundamental shift in the role of patients and families relative to researchers, clinicians, investigators, and biorepositories. Patients with rare and ultra-rare genomic disorders are increasingly unwilling to accept the message that little is known about their disorders, no treatments are validated, and too few other patients share the disorder to support a research program. Instead these patients are using social media and other tools to find each other, build a research cohort, form alliances with researchers and clinicians, assemble a biorepository, and attempt therapeutic interventions based on genomic and proteomic analysis. Patients and families are driving research to speed knowledge aggregation, hypothesis testing, and benefit.

It is no accident that the federal Precision Medicine Initiative (PMI) aims to be patient-centered and prominently relies on patient advocacy organizations. Genomic knowledge, when openly shared with the individuals sequenced and their loved ones, is a powerful tool. It allows people to more deeply understand the etiology and nature of their disorder; to find others with the same genomic variants; to build research resources such as repositories and data banks; to negotiate with researchers on the terms of access to these resources; and to advocate for research dollars at NIH and private foundations, with funding "angels," and at pharmaceutical companies.

Patients and families are building organizations such as NGLY1.org, the Phelan-McDermid Syndrome Foundation, and the Multiple Myeloma Research Foundation (MMRF), creating what the MMRF website calls "a new patient-centered model of collaboration." The pioneers of precision medicine are harnessing the power of accessible genomics, mobile health tools for phenotypic measurement, and open-source analytic platforms, together with social media and linked communities to drive progress. Organizations such as SAGE Bionetworks are using open access tools to enable massive sharing of omics data to speed drug discovery. We Are Curious is helping people gather data on themselves and share that data with others to drive health insights.

This fundamental shift is a deep challenge to "business as usual" in biomedical research, translational genomics, and specimen and data assemblies. Major changes are now required to meet the needs of patients and families, effectively partner with them in research and care, and optimize progress. Much as AIDS and breast cancer advocacy organizations drove significant changes in the process of drug discovery, testing, and approval, patient-driven precision medicine in the era of accessible informatics and computer tools is driving a reorganization of genomic research, implementation, and care.

Pivotal to this revolution is the formation of cohorts, collection of data and specimens, and their use to drive discovery and progress. Yet the history of biorepositories is built largely on rejecting direct engagement with the individuals whose material and information are the contents. This chapter traces that history and the resultant policies and practices avoiding accountability to and control by these individuals. We then address the revolution under way powered by patients and families. We suggest that understanding this revolution requires looking at a century of transition, from early-twentieth-century paternalism to twenty-first-century goals of partnership.

## The History of Excluding Patients from Biospecimen Research

The history of accumulating specimen collections for study is an ancient one. Specimen collections range enormously from an individual physician's collection of specimens in a freezer to large-scale collections (De Souza and Greenspan 2013; IOM 2012). The heterogeneity has posed a major challenge to the analysis of these collections and even development of agreed terminology. Terms vary, including "biobank" (a term that may be used to encompass collections of specimens and data), "biorepository,"

"DNA databank," and "biospecimen collection." Even efforts to enumerate US human biorepositories have proved complex, with researchers at RAND, and more recently Henderson and colleagues, making progress (Eiseman and Haga 1999; Eiseman et al. 2003; Henderson et al. 2013; Edwards et al. 2014). Countries have now created population-wide repositories such as UK Biobank to generate the statistical power and sample diversity to address fundamental questions in epidemiology, genomic discovery, and treatment outcomes. With the 2015 announcement of the PMI, the US has joined this effort.

The modern development of computer informatics, robotic specimen retrieval, and a sophisticated science of biospecimen preservation and analysis has transformed the practice of assembling, maintaining, and utilizing collections. The International Society for Biological and Environmental Repositories (ISBER) was founded in 2000, publishing recommended best practices in *Cell Preservation & Technology*. In 2005, the National Cancer Institute (NCI) at the National Institutes of Health (NIH) formed the Office of Biorepositories and Biospecimen Research (OBBR), which later became the still-operating Biorepositories and Biospecimen Research Branch (BBRB). In 2007, 2011, and then 2016, this office published *Best Practices for Biospecimen Resources* (NCI 2016). Publication of quality standards and plans for biorepository certification testify to the growing maturity of the field and its importance in driving research globally.

Along with the evolution of modern biorepositories came the need to determine how specimens would be obtained, whether consent was needed from source individuals, what rights they retained over their specimens and data, and what rules would govern specimen use. In 1997, the DHHS Office for Protection from Research Risks (OPRR), the predecessor to the Office for Human Research Protections (OHRP), issued guidance obviating the need to seek consent from source individuals or even inform them when specimens and data collected for clinical use were subsequently archived and used in research, as long as specimens and data were de-identified and the investigators could not readily re-identify the source. The Office for Protection from Research Risks (1997) ruled that that "human subjects" are not involved when the material "was collected for purposes other than submission to the Repository," and the "material is submitted to the Repository without any identifiable private data or information." This was reaffirmed by the National Bioethics Advisory Commission (NBAC) shortly thereafter: "Federal regulations governing human subjects research ... that apply to research involving human biological materials should be interpreted ... [such that] Research conducted with unidentified samples is not human subjects research and is not regulated by the Common Rule" (NBAC 1999). OHRP again confirmed this approach in 2008 (OHRP 2008a,b). Each of these authorities further agreed that even when samples could be re-identified, so that investigators working with those samples were conducting human subjects research, the requirement to obtain informed consent could be waived by IRBs under some circumstances.

This prioritization of biospecimen collection and use over direct accountability to individual sources and patient decisional control has had complex roots. Strong arguments have been made that access to de-identified specimens as well as de-identified data is important to fuel research, quality oversight of health-care institutions, and public health surveillance. Yet there is also a long history of regarding specimens removed from the body for clinical purposes as material over which the source individual has no rights and in which the individual has no interests (Brothers 2011; Brothers and Clayton 2010). The Supreme Court of California famously took that position in the 1990 case of *Moore v. Regents, University of California* (793 P.2d 479 (Cal. 1990)).

Twenty years later, however, the landscape had begun to change, as evidenced by the reaction to Rebecca Skloot's book on the creation of the HeLa cell line from cancer cells removed in 1951 from Henrietta Lacks (Skloot 2010). These cancer cells were used to create a cell line for research, but without notice to the patient or her consent. As Skloot recounts, the failure to inform the patient and seek her consent was in keeping with the prevailing rules and norms surrounding specimen collection and use. Indeed, if the cells were removed for treatment purposes and then de-identified before use, these actions would still technically comply with current rules. In the case of the HeLa cells, however, the source individual's identity was not protected. Indeed, her family members were contacted later in an effort to recruit them for genetic analysis. In 2013, when researchers set out to publish the full genomic sequence of a strain of HeLa cells, in a changed public environment more sensitive to privacy risks and the disturbing health disparities and racism that Skloot's book revealed as part of the HeLa story, concern finally became acute (Hudson and Collins 2013). Ultimately, NIH reached a historic agreement with the Lacks family and set up a process for their involvement in research decisions involving the cell line (National Institutes of Health 2013)—recognizing through these actions that the interests of a source individual and her family do not end once the cells are separated from her body.

NIH has argued that the agreement with the Lacks family is unique and should not be construed as precedent. However, they also recognized the inevitable importance of the agreement, expressing the hope that the agreement would "spur broader discussions regarding consent for future use of biospecimens, with a goal of fostering true partnerships between researchers and research participants" (Hudson and Collins 2013). The Lacks case and agreement have indeed become an important part of a seismic shift in attitudes toward what researchers owe to individuals whose specimens and data fuel research. When the Havasupai tribe sued Arizona State University for research analyses of their materials that went beyond the limits mutually agreed to (Mello and Wolf 2010), they were asserting a continuing interest over research uses of their materials, harm to those interests, and persisting rights to hold the researchers accountable. The case was settled on terms favorable to the tribe, including return of their specimens (Harmon 2010).

In 2009, parents sued the Minnesota Department of Health (*Bearder v. Minnesota*, 806 N.W.2d 766 (Minn. 2011)) for retaining and performing research on Guthrie cards with dried blood spots generated in newborn screening. The parents claimed a privacy violation in the retention and research use of these cards without notice or consent, even though the cards were de-identified. The Department of Health argued that retention and use facilitated quality control and development of the newborn screening program, as well as research use. But in 2011 the Minnesota Supreme Court ruled in the parents' favor in the *Bearder* case, leading to the ordered destruction of about a million Guthrie cards (Olson 2014). Parents in Texas brought a similar case (Lewis et al. 2011).

Growing source concern over the use of biospecimens and the potential for re-identification and privacy risks (Grajales et al. 2014) has been only one manifestation of a larger shift in attitudes toward the researcher-participant relationship, responsibilities toward research participants and specimen sources, and the duties of biorepositories. That shift is manifested by the turn toward various forms of community-based participatory research (CBPR), some involving deep engagement with and partnership with the community. It is manifest in an enormous literature that has arisen arguing that data and specimen collections used in research have duties of responsible custodianship and accountability to data and specimen sources (e.g., Hoffman 2016); many of those collections have set up governance mechanisms involving the source community. And the shift is evident in the recognition over the last decade that researchers as well as data and specimen archives may shoulder responsibilities to offer back to source individuals their health-relevant results and incidental findings (IOM 2014; Wolf et al. 2008).

All of these developments are propelling a fundamental shift toward treating "patients as partners in research rather than passive, disenfranchised purveyors of biomaterials and data" (Kohane et al. 2007). Yet resistance continues. Arguments that biobank research systems—conceived as the combination of primary collection sites, the central biobank, and secondary research sites—have obligations to clarify their policy on return of results, make that transparent to source individuals, and consider return of high-stakes individual results if feasible (Wolf et al. 2012) have met with some objection. Those involved in biobank research argue that addressing these issues will cost money, that their budgets are not designed for this, and that they are merely aggregating material for research and should be absolved of these duties (Bledsoe et al. 2013). Of course, all engagement with source individuals, research participants, and their communities requires time and resources. Indeed, fulfilling basic ethical requirements in research, such as seeking informed consent, costs money. To say that current business models for biobanks or biorepositories have failed to budget for source individual engagement and for addressing issues such as return of results is merely to restate the history of repositories and resist change (Wolf 2013). A more nuanced conversation might attempt to differentiate repositories by factors that should dictate a greater or lesser degree of involvement with source individuals (Wolf et al. 2012). But simply to resist any role other than aggregator and insist that source individuals remain "passive, disenfranchised purveyors of biomaterials and data" (Kohane et al. 2007) is to ignore a sea change in public attitudes and professional opinion.

By 2011, the NCI's Best Practices for Biospecimen Resources urged those resources to reveal policies on return of results and other issues to source individuals, and to involve them in governance. And in 2011, DHHS issued an Advance Notice of Proposed Rulemaking (ANPRM) offering for comment proposed changes to the Common Rule governing research with human participants. The ANPRM proposed requiring that "a person needs to give consent, in writing, for research use of their biospecimens" (Department of Health and Human Services 2011). The ANPRN took notice of fundamental changes under way in the conduct of research. Those included increasing use of patient data and biospecimens. Indeed, academic health centers increasingly used their own health-care systems as "living laboratories for research" (Kohane 2011) and were committed to developing learning health-care systems (Friedman et al. 2010). The necessary analysis of data and specimens called for clarity about what duties were owed to source individuals.

In 2014, Congress enacted the Newborn Screening Saves Lives Reauthorization Act, providing that "Research on newborn dried blood spots shall be considered research carried out on human subjects," thereby requiring consent as a general matter (Newborn Screening Saves Lives Reauthorization Act 2014). This signaled a major change in the approach to research with biospecimens. By late 2015, when agencies took the next step in the process of revising the Common Rule by issuing a Notice of Proposed Rulemaking (NPRM), the proposed requirement of consent to biospecimen use appeared to be part of a wider change in the role of source individuals and participants in research (Department of Homeland Security et al. 2015).

In 2016, NCI issued a revision to its Best Practices for Biorepository Resources, updating recommendations on topics including informed consent, return of results, and community engagement. The document urged that federal regulations be considered the floor, not ceiling—minimal requirements, rather than the limit of appropriate practices: "The NCI recommends seeking the informed consent of research participants who provide biospecimens and associated data whenever such consent is required by regulation, and also when consent is ethically appropriate and can practicably be obtained. Respect for individuals who have provided data and/or biospecimens for research is of paramount importance...." (section C.2.2.1, p. 35). NCI also endorsed the option of broad consent to unspecified future research uses (section C.2.3.9, p. 39).

In early 2017, DHHS and other agencies issued the final rule amending the Common Rule, the culmination of the ANPRM and NPRM process (Department of Homeland Security et al. 2017). The final rule retreated from the NPRM's proposal that research with biospecimens require source individual consent, even when the specimens were de-identified. Similarly, the rule superseded the Newborn Screening Act's provision that research on newborn dried blood spots was human subjects research, which would generally trigger consent requirements even if the blood spots were de-identified (Newborn Screening Saves Lives Reauthorization Act of 2014). While clearly recognizing the importance of specimen source choice and autonomy, the final rule sought to balance the importance of using biospecimens to make research progress and yield new therapies.

The final rule nonetheless made significant advances toward recognizing research participants and specimen sources as partners in research. The rule requires greater transparency in the informed consent process concerning future uses of biospecimens. Even when informed consent is not required, the rule creates an option for investigators to seek broad consent from sources for the use of their biospecimens in research. Indeed, the rule requires broad consent when identifiable biospecimens are stored for possible secondary research. The rule sets out required elements and additional elements of informed consent, as well as required elements of broad consent; in both contexts the rule clarifies the importance of addressing return of results. Importantly, the rule recognizes that requirements to seek consent may grow over time; with changing technology and growing use of whole genome sequencing, the effectiveness of de-identification may decline. The rule thus creates a process for periodic reconsideration in light of these advances, a process that may expand requirements for seeking consent even when biospecimens have been de-identified. Finally, the revised Common Rule recognizes that its requirements are a floor not ceiling, acknowledging that agencies and other authorities (including state, local, and tribal authorities) may add more requirements. The rule reiterates "the central tenet that participants should be active partners ... not merely passive subjects" (p. 7,152) and strives to respect public sentiment on biospecimen use. Based on public comments received on the NPRM, the rule seeks to reconcile "the need for obtaining consent before using ... biospecimens for research" (p. 7,168) and the need to facilitate research progress.

**The Patient and Family Revolution**

As both the NPRM and final rule note, the last twenty years have seen "a paradigm shift...in how research is conducted" (Department of Homeland Security et al. 2017, p. 7,151). As the NPRM recognized, "more people want to play an active role in research, particularly related to health, and they have different expectations than when the Common Rule was first established. A more participatory research model is emerging. ..." The NPRM saw these changes rooted in broader trends, including "widespread use of social media, in which Americans are increasingly sharing identifiable personal information ... including health-related information" (Department of Homeland Security et al. 2015, p. 53,938).

Shifting attitudes toward the role of source individuals in research on biospecimens as well as data is a function of multiple changes. These include new recognition of the privacy and informational risks associated with research on an individual's specimens and data, risks that have become more evident with demonstrations of re-identifiability. But the changes also include new empowerment of patients and families to act in response to illness and disability—by gathering data, finding others with like conditions, assembling cohorts for research and mutual support, conducting research, and partnering with academic and other investigators (Desmond-Hellman 2012). Many of these developments, though not all, rely on the capacity of individuals and communities to utilize technologies and software platforms including social media and mobile health devices (Fleurence et al. 2014).

As Might and Wilsey, fathers of children with an *NGLY1* gene, write, "Two new developments in genetics promise to dramatically shorten the time to reach a successful diagnosis: next-generation sequencing (NGS) and family engagement through social media. ... [There is] a new model for clinicians and researchers. In this model, families, patients, and scientists work jointly. ..." (Might and Wilsey 2014; Might 2012). They go on to highlight challenges raised by this model, including whether researchers should disclose to families results that are not fully established and retain significant uncertainty. True partnership in research and full respect for participant and family choice and control would seem to militate in favor of openness about research results.

Terry, who leads a confederation of patient advocacy groups, similarly emphasizes the importance of the shift toward patient engagement, shared control, return of results, and a relational instead of transactional model of biobanks. She points to online tools and platforms that facilitate patient control over sharing their own information, and decisions that they can make and modify on what research results to receive. "Perhaps the greatest tension for institutions," she writes, "will come from engaging participants as true participants in the process. Institutions and their IRBs ... are not built on a relationship-based engagement model. ... I have built a relational-trust model in which the biobank is steward for the participants in which the actual community owns and manages the biobank." (Terry 2012; Terry et al. 2007). Terry and Terry (2011) recommend that participants themselves own, control, and share data.

Vayena and colleagues consider issues raised by such research models that go beyond participant engagement to participant leadership of research (Vayena et al. 2015; Vayena 2014). They cite the well-known example of a trial of lithium initiated by patients with ALS and ultimately published in *Nature Biotechnology* (Vayena and Tasioulas 2013). They argue that participant-led research "promises to be a vital supplement to standard research: it can focus on conditions that are neglected by standard research ... and can draw on a broader range of data and deliver outcomes more rapidly" (Vayena et al. 2015; Schumacher et al. 2014). However, they acknowledge significant challenges raised by this citizen science, and argue that it "should meet the same high ethical and scientific standards expected of standard research." Yet they suggest that standard oversight mechanisms, such as IRB review, should not necessarily apply. Instead, ethics review by a committee of participants or through crowd-sourcing might be used. They urge the establishment of resources to facilitate participant-led research, such as "development of an online platform where ... activities may be publicly registered; the provision of scientific advice on research proposals through publicly funded panels of experts, operating at an international level; online tools, including scientific and ethical checklists of relevant considerations."

What emerges is a picture of diverse research models with some common characteristics. These models are characterized by deep engagement on the part of individuals whose specimens and data are used in research, to the point of research partnership or even leadership. Individuals may collect data, shape research agendas, vet research designs, and exercise governance control over projects and resources such as biorepositories. These models are further characterized by the open flow of data and results among those individuals, investigators, and the biorepository or other entity aggregating the material and data for research. This includes individuals' access to their own data and results. Indeed, a core ambition of many of these efforts is to seed rapid and wide data sharing across aggregators to share knowledge, increase analytic power, and speed therapeutic progress.

A third core feature is the innovative use of online and informatics tools to achieve core research purposes. These include building cohorts, raising money to fund research, recruiting participants, eliciting consent, and enabling participants to control access to and use of their materials and to change preferences dynamically in real time. Once the research is under way, these tools facilitate offering participants both individual and aggregate research results, communicating about new research opportunities, and building forums in which participants can share results and generate new ideas for research. The commoditization and open source licensing of technologies that can be used to store and manage access to and analyses of data of entire health-care systems (Kohane et al. 2012) have enabled patient advocacy organizations, large and small, to develop registries under patient control that previously were only affordable and manageable by a few expert research organizations.

**Evolving New Models for Research and Biospecimen Resources**

These new models of research, including biospecimen and data aggregation and sharing, pose a significant challenge to conventional research and biospecimen practices characterized by researcher and biobank control. Efforts to advance research and biorepository design have thus far been only partly successful in meeting these challenges. For example, the last decade has seen tremendous advance toward sharing individual and aggregate results with participants, but resistance to extending return of results into genomic research involving biobanks (Bledsoe et al. 2013; Wolf 2013) and remaining uncertainty about whether research-grade information should be offered before clinical confirmation.

Another domain of debate is whether consent should be required for the use of clinically derived biospecimens in research. It is difficult to see how omitting transparency about research use of clinical specimens and conducting research with no notice, consent, and reporting is compatible with the emerging models of engaged research (Kaye et al. 2012). Yet when the ANPRM and then NPRM proposed changes to the Common Rule that would require rudimentary "broad" consent to the research use of such specimens, a firestorm of debate emerged. Though some complained that the proposal failed to go far enough in seeking truly informed consent (O'Connor 2013; Vayena et al. 2013), the dominant objection was that seeking consent at all went too far and would hamper research as well as biorepositories. Thus, the final rule abandoned a blanket approach of treating research with non-identified specimens as human subjects research and requiring at least some form of consent.

Throughout, the question has persisted among many researchers and their institutions of how much to trust research participants and specimen sources to serve as partners in the research process, with a stake in both protections and progress (Janssens and Kraft 2012). Though research and repositories depend on these individuals, recruitment for research participation remains difficult (especially among traditionally underrepresented minorities), and individuals burdened with or at risk of disease actually have the greatest stake in supporting research advances and unique insights, debate continues (Anderson et al. 2012).

Progress at this point requires recognizing that the transformation under way is profound. A century ago, Justice Benjamin Cardozo launched the revolution that would eventually create the requirement of "informed consent." In deciding the *Schloendorff* case, Cardozo penned what remains the most famous declaration in US health law jurisprudence: "Every human being of adult years and sound mind has a right to determine what shall be done with his own body" (*Schloendorff v. Society of New York Hospital*, 105 N.E. 92 (N.Y. 1914).). As is chronicled by Katz (2002), it took decades for American judges to muster the courage to make this revolution real by facing down resistance to shared decision making and the centrality of patients' values in making decisions about health care.

A hundred years after *Schloendorff*, human beings are again seeking the right to determine what shall be done with their bodies—now in the form of their genetic sequences, their organs and tissues, and their data. To be sure, Cardozo ventured onto the then-thin ice of patients' rights by relying on the right to refuse unwanted physical intrusion. That was the first tentative step. But patients' rights grew in the twentieth century to embrace rights of physical and informational privacy, as well as the right to consent to or refuse participation in research.

One might object that the *Schloendorff* revolution was about consent to clinical care, while the new models of research and biorepository assembly and governance are about research. But that would underestimate the transformation under way. With the rise of translational research, the conventional line drawn between research and clinical care has blurred (Kohane 2015; Wolf et al. 2015). Indeed, in translational genomics, research may yield increasing certainty about some variants that then move into clinical use (for example, in cancer care), while clinical sequencing generates new questions about novel variants or poorly understood associations that require further research. The progression is not a one-way street but a multi-directional process yielding knowledge synthesis over time (Wolf et al. 2015; Khoury et al. 2012).

Seen this way, the current revolution in research, including the rules for biospecimen and data collections, is the next logical step in a revolution that began a century ago. The first revolution was confined to clinical care and the right of each individual to control the physical self. A century later, the revolution extends to the full translational cycle embracing research and clinical care, recognizes decisional and privacy rights to control not only the body but also one's data and specimens, and reaches beyond lone individuals to link communities to form cohorts, share data, and seek knowledge. The full power of this second revolution and the full extent of the challenges it poses are only coming into view. This is part of the promise of a range of new efforts including the PMI—their capacity to evolve based on an underlying commitment to deep engagement, return of individual and cohort results, dynamic consent extending to use of specimens and data, shared data, transparency, and accountability (White House 2015; Precision Medicine Initiative 2015). This type of evolution promises to involve participants themselves in the development of the ground rules for this kind of twenty-first-century research.

## Conclusion

When Katz looked back on the first revolution, the one set in motion by Justice Cardozo's declaration that even patients had decisional rights, he despaired. Judicial willingness even to begin developing the doctrine of informed consent took decades to emerge, with little progress between 1914 and the 1950s. Once judges got serious, resistance was tremendous to changing medical practice to embrace the importance of patient decision making and centrality of patient values (Katz 2002).

Alexander Capron urged in his foreword to Katz's book that we await the next generation of physicians to see transformative change. Wolf went further, arguing that it would be the next generation of patients who drove the next advance, as attitudinal data suggested that younger patients wanted more involvement in medical decision making and greater control (Wolf 2006). With the generational shift now well under way, technology has accelerated the empowerment of patients and research participants by enabling online communities, wide information-sharing, and purposeful collaboration.

The change is at hand. Innovations—from Patients Like Me, to Sage Bionetworks, to the patient-driven PXE International Blood and Tissue Bank, to the PMI design documents and trust principles—attest to it. This second revolution extends further than the first—beyond the body to the data and specimens generated. It reaches beyond clinical care to the full translational process. And it goes beyond consent to bodily invasion to partnership in research design, dynamic consent over time, and open access to data and results. Recognizing this as a second revolution places today's debates on research design, biospecimens, and data in the context of a century-long evolution from paternalism to partnership.

## Note

1. Preparation of this chapter was supported by National Institutes of Health (NIH), National Human Genomic Research Institute (NHGRI) and National Cancer Institute (NCI) grant 1R01HG008605 (Wolf, Clayton, Lawrenz, PIs); NIH, NCI and NHGRI grant 1R01CA154517 (Petersen, Koenig, Wolf, PIs); NIH grant U01HG007530 (Kohane, PI); NIH grant P50MH106933 (Kohane, PI); and NIH Common Fund U54HG007963 (Kohane, PI). The authors additionally participate in the Clinical Sequencing Exploratory Research (CSER) Consortium supported by NHGRI and NCI. All views expressed are those of the authors and do not necessarily reflect the views of NIH, NCI, NHGRI, or the CSER Consortium. Thanks to Luke Haqq and Chris Brown for research assistance.

## References

Anderson, Nicholas, Caleb Bragg, Andrea Hartler, and Kelly Edwards. 2012. Participant-centric initiatives: Tools to facilitate engagement in research. *Applied & Translational Genomics* 1: 25–29.

Bledsoe, Marianna J., Ellen Wright Clayton, Amy L. McGuire, William E. Grizzle, P. Pearl O'Rourke, and Nikolajs Zeps. 2013. Return of research results from genomic biobanks: Cost matters. *Genetics in Medicine* 15 (2): 103–105.

Brothers, Kyle B. 2011. Biobanking in pediatrics: The human nonsubjects approach. *Personalized Medicine* 8 (1): 71–79.

Brothers, Kyle B., and Ellen Wright Clayton. 2010. "Human non-subjects research": Privacy and compliance. *American Journal of Bioethics* 10 (9): 15–17.

Department of Health and Human Services. 2011. Advance notice of proposed rulemaking (ANPRM). Human subjects research protections: Enhancing protections for research subjects and reducing burden, delay, and ambiguity for investigators. *Federal Register* 76 (143): 44,512–44,531.

Department of Homeland Security et al. 2015. Notice of proposed rulemaking (NPRM). Federal policy for the protection of human subjects. *Federal Register* 80 (173): 53,933–54,061.

Department of Homeland Security et al. 2017. Federal policy for the protection of human subjects. *Federal Register* 82 (12): 7,149–7,274.

Desmond-Hellman, Susan. 2012. Toward precision medicine: A new social contract? *Science Translational Medicine* 4 (129): 1–2.

De Souza, Yvonne G., and John S. Greenspan. 2013. Biobanking past, present, and future: Responsibilities and benefits. *AIDS* 27 (3): 303–312.

Edwards, Teresa, R. Jean Cadigan, James P. Evans, and Gail E. Henderson. 2014. Biobanks containing clinical specimens: Defining characteristics, policies, and practices. *Clinical Biochemistry* 47 (4–5): 245–251.

Eiseman, Elisa, and Susanne B. Haga. 1999. *Handbook of Human Tissue Sources*. RAND Corporation.

Eiseman, Elisa, Gabrielle Bloom, Jennifer Brower, Noreen Clancy, and Stuart S. Olmsted. 2003. *Case Studies of Existing Human Tissue Repositories: "Best Practices" for a Biospecimen Resource for the Genomic and Proteomic Era*. RAND Corporation.

Fleurence, Rachael L., Anne C. Beal, Susan E. Sheridan, Lorraine B. Johnson, and Joe V. Selby. 2014. Patient-powered research networks aim to improve patient care and health research. *Health Affairs* 33 (7): 1212–1219.

Friedman, Charles P., Adam K. Wong, and David Blumenthal. 2010. Achieving a nationwide learning health system. *Science Translational Medicine* 2 (57): 57cm29.

Grajales, Francisco, David Clifford, Peter Loupos, Sally Okun, Samantha Quattrone, Melissa Simon, Paul Wicks, and Diedtra Henderson. 2014. *Social Networking Sites and the Continuously Learning Health System: A Survey*. National Academy of Medicine.

Harmon, Amy. 2010. Indian tribe wins fight to limit research of its DNA. *New York Times*, April 21.

Henderson, Gail E., R. Jean Cadigan, Teresa P. Edwards, Ian Conlon, Anders G. Nelson, James P. Evans, Arlene M. Davis, Catherine Zimmer, and Bryan J. Weiner. 2013. Characterizing biobank organizations in the US: Results from a national survey. *Genome Medicine* 5 (1): 3. doi:10.1186/gm407.

Hoffman, Sharona. 2016. Citizen science: The law and ethics of public access to medical big data. *Berkeley Technology Law Journal* 30 (3): 1741–1805.

Hudson, Kathy L., and Francis S. Collins. 2013. Biospecimen policy: Family matters. *Nature* 500 (7461): 141–142.

Hudson, Kathy L., and Francis S. Collins. 2015. Bringing the Common Rule into the 21st century. *New England Journal of Medicine* 373 (24): 1–4.

IOM (Institute of Medicine). 2012. *Future Uses of the Department of Defense Joint Pathology Center Biorepository*. National Academies Press.

IOM. 2014. *Assessing Genomic Sequencing Information for Health Care Decision Making: Workshop Summary*. National Academies Press.

Janssens, A. Cecile, and Peter Kraft. 2012. Research conducted using data obtained through online communities: Ethical implications of methodological implications. *PLoS Medicine* 9 (10): 1–4.

Katz, Jay. 2002. *The Silent World of Doctor and Patient*. Johns Hopkins University Press.

Kaye, Jane, Liam Curren, Nick Anderson, Kelly Edwards, Stephanie M. Fullerton, Nadja Kanellopoulou, David Lund, et al. 2012. From patients to partners: Participant-centric initiatives in biomedical research. *Nature Reviews Genetics* 13 (5): 371–376.

Khoury, Muin J., Marta Gwinn, M. Scott Bowen, and W. David Dotson. 2012. Beyond base pairs to bedside: A population perspective on how genomics can improve health. *American Journal of Public Health* 102 (1): 34–37.

Kohane, Isaac S. 2011. Using electronic health records to drive discovery in disease genomics. *Nature Reviews Genetics* 12 (6): 417–428.

Kohane, Isaac S. 2015. Ten things we have to do to achieve precision medicine. *Science* 349 (6243): 37–38.

Kohane, Isaac S., Kenneth D. Mandl, Patrick L. Taylor, Ingrid A. Holm, Daniel J. Nigrin, and Louis M. Kunkel. 2007. Reestablishing the researcher-patient compact. *Science* 316 (5826): 836–837.

Kohane, Isaac S., Susanne E. Churchill, and Shawn N. Murphy. 2012. A translational engine at the national scale: Informatics for integrating biology and the bedside. *Journal of the American Medical Informatics Association* 19 (2): 181–185.

Lewis, Michelle H., Aaron Goldenberg, Rebecca Anderson, Erin Rothwell, and Jeffrey Botkin. 2011. State laws regarding the retention and use of residual newborn screening blood samples. *Pediatrics* 127 (4): 703–712.

Mello, Michelle M., and Leslie E. Wolf. 2010. The Havasupai Indian Tribe case: Lessons for research involving stored biologic samples. *New England Journal of Medicine* 363 (3): 204–207.

Might, Matthew. 2012. Hunting down my son's killer. http://gizmodo.com/5914305/hunting-down-my-sons-killer.

Might, Matthew, and Matt Wilsey. 2014. The shifting model in clinical diagnostics: How next-generation sequencing and families are altering the way rare diseases are discovered, studied, and treated. *Genetics in Medicine* 16 (10): 736–737.

National Bioethics Advisory Commission (NBAC). 1999. *Research Involving Human Biological Materials: Ethical Issues and Policy Guidance.*

National Cancer Institute (NCI), Office of Biorepositories and Biospecimen Research. 2011. *NCI Best Practices for Biospecimen Resources.* http://biospecimens.cancer.gov/bestpractices/2011-NCIBestPractices.pdf.

National Cancer Institute (NCI), Office of Biorepositories and Biospecimen Research. 2016. *NCI Best Practices for Biospecimen Resources.* http://biospecimens.cancer.gov/bestpractices/2016-NCIBestPractices.pdf.

National Institutes of Health. 2013. Advisory Committee to the Director, HeLa Genome Data Access Working Group. *Background.* http://acd.od.nih.gov/hlgda.htm.

Newborn Screening Saves Lives Reauthorization Act. 2014. H. R. 1281, 113th Congress, Pub. L. 113–240, Section 12. https://www.govtrack.us/congress/bills/113/hr1281.

O'Connor, Dan. 2013. The apomediated world: Regulating research when social media has changed research. *Journal of Law, Medicine & Ethics* 41 (2): 470–483.

Office for Protection from Research Risks. 1997. *Issues to Consider in the Research Use of Stored Data or Tissues.* http://www.hhs.gov/ohrp/policy/reposit.html.

OHRP (Office for Human Research Protection). 2008a. *Guidance on Research Involving Coded Private Information or Biological Specimens.* http://www.hhs.gov/ohrp/policy/cdebiol.html.

OHRP. 2008b. *Guidance on Engagement of Institutions in Human Subjects Research.* http://www.hhs.gov/ohrp/policy/engage08.html.

Olson, Jeremy. 2014. Minnesota must destroy 1 million newborn blood samples. *Star Tribune*, January 14.

Precision Medicine Initiative. 2015. *The Precision Medicine Initiative Cohort Program—Building a Research Foundation for 21st Century Medicine.* https://www.nih.gov/precision-medicine-initiative-cohort-program.

Schumacher, Kurt R., Kathleen A. Stringer, Janet E. Donohue, Sunkyung Yu, Ashley Shaver, Regine L. Caruthers, Brian J. Zikmund-Fisher, Carlen Fifer, Caren Goldberg, and Mark W. Russell. 2014. Social media methods for studying rare diseases. *Pediatrics* 135 (5): e1345–e1353.

Skloot, Rebecca. 2010. *The Immortal Life of Henrietta Lacks.* Crown.

Terry, Sharon F. 2012. The tension between policy and practice in returning research results and incidental findings in genomic biobank research. *Minnesota Journal of Law, Science & Technology* 13 (2): 691–736.

Terry, Sharon F., and Patrick F. Terry. 2011. Power to the people: Participant ownership of clinical trial data. *Science Translational Medicine* 3 (69): 1–4.

Terry, Sharon F., Patrick F. Terry, Katherine A. Rauen, Jouni Uitto, and Lionel G. Bercovitch. 2007. Advocacy groups as research organizations: The PXE International example. *Nature Reviews Genetics* 8 (2): 157–164.

Vayena, Effy. 2014. The next step in the patient revolution: Patients initiating and leading research. *British Medical Journal* 349:g4318. doi:10.1136/bmj.g4318.

Vayena, Effy, and John Tasioulas. 2013. The ethics of participant-led biomedical research. *Nature Biotechnology* 31 (9): 786–787.

Vayena, Effy, Anna Mastroianni, and Jeffrey Kahn. 2013. Caught in the Web: Informed consent for online health research. *Science Translational Medicine* 5 (173): 1–3.

Vayena, Effy, Marcel Salathé, Lawrence C. Madoff, and John S. Brownstein. 2015. Ethical challenges of big data in public health. *PLoS Computational Biology* 11 (2): e1003904.

The White House. 2015. *Precision Medicine Initiative: Privacy and Trust Principles.* https://www.whitehouse.gov/sites/default/files/microsites/finalpmiprivacyandtrustprinciples.pdf.

Wolf, Susan M. 2006. Doctor and patient: An unfinished revolution. *Yale Journal of Health Policy, Law, and Ethics* 6 (2): 485–500.

Wolf, Susan M. 2013. Return of results in genomic biobank research: Ethics matters. *Genetics in Medicine* 15 (2): 157–159.

Wolf, Susan M., Frances P. Lawrenz, Charles A. Nelson, Jeffrey P. Kahn, Mildred K. Cho, Ellen Wright Clayton, Joel G. Fletcher, et al. 2008. Managing incidental findings in human subjects research: Analysis and recommendations. *Journal of Law, Medicine & Ethics* 36 (2): 219–248.

Wolf, Susan M., Brittney N. Crock, Brian Van Ness, Frances Lawrenz, Jeffrey P. Kahn, Laura M. Beskow, Mildred K. Cho, et al. 2012. Managing incidental findings and research results in genomic research involving biobanks and archived datasets. *Genetics in Medicine* 14 (4): 361–384.

Wolf, Susan M., Wylie Burke, and Barbara A. Koenig. 2015. Mapping the ethics of translational genomics: Situating return of results and navigating the research-clinical divide. *Journal of Law, Medicine & Ethics* 43 (3): 486–501.

# 17 Informing the Public and Including It in Discussions about Biospecimens

Jane Perlmutter and Heide Aungst

As a long-term cancer survivor and patient advocate, I recognize the importance and difficulty of the issues raised in this volume. Although people find their way to advocacy through different life events and motivations, we are united in the mission to rapidly improve the lives of patients. Research advocates in particular try to accomplish this by ensuring that patients' voices are heard and impact research. But as fields like genomics and technologies used in research rapidly evolve, consent procedures shift, and privacy concerns grow, researchers often retreat into their own labs, immersed in data, forgetting about the patients in whose name the research is being conducted. Some may add token representation of patient advocates, but many leave out our voices altogether. Many patients and researchers accept this as status quo. I find it disturbing. Therefore, I am grateful for the opportunity to provide the voice of a patient advocate here in the discussion about biobanking.

—Jane Perlmutter

Scholars debate exactly when "patient advocacy" began as a formal endeavor but agree that it evolved through several stages to where it is today; indeed, some enlightened organizations already give patients a voice in research efforts. A quick look at the history of patient advocacy sets the foundation for the current era of genomics and biospecimen repository research—weighed down by concerns of privacy, consent, ownership, data sharing, and returning results—complex issues that can be difficult for the average patient to grasp and were never part of the original awareness, self-help, and fundraising landscape of advocacy.

One of the first health advocacy organizations began in 1913 as the American Society for the Control of Cancer, the predecessor of the American Cancer Society (Perlmutter, Bell, and Darien 2013). By the 1930s, organizations such as the March of Dimes began raising funds earmarked for disease-specific research, in that case to eradicate polio (ibid.). In the 1950s, people with chronic diseases first formed organizations for mutual support

(Aymé, Kole. and Groft 2008) or gathered to promote a "self-help" philosophy, "in which self-determination and patient empowerment were important social drivers" (Koay and Sharp 2013). Others believe patient advocacy originated in the 1960s alongside other grassroots efforts, including environmentalism and consumer rights (Koay and Sharp 2013). Most agree that the formation of the Patient-Centered Outcomes Research Institute, an independent nonprofit, nongovernmental organization authorized by Congress in 2010, has been another major milestone in advancing patient engagement and involvement.

However inexact its origins, patient advocacy changed dramatically in the 1980s with the HIV/AIDS movement. Patients moved from seeking support to becoming disease experts capable of challenging the system, especially to expedite drug trials, and "for the first time, the activist/expert held a distinct role in what was funded and how research was done" (Perlmutter, Bell, and Darien 2013).

The HIV/AIDS movement that cemented the role of patient advocate, also led to the understanding among researchers about the critical need to establish biospecimen repositories. While specimens had been kept in pathology labs for research purposes dating back a century or more, one of the first disease-specific biospecimen repositories established was the AIDS Specimen Biobank (ASB) at the University of California San Francisco in 1982 (De Souza and Greenspan 2013). More than thirty years later, the ASB remains an important resource for researchers around the world. The intersection of the patient advocate's role in biospecimen repository research, although firmly rooted and continuing to evolve, must work to balance researchers' priorities with the compelling needs of patients. The complexity of issues around genomic research demand, and will benefit from, an active collaborative effort and dialogue between researchers and patient advocates.

Patient advocates are encouraged to contribute not only their specimens to research but also their voice to the research process. A patient advocate serves a role that is different from the general public or the average patient, for reasons enumerated. And the voice of the patient advocate is important in promoting biospecimen research, especially for specific diseases and drug development in areas with active patient advocacy groups.

The public has limited knowledge and generally few opinions on such topics as biospecimen repositories specifically, and research more generally. Nevertheless, the general public plays a pivotal role in donating to biospecimen repositories. Rivera and Aungst (chapter 6) have reviewed the sizable literature of studies that have asked individuals about their motivations to donate biospecimens for research. While altruism and contributing to science stand out as admirable reasons to donate, the literature has limitations. Many of the scenarios presented to individuals in the context of a research survey are hypothetical, and may not comport to how a person would react in a "real" situation, especially if they had recently received a serious medical diagnosis. Further, many of the published studies have not focused on participants in clinical trials with specific diseases in search for a cure many of whom hope for significant personal impact of the research.

The role of "patient" differs from that of the "public" in that a patient is immersed in the medical system. Patients, however, may be traumatized because of a recent diagnosis. They aren't yet fully educated on their disease, its process, or their personal prognosis. They may never transcend from patient to patient advocate because of the emotional toll of fighting a very personal battle with their disease. But those who do get past their diagnosis and are able to view the larger picture can become true patient advocates individually and/or strong contributors to patient advocacy organizations.

Patient advocates think beyond their personal situations and become educated not only about their disease but also about all aspects of discovery, prevention, treatment and cure, including research, ethics, and funding—and many attain a deeper understanding of genomics, biospecimens, and repositories. The result is a perspective beyond that of the general public or a specific patient and one that can inform researchers, funders, and the research process. A patient advocate is a liaison between both sides of the research equation: understanding, communicating, and collaborating what patients want and need from researchers and what researchers want and need from patients. Patient advocates are often willing to "rally the troops" to help researchers get what they need to conduct research—whether more patients or more biospecimens or more funding—but they won't hesitate to vocalize their positions if they feel researchers are headed in a direction that doesn't value or ultimately won't help patients.

**Patients' Understanding of Biospecimen Research**

Perhaps the best role for advocates is to become true partners alongside the researcher, to recognize communication and educational gaps between researchers and subjects. This happens frequently, and a typical scenario can be described: Researchers know at the beginning of a study what they need from participants and why. Yet, their hypotheses and the questions they choose to ask may not be the most central to patients. For example, the movement toward patient-centered drug development (http://www.fda.gov/downloads/AboutFDA/CentersOffices/CDER/UCM310754.pdf) has resulted in the recognition that clinical trials often address endpoints that are only of secondary importance to patients. Further, too often, researchers do not clearly communicate their goals or explain the research to participants and instead, make assumptions about their knowledge base. As is discussed extensively elsewhere in this volume, the story of Henrietta Lacks is one of the best-known cases of misunderstanding between researchers and patients and their families. When contacted by researchers, Henrietta's husband thought "cell" meant jail cell, implying that Henrietta might be being kept somewhere (Skloot 2010, 183). The next generation, Henrietta's children, admitted to not knowing what a cell was (ibid., 162). The patient advocate can help bridge this gap.

It is especially important that researchers develop educational programs so that people with low health literacy—and who are most vulnerable—understand biospecimen donation. The biospecimen supply should be diverse and representative to ensure adequacy of disease studies and generalizability in the drug development process. But diversity alone is insufficient: ideally everyone should understand the societal importance of contributing to scientific research. The educational program could profitably begin with a K–12 curriculum—and extend beyond commonly taught middle school Mendelian genetics. Beyond classroom education, organizations that fund and use biospecimen research have a further responsibility to invest significant resources educating the public about the issues that could affect them when they donate biospecimens. If the public understood biomedical research and the importance of biospecimen donation, a person facing a health crisis would be better informed and have the relevant background to process a request to donate a biospecimen.

**Donor and Patient Education**

Patient education must begin with a foundational understanding of terms used in research overall and in biospecimen repository research in particular. Even terms common to researchers like "biobank" or "biospecimen" may be unfamiliar to the public, regardless of educational level. In fact, in one study, 67 percent of participants had never heard the word "biobank" (Rahm et al. 2013). In another, 46 percent had never heard the word and, of the remaining individuals, 45 percent had heard it, but admitted they did not know what it meant (Tupasela et al. 2009). Furthermore, the vocabulary specific to biospecimen research must be clearly understood. One example is understanding the difference between genetics and genomics. The study of genetics examines single genes, so genetic tests assess heritable characteristics of an individual, and, therefore, could impact genetically related family members. Genomics, on the other hand, investigates the entire genome or exome of a person or, in oncology, the genome of a tumor. Rather than single genes, genomics also looks at gene function—how the genome works, including gene regulation and expression. As genomic research explodes, this distinction is important to biospecimen donors who want to fully understand a study they might participate in.

Education outside of a specific research protocol is critical and should be ongoing, not only in schools and colleges, but with general open-to-the-public educational sessions, as well.

**Perception vs. Reality**

Education is not needed just around the fundamental purpose of biospecimen repository donation for research, but on the perception vs. reality of research initiatives, as well as on the potential of individual benefit from research.

One of the most important areas where education is needed to promote patients giving to biospecimen research is the Precision Medicine Initiative (PMI), where there has been a fundamental misunderstanding by many in the public as to what is likely to be achieved in the near term. A critical part of the PMI is to collect biospecimens from one million American volunteers to improve disease treatment and prevention by understanding individual variability through genomics. A PMI Working Group has recommended that a biospecimen repository be in place before individuals are recruited for the cohort and that a clearer understanding be developed as to which biospecimens to collect—that is, not only blood but also possibly other biospecimens, such as microbiome or hair and nail clippings. Researchers will use these biospecimens to better understand a depth of influences on human health, including biomarkers for disease risk, the impact of mutations, gene-environment interactions, and pharmacogenomics—understanding the efficacy and safety of therapeutics on the individual (White House Press Office 2015).

For a million biospecimen donors to volunteer for PMI, they must understand the reality of the initiative. However, it already has generated some misperceptions. For almost two decades, the concept of personalized medicine has been heralded as the coming revolution in medicine that will have a huge impact on the health of all Americans. When the media touts that precision medicine is right "around the corner" there can be a vast distinction between what a patient believes and what a scientist believes "right around the corner" actually means. When researchers say a discovery will be "in the clinic" within two years, they may mean it will be in Phase I clinical trials in two years—not widely available. It is likely, however, that a patient would interpret "right around the corner" to mean that the discovery will be available for everyone from their doctor in two years. Moreover, the researcher, but not the patient, is aware that the drug discovery cycle takes eight to 20 years, with only about 8 percent of drugs making it to market (http://www.fda.gov/downloads/AboutFDA/CentersOffices/CDER/UCM310754.pdf). There is no intention to mislead patients; researcher are simply unaware of the vast differences in their world views.

**Donation Utility**

While education is an important foundation, patients also need to understand the utility and purpose of donation. Patients may falsely believe they will benefit directly from donating a specimen. This therapeutic misconception occurs, in part, because the request for donation may take place during clinical procedures. Clinicians and researchers add to the confusion by not being clear about whether test results will be returned to patients and/or their health care providers, or by giving exaggerated hope to patients about the likely personal benefits of participating in research.

Even when tests will be conducted or results will be returned, there may be misunderstandings as to what the information means. Too often, patients believe that all tests elicit precise and actionable data. This is often not the case. Two different examples where misconception exists include newborn screening which often indicates a need for further testing rather than a definitive diagnosis, and genomic sequencing which often identifies mutations or sequence variations that are not yet understood. Widespread innumeracy of the American public (Paulos 1988 [2001]; Hacker 2016) is a further impediment to public understanding of the meaning of test results, given the probabilistic characteristic of all tests. Even for individuals with a good understanding of probability, it is important—as addressed earlier with the vocabulary of research—to educate patients on the glossary of terms used to characterize tests. For example, *sensitivity* measures the proportion of positives that are correctly identified as such, while *specificity* measures the proportion of negatives that are correctly identified as such. Further, it is important to understand that a test may be flawed because it is not *analytically valid*—accurate in detecting the specific entity that it was designed to detect—or because it is not *clinically valid*—not accurate for a specific clinical purpose, such as diagnosing or predicting risk of a disorder. A third problem is that a test might not have *clinical utility*—that is, there is a lack of evidence that use of the test results in improved measurable clinical outcomes. Finally, none of these factors is black and white. As more data become available, a test can be better assessed on all of these factors. As a consequence, new tests are often incompletely evaluated, but the average patient does not grasp these limitations.

To complicate matters further, the regulations surrounding tests are inconsistent and often misunderstood (Institute of Medicine 2015; Lyman and Moses 2016). The FDA regulates some tests, but others, especially those sent to a single laboratory, often are not regulated. However, the Centers for Medicare and Medicaid Services (CMS) and most health insurers only will reimburse tests conducted in Clinical Laboratory Improvement Amendment (CLIA)-certified labs. With technology rapidly evolving, a Center for Disease Control (CDC) working group issued guidelines for next-generation sequencing in an effort to ensure quality of results across research and clinical laboratories (Gargis et al. 2012) and the following year, the American College of Medical Genetics issued its own set of laboratory standards for next-generation sequencing (Rehm et al. 2013). Patients should understand that when they donate tissue for research, research labs are not required to be, and are often not CLIA certified. This limitation in standards allows research to progress rapidly, but has implications for the accuracy and consistency of the tests and on the ability of the researchers to provide results to patients and/or their clinicians. Within the Department of Health and Human Services, the rules for returning results to patients conflict between CLIA policy and HIPAA (Health Insurance Portability and Accountability Act) rules. According to the website of the DHHS, "certain laboratories in HIPAA covered entities that process research results have a legal responsibility to provide the results to research subjects upon request, although CLIA does not allow returning non-CLIA lab results for a treatment purpose, because of legitimate and long-standing concerns for the validity, reliability and accuracy of results generated in non-CLIA-certified laboratories" (http://www.hhs.gov/ohrp/

sachrp-committee/recommendations/2015-september-28-attachment-c/#). Therefore, if results are returned to patients during a research project, the test may need to be repeated in CLIA-certified laboratory to confirm the results.

**Empowered Patient Involvement**

In recent years, the government has recognized the importance of patients involvement in research initiatives by valuing their contributions to programs, including PMI, the establishment of the Patient-Centered Outcomes Research Institute (PCORI), and the FDA's Patient-Focused Drug Development Program (http://www.fda.gov/downloads/AboutFDA/CentersOffices/CDER/UCM310754.pdf). Beyond government programs, two outstanding examples of empowered breast cancer patient advocates' roles in biospecimen repositories are the Komen Normal Breast Tissue Bank (http://komentissuebank.iu.edu) and Inflammatory Breast Cancer Foundation Tissue Bank (http://www.ibcresearch.org/biobank/). In both of these cases, researchers identified the need for biospecimens as an important priority, and discussed this with patient advocates with whom they regularly collaborated. The advocates then raised the funds to start biospecimen repositories and played key role in their governance. The advocates also helped educate their constituents about the importance of biospecimen donation.

In addition to direct benefit to the research, public engagement increases transparency, which in turn, increases public trust and support for medical research. Many initiatives value and empower the patient—but token patient involvement is still all too common in research.

True empowerment of patient advocates means treatment as full collaborators. This may mean establishing training for researchers to understand exactly how to engage, respect, and partner with patient advocates. Some ways of empowering advocates includes involving them early in the research planning process—especially ideal for institutions just launching a biospecimen repository—and continuing to involve them as part of the research team, including in conference presentations and invitations to be coauthors on professional publications. Token involvement is asking only one or two advocates to participate and rarely asking for or carefully considering their opinions. Empowered involvement means they are equal collaborators at all stages of the research process, from planning through publication and public dissemination.

Empowerment of patient advocates and advocacy organizations needs to go beyond numbers in individual research projects, although numbers and diversity of patient advocate involvement is important. Empowering patients needs to be valued to penetrate the barriers of policy makers' agendas. Too often in the political arena, an unconscionable paucity of patients or their advocates exists. However, some strong countervailing examples have evolved. One is in the rare disease community in which more than 1200 patient organizations are connected to a larger, umbrella organization such as the Genetic Alliance or National Organization for Rare Disorders (NORD) (Aymé, Kole and Groft 2008). These organizations support biospecimen research to better understand their diseases. This involvement provides one strong patient advocacy voice that can shape policies and, ultimately, positively impact the health of many people.

That voice of patient advocacy is also essential in setting policy about research such as discussed elsewhere in this volume. Ethicists and legal scholars may be in the best position to frame many of the relevant questions, but lay people should fill the gaps, add to the discussion, and help evaluate the priorities and trade-offs that impact policy—such as carefully assessing the balance between autonomy and public good. Admittedly, the professionals already involved in these discussions have been—or will someday be—patients themselves who may be asked to donate their biospecimens for research. Nevertheless, they probably will see things differently at that point, since they already have significant depth of knowledge on the subject. Note, however, that it is not easy to predict one's reaction to becoming a patient. Rebecca Dresser captures bioethicists' experiences as they become patients in her book *Malignant: Medical Ethicists Confront Cancer*. Most of the bioethicists—cancer survivors or caregivers of cancer patients—agreed that they expected their professional experiences to help them deal with their experiences as a patient. Yet this did not happen. Rather, in reverse, they found that their patient experiences profoundly influenced their subsequent professional engagements. This change captures why patients and their advocates need to be at the table in research and in setting policy. Despite some professionals' assumptions, the medical world looks different when the reality and urgency of being a patient is experienced.

## Informed Consent

Public education about—and understanding of—research is just the beginning of valuing the patient in the research process. As the patient enters the agreement to participate in research, it is important to fully understand all aspects of the informed consent before signing it. Current approaches to consenting patients to research or biospecimen donation often focuses on obtaining a signed *document*, rather than assessing if the patient or donor truly understands and, therefore, can consent to the research (Henderson 2011). A recent Clinical Trials Transformation Initiative (CTTI) project recommended more focus be placed on the informed consent *process*, one that is highly interactive and tailored to the individual patient (http://www.ctti-clinicaltrials.org/what-we-do/projects/informed-consent/products). CTTI's recommendations include a tool to assist those requesting consent to follow such a process.

Whenever researchers collect data with the goal of general knowledge acquisition, potentially resulting in publication, it is considered research and consent is required. Patients and participants may want to know, understand, and potentially consent. At a minimum, patient representatives and their advocates should be involved at the outset in determining whether and how consent should be obtained.

Four points should be especially clear when consenting people to donate biospecimens for research:

- Their participation is totally voluntary.
- They may withdraw their consent at any time.
- The risks and benefits of donation are articulated.
- Signing a consent document does not nullify the right to sue a sponsor.

Different institutions write their own informed consents and some use burdensome, academic language, while others make the consent easier to read and to understand. An ideal informed consent for biospecimen donation would cover these topics:

- Privacy and protection of donors' information.
- Whether and how biospecimens and data will be shared with other researchers.
- Whether, when, and how aggregate and/or individual results will be provided to donors. (The current regulations require the informed consent document to include a statement regarding whether clinically relevant research results, including individual research results, will be disclosed to subjects, and if so, under what conditions.)

Each donor is likely to have his or her own preferences on these issues, and each study may take different approaches to dealing with them, sometimes even allowing individual donors to select among several options.

However, whatever options are offered should apply consistently to individuals at all sites in a multisite study. As is discussed in other chapters in this volume, there are many legal and ethical issues associated with each of these three topics, but from the patient advocate perspective, the most critical aspect is that research participants understand all details of what they are signing.

**Privacy and Protection of Donors' Information**

Privacy issues associated with donors' biospecimens and resulting data need to be discussed and disclosed in the informed consent. Some studies show that concerns about lack of privacy is associated with lower willingness to donate biospecimens (Gaskell 2013). However, while privacy is sometimes a concern of potential biospecimen donors, in this age of the Internet, social media, and common passive collection of non-medical data, much of the public is used to a lack of privacy and willing to trade-off some of their privacy for other conveniences. A variety of research studies (e.g., Lee 2012; Hensley 2014) have found that between about 10 and 25 percent of patients are concerned about the privacy of their medical data. Other studies have found that even lower percentages of people are disinclined to donate biospecimens, especially among those facing serious health problems (Baer, Smith, and Bendell 2011). Even with low numbers impacted by fear of loss of privacy, this is another area where education and clear communication will help people understand consent in biospecimen donation.

Although a significant effort is made in informed consent documents to inform patients about the protection of their privacy, terminology associated with privacy—as with general research terms and testing terms discussed earlier—is often confusing and rarely standardized; terms should become part of the general vernacular, and education provided, so that individuals are prepared to make informed decisions when asked to donate biospecimens. For example, even researchers often confuse the terms "anonymized" and "de-identified" as being synonymous. Anonymized (sometimes called anonymous) data cannot, in theory, be traced back to the donor because all identifiers have been destroyed. De-identified data, on the other hand, have had identifiers removed, so the researcher does not know the identity of the donor, but identifying codes are maintained independently and therefore the donor can theoretically be known. As Rothstein (2010) writes, "Deidentification exists on a continuum." Today, genomic research poses even more questions, since DNA is itself an identifier—and can sometimes be used along with public databases to tie back to specific individuals (Rothstein 2010). In fact, de-identified specimens can hurt some genetic research. When Chen et al. (2016) discovered thirteen people who genetically should have experienced a devastating childhood disease but didn't, he could not re-identify them to learn about possible protective agents within their genomes that could have been further studied to develop therapies.

The consent should also be clear about ownership and future use of a donors' biospecimens, including potential commercial use. What patients do not recognize is that once the material is taken from them, they no longer own their cells or tissues (Minkel 2006). Given that research findings that used these specimens can lead to profitable products, such as drugs or diagnostic tests, this issue of ownership is not a trivial issue, and may be of interest to potential biospecimen donors. Sometimes, but not always, informed consent documents state clearly that the donor will not profit from such products. Informed consent documents should clearly address issues of ownership and potential for compensation or payment. Also of interest to potential biospecimen donors is whether or not their tissues can be used for future research—and, if so, the parameters for that use, and whether biospecimen donors can direct limitations of use (e.g., "these specimens may only be used for cystic fibrosis research.")

Biospecimens can often be used for many types of research—and the demand for specific types of biospecimens can be significant. Patient advocates, although not necessarily those who donated the biospecimens, should be part of these decisions, while collaborating with researchers. Some patient advocacy groups play an important role in making such decisions (Landy et al. 2011).

The question of sharing is somewhat different when it comes to the data generated from the tissue and not the tissue itself. These data can be re-analyzed and/or combined with other data to lead to new useful information in an unlimited number of ways. Thus, many biospecimen donors, if they understood these issues, would want their data to be made widely available, albeit in a de-identified form. Because of the competitiveness of science, however, this sharing is not always the practice. Advocates want to "free the data" so as to have more power in numbers for answers on diseases. Some agencies that fund biospecimen research require open access to data; yet no provisions or resources for doing so are provided, and in reality much data is not shared, slowing progress. Data sharing is one of the key goals of the Cancer Moonshot initiative led by former Vice President Joe Biden.

The recently finalized revision of the Common Rule does not provide any restrictions (e.g., patient consent) on use or sharing de-identified biospecimens or data. That said, many patients and advocates may opt to exercise more control over use of their biospecimens and data, whether de-identified or not. The revised Common Rule does not preclude researchers from providing such choices.

### Providing Aggregate and/or Individual Results to Donors

The consent should clarify what types of information the donor should expect to receive about the research results in general, and their individual biospecimens and tests in particular, as well as the likely timing of receiving such information. There is a wide range of acceptability in these areas—and what works best for each patient/donor depends on their own personal values. Ideally, patients would be asked to indicate their preferences and the study would make the process of return clear (e.g., whether results would be returned through the donor's clinician or delivered by a trained genetic counselor working with the research team).

Many patients believe that they are entitled to the results of their data regardless of the research focus (Daniel and Haga 2011). Researchers differ in their beliefs about this issue, and those who agree with donors' rights in this regard are often perplexed about which data to share with patients and how to best do so. For example, should only currently actionable findings be shared? What happens to incidental findings which might become significant down the road? Should the information be shared directly with the donor or their clinician or both simultaneously? Who is responsible for explaining the meaning of the information, and how can that responsibility be effectively and efficiently carried out? When the information is genetic, is there any responsibility to explain the implications of the finding to genetically related family members—or does the patient's right to privacy and autonomous decisions override that? The answers to these questions are evolving and the engagement of a diverse group of patient advocates in the discussion is advised.

Ideally, donors who have opted to receive results should, and should be given the opportunity to turn down receiving results when they become available, since circumstances and opinions may shift. Some results could have implications for life or long-term care insurance—or simply to the emotional status of the donor, such as if they were to discover they carried a gene for early onset Alzheimer's Disease. Processes also must be clear about how to deal with incidental findings that predict adult-onset conditions in a pediatric population. Adequate support must be provided to interpret results and educate the patient about what they mean and this support should be tailored to the donor's level of health literacy. The CR Final Rule does not require that either aggregate or individual results be provided to research participants but does require that informed consent documents include a statement regarding whether and under what circumstances clinically relevant research results, including individual research results, will be disclosed to subjects. Further, in the limited setting of a broad consent for future use of biospecimens, the CR Final Rule specifies that "unless it is known that clinically relevant research results, including individual research results, will be disclosed to the subject in all circumstances, a statement that such results may not be disclosed to the subject must be included in the broad consent."

## Summary

To make advancements in health, especially at the genomic level, researchers must value patient advocates and participants at every point of the research process—from including their input into policies, respectfully soliciting their donation, maximizing its value, and helping them understand what is being learned and how it may impact them and future patients. The following three themes cut across all topics in this chapter.

*Better Public Education about the Research Process in General, and Biospecimen Research in Particular* Currently the potential of precision medicine is celebrated in the public press, but there is limited understanding of the research process or any realistic understanding of about the anticipated timeframe for realization of results. This will lead to public frustration, and reduction of support of both research funding and tissue donation.

*A Voice at the Table* Patient advocates must be involved in discussions of issues about biospecimen research, including those discussed in this volume. These individuals should have a strong foundation in the relevant research and continue to have broad two-way conversations with the diversity of the patients they represent.

*Respect for Donors* When someone donates a sample, the purpose must be made clear so as not to blur the boundary between clinical and research use. It must not only be made clear what the specimen will be used for immediately, but also what will happen to it in the future, and whether the donor has any say about allowable research, withdrawal of the samples, and receiving results. Finally, the value of their donation for the greater good of all should be acknowledged and appreciated.

In this day of the genomic revolution, researchers must consider biospecimen donors, just as with all patients, valued collaborators and partners in the research enterprise.

## References

Aymé, Ségolène, Anna Kole, and Stephen Groft. 2008. Empowerment of patients: Lessons from the rare diseases community. *Lancet* 371 (9629): 2048–2051.

Baer, A. R., M. L. Smith, and J. C. Bendell. 2011. Donating tissue for research: Patient and provider perspectives. *Journal of Oncology Practice* 7 (5): 334–337.

Chen, Rong, Lisong Shi, Jörg Hakenberg, Brian Naughton, Pamela Sklar, Jianguo Zhang, Hanlin Zhou, et al. 2016. Analysis of 589,306 genomes identifies individuals resilient to severe Mendelian childhood diseases. *Nature Biotechnology* 34 (5): 531–538.

Daniel, J. O., and S. B. Haga. 2011. Public perspectives on returning genetics and genomics research results. *Public Health Genomics* 14: 346–355.

De Souza, Yvonne G., and John S. Greenspan. 2013. Biobanking past, present and future. *AIDS (London, England)* 27 (3): 303–312.

Dresser, Rebecca. 2012. *Malignant: Medical Ethicists Confront Cancer*. Oxford University Press.

Free the Data. http://www.free-the-data.org.

Gargis, Amy S., Lisa Kalman, Meredith W. Berry, David P. Bick, David P. Dimmock, Tina Hambuch, Fei Lu, et al. 2012. Assuring the quality of next-generation sequencing in clinical laboratory practice. *Nature Biotechnology* 30 (11): 1033–1036.

Hacker, Andrew. 2016. *The Math Myth and Other STEM Delusions*. New Press.

Hallowell, N., A. Hall, C. Alberg, and R. Zimmern. 2014. Revealing the results of whole-genome sequencing and whole-exome sequencing in research and clinical investigations: Some ethical issues. *Journal of Medical Ethics* 41 (4): 317–321.

Henderson, Gail E. 2011. Is informed consent broken? *American Journal of the Medical Sciences* 342 (4): 267–272.

Hensley, Scott. 2014. *A Worry in theory, medical data privacy draws a yawn in practice*. National Public Radio.

Huckman, Robert S., and Mark A. Kelley. 2013. Public reporting, consumerism, and patient empowerment. *New England Journal of Medicine* 369 (20): 1875–1877.

Institute of Medicine. 2015. Policy issues in the development and adoption of biomarkers for molecularly targeted cancer therapies.

Klein, Daniel B., and Alexander Tabarrok. 2016. Is the FDA safe and effective? FDAReview.org, a Project of The Independent Institute. http://www.fdareview.org/.

Koay, Pei P., and Richard R. Sharp. 2013. The role of patient advocacy organizations in shaping genomic science. *Annual Review of Genomics and Human Genetics* 14 (1): 579–595.

Landy, David C., Margaret A. Brinich, Mary Ellen Colten, Elizabeth J. Horn, Sharon F. Terry, and Richard R. Sharp. 2011. How disease advocacy organizations participate in clinical research: A survey of genetic organizations. *Genetics in Medicine* 14 (2): 223–228.

Lee, Christoph I., Lawrence W. Bassett, Mei Leng, Sally L. Maliski, Bryan B. Pezeshki, Colin J. Wells, Carol M. Mangione, and Arash Naeim. 2012. Patients' willingness to participate in a breast cancer biobank at screening mammogram. *Breast Cancer Research and Treatment* 136 (3): 899–906.

Lyman, G. H., and H. L. Moses. 2016. Biomarker tests for molecularly targeted therapies: Laying the foundation and fulfilling the dream. *Journal of Clinical Oncology* 34 (17): 2061–2066.

Minkel, J. R. 2006. Uninformed consent. *Scientific American* 295 (4): 22–24.

Paulos, John Allen. 1988. *Innumeracy: Mathematical Illiteracy and Its Consequences.* Hill and Wang. [reprinted 2001]

Perlmutter, J., S. K. Bell, and G. Darien. 2013. Cancer research advocacy: Past, present, and future. *Cancer Research* 73 (15): 4611–4615.

Rahm, Alanna Kulchak, Michelle Wrenn, Nikki M. Carroll, and Heather Spencer Feigelson. 2013. Biobanking for research: A survey of patient population attitudes and understanding. *Journal of Community Genetics* 4 (4): 445–450.

Rehm, Heidi L., Sherri J. Bale, Pinar Bayrak-Toydemir, Jonathan S. Berg, Kerry K. Brown, Joshua L. Deignan, Michael J. Friez, Birgit H. Funke, Madhuri R. Hegde, and Elaine Lyon. 2013. ACMG clinical laboratory standards for next-generation sequencing. *Genetics in Medicine* 15 (9): 733–747.

Rothstein, Mark A. 2010. Is deidentification sufficient to protect health privacy in research? *American Journal of Bioethics* 10 (9): 3–11.

Skloot, Rebecca. 2010. *The Immortal Life of Henrietta Lacks.* Crown.

Tupasela, A., S. Sihvo, K. Snell, P. Jallinoja, A. R. Aro, and E. Hemminki. 2009. Attitudes towards biomedical use of tissue sample collections, consent, and biobanks among Finns. *Scandinavian Journal of Public Health* 38 (1): 46–52.

Vrijenhoek, Terry, Ken Kraaijeveld, Martin Elferink, Joep De Ligt, Elcke Kranendonk, Gijs Santen, Isaac J. Nijman, et al. 2015. Next-generation sequencing-based genome diagnostics across clinical genetics centers: Implementation choices and their effects. *European Journal of Human Genetics* 23 (9): 1270.

White House Press Office. 2015. Fact Sheet: President Obama's Precision Medicine Initiative. January 30. https://www.whitehouse.gov/the-press-office/2015/01/30/fact-sheet-president-obama-s-precision-medicine-initiative.

# 18 Investigator's Commitment during the Consent Process for Biospecimen Research

Erin Rothwell and Erin Johnson

It has been noted by many scholars that the current informed consent process is seriously flawed (Emanuel et al. 2008). Clinical informed consent for study-specific research aggregated over 45 years (1961–2006) found that only 54 percent of the research subjects understood the aims of the research, 50 percent understood randomization and the risks of the research, 47 percent knew participation was voluntary, 44 percent were aware that they could withdraw from the study, and 57 percent understood the benefits of the research (Falagas et al. 2009). With the added complexities of communicating what is "unlimited future" research through broad consent—now an option for research with identifiable biospecimens under the revised Common Rule finalized in January 2017—and the lack of a relationship between investigators and research subjects, these challenges are magnified.

Efforts to promote comprehension among donors during the consent process should continue, but efforts should also include methods to improve transparency, trust, and partnership of investigators with donors (Grady et al. 2015). This will be more important among diverse populations in garnering participation with broad consent for future research. In chapter 8 of this book Grady and her co-authors expand upon reported outcomes from an National Institutes of Health (NIH) Clinical Centers' Department of Bioethics workshop about broad consent for research with biological samples. Conclusions from this workshop suggested that broad consent communicates respect for donors, allows donors control of whether or not samples are used in research, allows donors to determine if the risks and benefits are acceptable, gives donors the opportunity to determine whether or not they want to contribute to the goals of the research, and makes future decisions about research with biospecimens transparent. However, most biospecimens are obtained within the clinical context and some scholars believe that broad consent for research with biospecimens obtained in the clinical setting will not promote these goals (Botkin 2015). Other competing demands in the clinical setting, such as high patient volume and more immediate clinical concerns, limit available time for research and create further challenges to obtaining informed consent.

Another challenge to broad consent includes limited public knowledge about what is research. The evidence clearly supports that much of the public does not have a reasonable understanding about what type of research is conducted with biospecimens (Lemke et al. 2010; Bates et al. 2005). For example, in our research on parental understanding about research with residual newborn screening specimens, we found that many participants had questions about the definition of research with biospecimens (Rothwell, Anderson, Swoboda, et al. 2012; Rothwell, Anderson, Goldenberg, et al. 2012). Further, some scholars argue that broad consent is not "informed" consent because, along with other ethical concerns, the consent cannot explain future research and the use of the biospecimens in that context (Caulfield, Upshur, and Daar 2003; Hofmann 2009).

However, communicating to research participants about how investigators will use biospecimens *responsibly* may help promote transparency and trust. In that the specific elements of future research cannot be described, it may be important to communicate the oversight, process, and responsibility executed by the investigator (and the institution) as a way of imparting trust in the system. Investigators are the ones accessing and continually using biospecimens. Premises that support broad consent for biospecimen research include that personal information will be handled safely, individuals can withdraw consent, and new studies will be reviewed by an institutional review board to provide timely oversight to the research (Hofmann 2009). Investigators manage many of these responsibilities. It may be prudent to focus education efforts for research participants on how investigators will adhere to an ethical code of conduct in any future research activity. The research described here was conducted to explore investigator acceptance of an investigator oath and, if acceptable, what components should be included in this oath. Interviews were conducted with investigators to identify attitudes and opinions about broad consent and to explore the acceptability of an oath by investigators given to research subjects along with the consent form. Further, the initial items of an investigator oath were edited and revised through an e-Delphi method among national experts in biobanking, informed consent, and/or research ethics.

## Methods and Materials

Investigators who use biospecimens in their research and have active studies with the Utah Genome Project were identified and contacted through email. After the first 4 individuals were interviewed, each of them were asked to volunteer an additional 3–5 individuals who may be appropriate for an interview, and this was repeated one more time after the second round of interviews. This snowball technique resulted in eighteen interviews. Interviews took place during the month September 2015 immediately following the issuance of the Notice of Proposed Rulemaking (NPRM) on September 8, 2015; the NPRM would have required at least broad consent for research with biospecimens regardless of identifiability, whereas the final rule offers it as an option, alongside the status quo of stripping existing biospecimens of identifiers.

The purposes of the interviews were to garner initial impressions about broad consent for future research and to assess initial reactions to the development of an investigator oath. The following questions were asked during the interviews:

1. Can you tell me how you have used biospecimens in your research?
2. What are your opinions about open broad consent for future unlimited research?
    a. Can you explain what type of consent, if at all, have you used most frequently for biospecimen research?
3. The FDA has an investigator commitment oath/letter that is required. What are your thoughts about a commitment oath for investigators who use biospecimens from broad consent? This would be given to research subjects who sign a broad consent.
    a. What should be included in an investigator oath?
    b. Why should an investigator oath not be used?
    c. What type of perception would research subjects have if they received an investigator oath that was provided while they contemplated broad consent for future unlimited research?
4. In your opinion, why or why not is broad consent for future unlimited research beneficial for research?
5. In your opinion, why or why not is an investigator oath for a broad consent beneficial for research? How would it be different? How would it be the same?

On the basis of data from the interviews, statements to be included in an investigator oath were generated; the statements were then sent to experts in biobanking, research ethics, and/or informed consent; an e-Delphi method was used. Note that this is the first step in a larger study. The next step in the planned study will assess and revise the preliminary results reported here in a more rigorous evaluation and randomized study.

Using a qualitative descriptive approach, semi-structured interviews were conducted to gather initial impressions and similarities across the topics of interest. Kvale (1983) defines the qualitative interview as "an interview, whose purpose is to gather descriptions of the life-world of the interviewee with respect to interpretation of the meaning of the described phenomena." The interviews were completed in person and by telephone. Research evidence supports that data obtained by telephone and data obtained from face-to-face interviews do not differ significantly; data of these two kinds are acceptable to use and are interchangeable for purposes of qualitative descriptive research (Sturges and Hanrahan 2004).

This study was deemed to be exempt from the University of Utah institutional review board because no identifiers of the participants were retained. The interviews lasted between 15 and 50 minutes, with an average of 20 minutes. The interviews were not recorded, but notes were taken during the interviews and summaries were written immediately after the completion of each interview. The reason audio recording was not done is that the interviewer was known to some of the interviewees, and the discussion of an investigator oath in itself may raise questions about the ethical conduct of researchers. Therefore, to promote a more open conversation, audio recordings and data about individual characteristics of the interviewees were not collected. In anticipation of the lack of audio recordings, the number of questions was limited to ensure adequate time for sufficient note taking.

The qualitative software program ATLAS.ti (Friese 2013) was used to analyze the interview notes and summaries. A content analysis was conducted to identify the common codes across all of the interviews. The codes were developed from reading of the interview summaries and interview questions. The codes were systematically applied to the data and then categorized together based on degree of similarity. This is the same approach to data analysis used in our previous research (Rothwell 2010; Rothwell, Anderson, and Botkin 2010; Rothwell et al. 2011).

For the e-Delphi study component, a survey was emailed to Delphi panel members in two rounds. A list of 25 national experts in biobanking, informed consent, and research ethics were identified from the published literature. Respondents were asked to score the importance of each statement on a five-point Likert scale ranging from extremely important to not important at all. Respondents were also asked to provide feedback on each statement in text form. The scores and edited statements from the first round were sent to the Delphi panel for the second round. Consensus was defined as having a median score of 4.0 (moderately important) or higher. Out of the 25 invitations, 17 experts responded and provided feedback. During the second round, 12 out of the 17 experts responded.

## Results

Five major categories were identified from the analysis of the interview data: ever-changing purpose of consent, awareness and accountability, inconsistent nature of biospecimen research, similarities and differences with clinical care, and components of the investigator oath. These categories are described below and supported with representative quotes of the interviewees. The statements of the investigator oath are provided in table 18.1.

Table 18.1
Statements in an investigator oath and mean and median scores

|  | Mean | Median | SD |
|---|---|---|---|
| to adhere to all privacy, ethical and confidential safeguards that are in place for this research. | 4.83 | 5.00 | .577 |
| to engage in thoughtful and well-designed research | 4.58 | 5.00 | .793 |
| to seek approval for my research from appropriate institutional, ethics, or governing committees before beginning research. | 4.42 | 5.00 | .900 |
| to not release the biospecimen/data to anyone who does not have appropriate approval and who would not adhere to the ethical standards in this oath. | 4.42 | 5.00 | .900 |
| to recognize your donation as a limited resource and to use it wisely and carefully. | 4.42 | 4.50 | .669 |
| to respect your donation for research. | 4.33 | 5.00 | .888 |
| to remember that there are some risks with research and to seek to minimize these risks. | 4.08 | 5.00 | 1.165 |
| to be open and transparent about my research. | 4.17 | 4.50 | 1.030 |
| to use the donation only in research to improve the well-being of our society. | 4.17 | 4.50 | 1.030 |
| to publish results for the advancement of future research efforts. | 4.00 | 4.00 | .775 |
| to seek help when needed if I am unsure about the integrity of the research or the welfare of the donors. | 3.92 | 4.00 | 1.084 |

### Ever-Changing Purpose of Consent

Many of the interviewees were supportive of broad consent for future research, but also identified that advancements in technology and genetics change rapidly and that therefore interpretations of consent forms may change. For example, some investigators indicated that institutional review boards (IRBs) consistently change their interpretations regarding the nature of genetic research. This may influence how the written text or its interpretation in a broad consent may change over time. One participant reported, "We wrote in the consent form 'researching genes' but then when we wanted to use the specimens a few years later again the IRB did not think sequencing genes was researching genes." Some participants also questioned how someone could comprehend "unlimited" research and whether broad consent was really "informed" consent. One participant commented "It's scary because it is unlimited, anything, how do you communicate that?" Another asserted "No way broad consent equals informed consent."

Interviewees also mentioned concerns about implementation of broad consent within the clinical setting. The most common concern was practical: how would broad consent be implemented in a clinical setting. Half of the interviewees viewed broad consent as a mechanism to protect institutions or investigators and half of the interviewees thought it was more to help protect research subjects. At the minimum, the interviewees stated it would raise awareness among research subjects. The following responses are representative:

Is broad consent trust in the regulatory process? Is it trust in the investigator? Is it just to let the participant to say no?

Broad consent is designed to the save the investigator. Patients are less likely to consent. Broad consent is both attractive and scary.

If anything, broad consent raises awareness and gives options to seek more information.

### Awareness and Accountability

All of the interviewees were supportive of an investigator oath, but there were also concerns. In support of the oath, many of the interviewees said it would bring awareness to "research subjects" and the importance of the research. One said "Have to let them know why you are doing this and it is for the right reasons." A few interviewees said it would help with trust because it would communicate respect to individuals through awareness and choice, but there was an equal number of interviewees who were unsure it could help with trust ("It will bring awareness so that helps with trust, I guess"; "It couldn't hurt but not sure it will help with trust").

Several interviewees also noted that the oath might be most beneficial to investigators because it would provide accountability. Many of the interviewees mentioned they had seen questionable behavior of an investigator sometime in their career ("too many [investigators] have done things that are wrong and those are the ones we know about"). Interviewees also mentioned investigators are "trying to do the right thing." One interviewee mentioned an unwillingness to release specimens donated by patients to a central repository because the interviewee could "not vouch for other investigator behavior." Some interviewees suggested that, instead of or in addition to an investigator oath, "periodic review of investigators by the institution would be good and more effective route" to ensure ethical conduct. Other suggestions that could accompany an investigator oath was education of the investigators ("What are ethical behaviors beyond not sharing specimens without an MTA?"; "There appears to be a disconnect with investigators. It would be good to know what is expected behavior or protocol"). An oath would help institutions to be able "to drop the hammer on questionable behavior of investigators." Several interviewees stated that there was a perception of investigators to view biospecimens as only data points and not as donations from individuals; for example, one asserted that "investigators need to appreciate the donation." The interviewees believed that an oath would also help promote respect. The following two comments about respect are representative:

It starts with respect. Some investigators do not respect the donation as a contribution from participants.

It would respect the rights and welfare of the subjects.

Many of the concerns with an investigator oath included that it may generate unnecessary concern by donors. One representative quote included "We've been doing this for years and to have this oath all of sudden would raise suspicions." Other concerns were that it is not possible to have trust without a relationship and the interviewees were unsure if an oath could achieve trust. Finally, some saw the oath as "another hoop to jump through" to gain access to biospecimens.

### Inconsistent Nature of Biospecimen Research

One category of responses that emerged in response to the questions about broad consent was about how the type and quality of research is left to the individual investigator, the point being that there is no one way to conduct this type of research. Interviewees often used the expression "gray area"—"there are gray areas with research with biospecimens"; "it is such a gray area of research"; "a lot of independence [to make decisions about research]." When prompted to elaborate, some of the interviewees mentioned return of results. Investigators know that they should not return results, but they sometimes come across clinically actionable data. One representative statement was "For example, you know you do not return results, but it is conflicting when you know death could be involved." Another situation is when research subjects sign the consent forms and indicate particular preferences for return of results, but then they change their minds, and "[donors] have the option to change their minds about return of results."

Other inconsistencies with biospecimen research and broad consent for future research were related to rapid changes in technology. Interviewees discussed what was possible ten years ago, what was possible five years ago, and what is possible today. Continued change in technology made their research continually changing. One quote that captured this category was "Fluid is the definition of investigator research." Other interviewees mentioned DNA sequencing as inconsistent; one of them noted that a "need to stop and think about each case individually." One interviewee noted that how inconsistencies are managed in secondary research with biospecimens depends on the "integrity of the investigator."

**Similarities and Differences with Clinical Care**

Many of the interviewees made references to the similarities and differences between biospecimen research and clinical care ("I'm not a physician, I am a researcher"). In regard to donors' signing a broad consent, interviewees suggested that a relationship with a medical provider would be associated with a greater likelihood that the participant would consent to future research ("You need a personal relationship and that will only be associated with disease specific sub-specialties"). Similarly, many interviewees mentioned there is often no relationship between a research subject who donated biospecimens and an investigator who uses them for secondary research purposes ("Dealing with DNA, not patients"). Despite the lack of a clinical relationship with research subjects, there were several comments that research can have important health implications. However, these health implications are different than clinical care for patients in that the results of research are slower than those of clinical care: an intervention prescribed by the physician may appear to have more immediate effects, whereas "an investigator might have to wait five years to see the outcomes of their research that might yield life saving information." Another representative quote was "PhDs have no oath so they have to do what they 'think' or 'hope' is right."

**Components of the Investigator Oath**

Interviewees were asked what might be included in an investigator oath that would accompany a broad-consent form. (See table 18.1.) The most common suggestion was to increase transparency by explaining both that the investigator will store and use biospecimens and what happens after the biospecimens are used. The second most common suggestion focused on promoting respect of the donors and the promise to use the biospecimens wisely. The third most common category was the role of the donor in the process and methods of communicating general results back from this research. Several interviewees stated that donors may expect return of results or that their sample will be used for hundreds of studies. The need for clear communication as to what is anticipated to be returned is important. The fourth category was to reiterate safeguards and reassure donors how their privacy will be protected from outside entities. The other categories were to provide specific examples of the risks and benefits of the research, and to provide concrete, practical examples of the impact of research on society.

After the interviews were completed, statements to reflect what was reported in the interviews for an investigator oath were drafted. The draft for the initial investigator oath incorporated the categories mentioned above, the expectations of investigators laid out in the NPRM, and the elements of Form FDA 1572, the statement of the Investigator, as codified in FDA Code of Federal Regulations Part 312, subpart D—Responsibilities of Sponsors and Investigators ("FDA Statement of Investigator," http://www.fda.gov/downloads/AboutFDA/ReportsManualsForms/Forms/UCM074728.pdf). The FDA code states that the selected investigator for a drug trial must make a commitment to a number of ethical standards. The resulting investigator oath had seventeen statements. After the second round of feedback from the Delphi panelists, eleven statements remained. (See table 18.1.)

**Discussion**

On January 18, 2017, the US Department of Health and Human Services along with many other federal agencies announced changes to the Federal

Policy for Protection of Human Subjects (HHS 2017). One of the changes centered on improving the informed consent form to allow research participants to better understand the scope, risks, and benefits of the research. Another change (described more fully in chapter 1 of this volume) was the option for investigators to rely on broad consent for future research as an alternative to seeking IRB approval to waive the consent requirement. The type of consent for secondary uses of biospecimens studied in this chapter was a one-time broad consent for future research.

Much of the research on consent focuses on improving donor comprehension through multimedia tools, extended discussions, comprehension quizzes, and enhanced consent forms on comprehension (Palmer, Lanouette, and Jeste 2012; Nishimura et al. 2013). Yet little research has considered how ethical values may influence willingness to consent into a study. Although some research has identified trust with the clinician or the institution as an ethical value influencing consent decisions (Molyneux, Peshu, and Marsh 2005), transparency and trust of the *investigator* have not been explored. Broad consent for future research poses significant challenges to participant comprehension due to the unknown nature of how specimens are used by investigators. Given the independence of the investigators who use biospecimens in secondary research, it is surprising that very little research has assessed how to improve transparency and trust of investigators among research participants or patients who donate their biospecimens.

As the distance between research subjects and investigators who use biospecimens in secondary research expands, more emphasis on the ethical standards of investigators is needed. The concept of a one-time broad consent for future research—thereby reducing the burden of multiple subsequent consents each time a biospecimen is used—also increases investigator responsibility to manage biospecimens in an ethical manner without re-contacting or re-consenting donors about different uses. The preliminary study reported here identified concepts and concerns about the changing nature of informed consent and the role of the investigator during the consent process.

An investigator oath was well received by the interviewees in this study. Interestingly, whereas this research intended to explore how investigators perceived the impact of an investigator oath on research participants, the data highlighted the beneficial potential effect of such an oath on investigators themselves. The most common reasons in support of an oath included increased transparency and accountability of investigators. In addition, many interviewees agreed that transparency would be improved by giving research participants a copy of the investigator oath with a copy of the broad consent. However, there were concerns that broad consent for future research in the clinical context will be perfunctory and that only the institution will benefit from legal protections (Botkin 2015). An analogy is drawn to the Health Insurance Portability and Accountability Act authorization: patients are rarely engaged in a meaningful process that promotes comprehension when signing an HIPAA authorization in the clinical context. Yet the HIPAA had a significant impact on clinicians' behavior in requiring providers to take appropriate actions to protect patients' privacy, confidentiality, and rights. In a parallel manner, investigators may become more acutely aware of their responsibilities in secondary research uses of biospecimens.

The other most common reason provided by interviewees in support of an oath was accountability of investigators. Accountability was not anticipated as a reason for an oath, but almost of the interviewees mentioned it as an important outcome. First, an oath would help investigators have a better understanding about what is expected for ethical conduct of biospecimen research. Second, because the research would be conducted long after broad consent was given and because of the lack of relationship between investigators and donors, knowing that donors were aware of the investigator oath might impact investigator accountability. An investigator oath that improves transparency and accountability of investigators may, in turn, improve safeguards, privacy, and confidentiality of biospecimen research.

Finally, an investigator oath may cause an investigator to reflect upon the impact of their research on others. While biospecimen research has historically been thought to contribute to generalizable knowledge, it also has a potential for improving individual health outcomes. As mentioned by the interviewees, technology and genetic advancements with biospecimen research can have important implications for an individual's health, especially with next-generation sequencing.

Investigators need to recognize the growing importance of a partnership with donors and not merely see specimen donations as future data points detached from the individual. An oath may promote more respect of the donation by a research participant. As has been discussed elsewhere in this volume, there is a growing awareness of the personal connection individuals have with their own specimens, their DNA, and their potential contributions to science. Improving transparency may help promote trust with investigators for successful public support and high consent rates for broad consent for future research.

## Conclusion

The integrity of investigation and of the investigator is of paramount importance for research, scientific advancement, and public trust. Placing a stronger emphasis on the responsibilities of the future investigators during the consent process for broad consent for use of biospecimens in future research may help improve confidentiality, privacy, and safeguards associated with this research. We suggest that this can be achieved, in part, by a public commitment of the investigator through an investigator oath that is provided to research participants and potentially widely displayed throughout the institution, similar to the patient's bill of rights. A public, signed investigator oath may provide accountability for an investigator to adhere to those ethical standards.

Future research is needed not only to assess methods to improve research subject comprehension during the consent process but also to assess the impact on accountability, transparency, and trust of providing research participants with an investigator oath during the consent process for biospecimen research. In addition, research is needed to assess how expectations and assurances of investigators' behavior impact perceptions of the partnership with donors. While there is little to no interaction between investigators who use biospecimens and the donors of the biospecimens, an oath by investigators using biospecimens made publicly available may help promote the concepts of trust, accountability, transparency, and partnership. These commitments parallel, but do not displace, the importance of research participant comprehension during the consent process. When the specific elements of future research are unknown, communicating how investigators will adhere to regulatory and ethical standards may help to develop the partnership and support among those donating biospecimens.

## References

Bates, B. R., J. A. Lynch, J. L. Bevan, and C. M. Condit. 2005. Warranted concerns, warranted outlooks: A focus group study of public understandings of genetic research. *Social Science & Medicine* 60: 331–344.

Botkin, J. R. 2015. Crushing consent under the weight of expectations. *American Journal of Bioethics* 15 (9): 47–49.

Caulfield, T., R. E. Upshur, and A. Daar. 2003. DNA databanks and consent: A suggested policy option involving an authorization model. *BMS Ethics* 4: 1.

Emanuel, E., C. Grady, R. A. Crouch, R. K. Lie, F. G. Miller, and D. Wendler, eds. 2008. *The Oxford Textbook of Clinical Research Ethics.* Oxford University Press.

Falagas, M., I. Korbila, K. Giannopoulou, B. Kondilis, and G. Pappas. 2009. Informed consent: How much and what do patients understand? *American Journal of Surgery* 198: 420–435.

FDA. 2016. Electronic Code of Federal Regulations. http://www.fda.gov/downloads/AboutFDA/ReportsManualsForms/Forms/UCM074728.pdf.

Friese, S. 2013. ATLAS.ti 7 User Guide and Reference.

Grady, C., L. Eckstein, B. Berkman, D. Brock, R. Cook-Deegan, S. M. Fullerton, H. Greely, et al. 2015. Broad consent for research with biological samples: Workshop conclusions. *American Journal of Bioethics* 15 (9): 34–42.

HHS (US Department of Health and Human Services). 2013. Modifications to the HIPAA Privacy, Security, Enforcement, and Breach Notification Rules Under the Health Information Technology for Economic and Clinical Health Act and the Genetic Information Nondiscrimination Act; Other Modifications to the HIPAA Rules. *Federal Register* 78: 5565–5702.

HHS (US Department of Health and Human Services). 2015. NPRM for Revisions to the Common Rule.

Hofmann, B. 2009. Broadening consent—and diluting ethics? *Journal of Medical Ethics* 35 (2): 125–129. doi:10.1136/jme.2008.024851.

Kvale, S. 1983. The qualitative research interview: A phenomenological and a hermeneutical mode of understanding. *Journal of Phenomenological Psychology* 14: 171–196.

Lemke, A. A., W. A. Wolf, J. Hebert-Beirne, and M. E. Smith. 2010. Public and biobank participant attitudes toward genetic research participation and data sharing. *Public Health Genomics* 13 (6): 368–377.

Molyneux, C. S., N. Peshu, and K. Marsh. 2005. Trust and informed consent: Insights from community members on the Kenyan coast. *Social Science & Medicine* 61 (7): 1463–1473.

Nishimura, A., J. Carey, J. P. Erwin, J. C. Tilburt, M. H. Murad, and J. B. McCormichk. 2013. Improving understanding in the research informed consent process: A systematic review of 54 interventions tested in randomized control trials. *BMC Medical Ethics* 14: 28.

Palmer, B. W., N. M. Lanouette, and D. V. Jeste. 2012. Effectiveness of multimedia aids to enhance comprehension of research consent information: A systematic review. *IRB* 34 (6): 1–15.

Rothwell, E. 2010. Analyzing focus group data: Content and interaction. *Journal for Specialists in Pediatric Nursing* 25 (3): 202–214.

Rothwell, E., R. Anderson, and J. R. Botkin. 2010. Policy issues and stakeholder concerns regarding the storage and use of residual newborn dried blood samples for research. *Policy, Politics & Nursing Practice* 11 (1): 5–12. doi:10.1177/1527154410365563.

Rothwell, E., R. Anderson, M. Burbank, A. Goldenberg, M. H. Lewis, L. Stark, B. Wong, and J. R. Botkin. 2011. Concerns of newborn screening advisory committee members regarding storage and use of residual newborn screening bloodspots. *American Journal of Public Health* 101 (4): 2111–2116. doi:10.2105/AJPH.2010.200485.

Rothwell, E., R. Anderson, A. Goldenberg, M. H. Lewis, L. Stark, M. Burbank, B. Wong, and J. R. Botkin. 2012. Assessing public attitudes on the retention and use of residual newborn screening blood samples: A focus group study. *Social Science & Medicine* 74: 1305–1309. doi:10.1016/j.socscimed.2011.12.047.

Rothwell, E., R. Anderson, K. J. Swoboda, L. Stark, and J. Botkin. 2012. Public attitudes regarding a pilot study of newborn screening for spinal muscular atrophy. *American Journal of Medical Genetics* 161A (4): 679–686.

Sturges, J. E., and K. J. Hanrahan. 2004. Comparing telephone and face-to-face qualitative interviewing: A research note. *Qualitative Research* 4 (1): 107–118.

# 19 Biospecimen Repositories in the Era of Precision Medicine: Perspectives from a Biobanker "in the Trenches"

Quinn T. Ostrom and Jill S. Barnholtz-Sloan

The advent and growth of the use of high-throughput techniques for profiling of biospecimens for biomarker discovery and validation for precision medicine has made access to well-annotated biospecimen repositories and high-quality biospecimens critically important. Developing and maintaining these repositories is accomplished through collaboration among clinicians, researchers, and their institutions, and balancing the needs of and "value add" for these stakeholders is of utmost importance. Collaboration with clinicians, other clinical staff, and institutions is essential to obtaining informed consent, obtaining biospecimens with short ischemic time under strict standard operating procedures, and obtaining accurate and relevant clinical annotation.

Once biospecimen repositories are established, another challenge for the biobanker is developing standardized policies related to governance and distribution of biospecimens (both within and outside the institution). Involving both clinical and research stakeholders in the prioritization of projects obtaining access to biospecimens is necessary so that these biospecimens can be utilized in research with the largest potential benefit to participants. Having pre-established policies for transfer of biospecimens or biospecimen repositories between institutions is an important component for to ease the development of new collaborations. Creating a sustainable biobanking program can also be a challenge as the costs of consent, obtaining and processing biospecimens, storing biospecimens, and research on the biospecimens is expensive. New policies of the National Institutes of Health (2003, 2014) and of peer-reviewed journals related to public sharing of "omic" data generated by these stored biospecimens create new ethical challenges that must be addressed by the biobanker in developing informed consent documents, protocols, and standard operating procedures.

## Biobanking 101: The Consenting Process, Participant Information, and Data Sharing

There are two distinct groups of biospecimens that may be collected for biobanking: excess clinical biospecimens (also called "discarded" biospecimens), and biospecimens collected explicitly for research with informed consent. Excess clinical biospecimens are a portion of blood or tissue that was collected for clinical care where not all of the portions of biospecimens end up being needed for clinical care and hence can be utilized for research purposes (Riegman and van Veen 2011). These specimens typically have limited clinical annotation. The determination of "discarded" and use of these biospecimens for research is often covered under a hospital's policies and procedures, or under an institution-wide protocol.

Research biospecimens are those collected intentionally for the purposes of research, with informed consent. One of the most fundamental components of establishing a successful biospecimen repository is the development of a sustainable and thorough consent process. Many research institutions have relied on an "opt-out" consent for use for tissue considered "discarded," but with the increasing use of these biospecimens for omic studies an "opt-in" approach may be needed (Giesbertz, Bredenoord, and van Delden 2012; Pulley et al. 2010, Helgesson et al. 2007; Roden et al. 2008). Differentiating between excess clinical and research biospecimens is important for the purposes of biospecimen utilization, as each type of biospecimen is appropriate for specific types of investigations. Regardless of the type of biospecimen collected, biospecimen repositories must be established with well-defined standard operating procedures through a protocol approved by an institutional review board (IRB).

The ideal consent procedure for a successful biospecimen repository is one that is part of a streamlined process that dovetails with a participant's clinical needs. This can be achieved through collaboration between clinicians and researchers. Institutional support for this process is critical; both in terms of making time for the consent process during the participant's clinical encounter as well as providing research nurses and/or clinical research staff that are trained to carry out the process of informed consent.

Changes in NIH rules (National Institutes of Health 2003, 2014) and in many peer-reviewed journals' policies on the sharing of data (Alsheikh-Ali et al. 2011) require public sharing of data. Informing participants of these policies is required and must be introduced during the informed consent process. This may also involve providing information about the Genetic

Information Nondiscrimination Act (Pub.L. 110–233, § 122 Stat. 881, 2008) to potential participants.

When setting up a biospecimen repository it is important to determine whether or not the participants will be individually identifiable to researchers in the future. If they are to be identifiable then it is important to determine who will be allowed to gain access and use the identifiable information, and how this information will be able to be used. There are many benefits to having de-identified biospecimens (where any information that may potentially link the biospecimen to the individual it was obtain from is removed), including the potential for these biospecimens to be used under an exempt IRB protocol. It also removes an additional level of security from biospecimen repository record keeping. However, keeping a link to the participant of origin may be preferable, as it allows for continued collection of post-surgical and clinical outcomes data. It also allows for re-review of the medical records that may be necessary as new research questions are developed and pursued. Writing the ability to re-contact persons into your biobanking protocol also adds additional value to your repository, as it may allow for recruitment of the participants into future studies. For many projects, especially those focused on personalized medicine, retaining a link to the participant of origin enhances the availability and depth of data associated with the biospecimen. However, the number of individuals allowed access to identifiable information should be limited and each should be appropriately trained to handle these data in order to ensure security and confidentiality.

The CR Final Rule, issued on January 19, 2018, allows for the use of a "broad consent" for collection of biospecimens. A broad consent is a consent that requests the use of biospecimens and clinical data without a specific purpose in mind, but for a broad range of future studies and identifying information remains associated with their biospecimens and clinical data (Grady et al. 2015; Wendler 2013). This is in contrast to a "blanket" consent, where donors allow access to their biospecimens without any restriction, but often with the expectation that no identifying information be attached. The "broad consent" would take the form of an institution-wide informed consent, potentially presented to potential donors upon check in to the hospital system. Because of the potential expense and burden of obtaining informed consent from all potential donors entering a hospital, investigators have attempted to use different types of consent procedures, including online portals and tablets (Thiel et al. 2015). "Broad consent" would then allow for individual researchers to develop their own protocols—either exempt or non-exempt depending on the type of patient data required—to access these biospecimens and hence enhance translational research.

### Biobanking Governance and the IRB

A successful biospecimen repository requires engagement with the IRB at several points in the process. First, a biobanking protocol must be developed and approved by the IRB. Collaboration with the IRB allows the investigator to leverage the IRB's knowledge for the development of a comprehensive protocol and informed consent process that is accessible to all potential donors. This partnership between investigator and IRB helps to ensure the success of the biobanking process. The specific culture of each institution's IRB will drive many of the decisions an investigator will make in setting up a biospecimen repository. Once the protocol has been approved, the IRB will review it generally annually to determine that it is continuing to conform to ethical standards. Periodic review is important to biobanking, as "best practices" continue to change over time.

The IRB continues to be a partner after the biospecimen repository is established and has accrued sufficient samples to be utilized for research projects. IRBs will vary significantly on their determinations about the level of risk for secondary use projects (Goldenberg et al. 2015). Engaging with your IRB at all stages of your banking protocol will allow you to develop an understanding of the local IRB culture and thus to tailor your protocol and procedures. Consensus ethical guidelines for these repositories are still being developed, and as a result many IRB leaders are continuing to modify their policies and procedures about these projects (Rothwell et al. 2015).

### Assuring Quality in Biospecimen and Data Collection

When initiating a biospecimen repository, specific standard operating procedures must be developed and adhered to for collection, processing, and storage of biospecimens and for collection, coding, and storage of clinical data. Having a biospecimen repository of high-quality biospecimens with clinical annotations helps to ensure the success of a repository and for the repository's value for the research community. Many large established biospecimen repository operations—including the National Cancer Institute (Biorepositories and Biospecimen Research Branch 2014), the European Organization for Research and Treatment of Cancer (Mager et al. 2007; Morente et al. 2006), the UK Biobank (Peakman and Elliott 2008), and many others (Yong, Dry, and Shabihkhani 2014)—have made their standard operating procedures publicly available.

For many modern analytical techniques, tracking the *ischemic time* associated with each specimen is important for estimating the quality of the research biospecimen. The ischemic time is defined as the time period between when a specimen is removed from the body, and when it is "fixed," either by being frozen or exposed to a chemical. During the time that a tissue is removed from the body and not fixed, RNA, many proteins, and sometimes DNA degrade and are no longer as representative of the actual proteins, RNA, and DNA within the living cells. In the collection of both fresh frozen and formalin fixed paraffin embedded (FFPE) tissues, ischemic time can have a significant effect on biospecimen quality (Turashvili et al. 2012). For some studies and depending upon the product needed, the ischemic time of a tissue may not affect results significantly. DNA can be isolated from tissue that has had more than an hour of ischemic time without significant degradation. For RNA or protein-based studies, however, increasing ischemic time can cause significant increases in degradation (Hong et al. 2010; De Cecco et al. 2009). Standard operating procedures for biobanking generally recommends an ischemic time of less than 20 or 30 minutes (Mager et al. 2007; Shabihkhani et al. 2014), though the shortest time period possible is ideal. Fixation within the operating room or shortly after the specimen is transported to pathology can allow for assurance of a short ischemic time.

Obtaining high-quality biospecimens requires collaboration with clinicians. Strong engagement with pathologists that specialize in the specific tissue types to be collected is an important component of a successful biospecimen repository. Engagement and involvement of pathology is necessary for performing quality control, as well as facilitating access to discarded clinical biospecimens for research. For tissues that are collected specifically for research purposes, engagement with surgeons who obtain tissue in the operating rooms will help to decrease ischemic time, as they are the persons most able to minimize the time that between harvesting the biospecimen and tissue freezing or fixation in formalin. Engaging with clinicians requires balancing their clinical duties with their roles as researchers (Thasler et al. 2013; Tsikitis et al. 2013). The clinical needs of the participant and/or patient will always be primary and, as a result, may dictate the limits of biospecimen quality. For all specimens, it is important to communicate with the pathologist to confirm that all specimens needed to make an accurate clinical diagnosis are available before using any specimens (either "discarded" or consented research specimens) for research purposes.

Engaging with clinicians allows for the collection of more accurate biospecimen annotations. Tracking ischemic time is an important component of any biobanking protocol, and these professionals will be the best sources of information for when a biospecimen is removed and fixed. Many repositories may also want to collect biospecimens other than tissues including but not limited to blood, saliva, and urine. These biospecimens must be handled according to specific protocols in order to preserve the quality of DNA, RNA, and protein. Blood and saliva that will not processed immediately may be collected using commercially available tubes—specifically designed for this purpose—that contain compounds that will fix DNA and RNA for later extraction (Wahlberg et al. 2012). Blood may be separated into plasma, cells, or serum that can be easily stored to allow for later use. These allow the biospecimens to be stored at room temperature for a longer period of time, which may work well with the workflow of some biospecimen repositories. If specialized procedures are not used, biospecimens should be stored in a 4°C freezer as soon as possible. Urine should be frozen as soon as possible in order to maintain quality of protein, DNA, and RNA for future molecular analysis.

Once a biospecimen has been obtained and processed for storage, quality-control procedures should be performed. Depending on clinical needs and standard operating procedures, biospecimens may be evaluated for quality control before being or after fixation. Biospecimens may be evaluated using hematoxylin-and-eosin-stained slides made by the pathologist immediately before freezing or fixation, which would provide the most accurate structural view of the tissue. For biospecimens that are FFPE, it is a simple process to cut slides from the resulting block in order to verify tissue quality or to determine specific parts of a biospecimen that may be best suited for an individual project. For frozen biospecimens where a slide is not taken before freezing, it may be necessary to evaluate these for quality control after they have already been stored in a freezer but before use. The quality-control procedure may take the form of slicing a small section of tissue from the specimen in order to produce a slide with which the pathologist evaluates quality. Making quality-control slides from previous frozen tissue may produce artifacts in the tissue that inhibit the quality-control process, by producing structures in the tissue that may not be an accurate reflection of tissue quality; taking slides from the top and bottom of each biospecimen before freezing may decrease these artifacts.

## Disease-Team Collaboration for Prioritization of Biospecimen Repository Resources

Biospecimen repositories may take the form of private biobanks where the biospecimens have been collected solely for the use of one investigator or institution-wide biospecimen repositories, where any investigator within an institution has the opportunity to access these biospecimens. These specimens may also be made available to those beyond the institution through an approved request process. Biospecimens collected via the private model may also be available to other investigators, but this will often require scientific collaboration with the investigator managing the biospecimen repository. While a private model gives the investigator who has established the repository control over prioritization of projects utilizing these biospecimens, the investigator is usually responsible and accountable for funding their repository efforts.

Institutional biospecimen repositories are biobanks that have the financial support of the institution at large and may be collected via a biobanking core facility rather than by an individual investigator. This type of biospecimen repository may be less focused on a specific disease and, while casting a larger inventory, may lose the depth of clinical annotation that may be available in an individual repository. Having institutional support for banking efforts can help facilitate researcher-clinician collaboration, increasing the ability to collect appropriate clinical annotation hence increasing the long-term utility of banked specimens.

For either type of biospecimen repository, it is important to keep a well-organized inventory of collected biospecimens in order to facilitate utilization with up to date knowledge of the specific biospecimens available within the repository. It is important to have the ability to query this inventory in order to meet the needs of specific projects, as different research projects may have specific needs including participant demographics, biospecimen size, biospecimen heterogeneity, type of treatment the participant received, and clinical outcomes.

For both types of repositories, investigators must have well-defined procedures for investigator access to and utilization of biospecimens. The amount of effort that goes into collecting and annotating these biospecimens is significant, and as a result it is important to make sure that they are being used for the highest-priority projects. Individual institutions may have pre-determined policies about resource sharing and access to biospecimens. This may take the form of disease-focused teams that direct the research agenda for each disease and are thus the best authorities on which projects should take priority. After these determinations have been made, it is the responsibility of the biobanker (whether an individual or a core facility) to facilitate access for individual investigators. In addition to prioritization, this may also include signing a data-use agreement, a material transfer agreement, or other paperwork. A fee for service payment from the requesting investigator may also be required.

### Collaboration within and between Institutions

Science is becoming increasingly collaborative, and many investigators have multiple collaborators in multiple institutions. As a result, it is necessary to have a standard operating procedure for study prioritization and transfer of biospecimens between investigators and institutions.

Institutions probably have procedures for prioritization of use of stored biospecimens. In addition, it is important to ensure that any use of the stored biospecimens is governed by an IRB-approved protocol. Depending on the amount of annotation needed by the investigator, these may be ruled exempt or non-exempt. Investigators may choose to have collaborators or other utilizers of the biospecimens sign a data-use agreement, which may also be required by some institutions. This will specifically delineate the data that utilizers will have access to, as well as the appropriate uses of the data.

When biospecimens or associated data leave an institution, the standard mechanism is to prepare a material transfer agreement that will be signed by both institutions (Parodi et al. 2013; Hallmans and Vaught 2011). A material transfer agreement is a legal contract that governs the use of the exchange materials, defines the assignments of liabilities, and protects the interests of all involved institutions and investigators. The development of these agreements involves legal representation from both institutions, which must agree to the language of the agreement and sign on behalf of the institution. The International Society for Biological and Environmental Repositories (2012) has specific recommendations for what should be included in these agreements, although the final contents are at the discretion of participating institutions.

### Biospecimen Annotation and Longitudinal Collection Data of Outcomes Data

Collection of clinical outcomes data increases the usability and value of biospecimens, especially for studies of biomarkers of survival or response to treatment. The first step in collecting data of this kind is making sure that a request to collect this data prospectively is included in the informed consent document, and that each participant individually opts in to this active follow-up (i.e., they are contacted over time for status) or passive follow-up (i.e., their medical records as searched on a regular basis for status updates). Since not all donors that consent to biobanking may consent to this option, it is important to be able to track the detailed consent information for any participant included in the biospecimen repository, preferably in a database that includes consent information and is easily searchable. Depending on the specifics of the protocol guiding the biospecimen repository, as well as institutional policy, the informed consent may limit the investigator to collect donor information only from the location where the biospecimen was originally collected. Most donors will utilize health-care services at more than one institution or may move out of region, limiting the accuracy of follow-up information available to the investigator. Once information on a participant is no longer available, clinical annotation of the repository should note that this participant has been lost to follow-up. In the case of measuring outcomes data, failing to note this information may result in inflation of the participant's survival time or time to recurrence as the investigator does not have access to the date of death or recurrence via the institution's records.

Within the consent, it is important to note specifically what information the investigator will be collecting. The appropriate level of specificity of information collected from the clinical data varies depending on institutional policies. An investigator can use consent language that provides blanket access to the entirety of the donor's record at that institution. Without this type of a language, investigators may be limited as research questions change. For example, in a cancer-based study in which a donor has consented only to follow-up regarding their specific disease, an investigator may not be able to access information regarding future health conditions that may or may not be associated with their cancer diagnosis.

With biospecimen repositories that are institutional and individually identifiable, linkage of the repository to the Electronic Medical Record provides can provide cost effective clinical annotation with relatively high accuracy (Bowton et al. 2014). Regardless of whether this type of infrastructure exists, careful thought should be given to the knowledge level of the person responsible for abstracting the data. For some pieces of data (e.g., specific disease diagnosis based on lab results or interpretation of symptoms), it may be necessary for a clinician to review a participant's record. For other information, such as prescription drugs a participant may have taken, this information could be abstract from a well-trained non-clinical research professional. The individual responsible for abstracting this information has a significant effect on the accuracy and reliability of the data collected, and it is important to make sure that both of these features are being maximized.

### Working toward Long-Term Sustainability of Biobanking

Building a successful and useful biospecimen repository requires long-term commitment from individual investigators, as well as sustained institutional support. Initial costs for starting a biospecimen repository will be much higher than those for regular maintenance, and these initial costs are not typically supported by investigators' grants. Development of these biospecimen repositories within institutions is an important foundation on which to build grant submissions by demonstrating an institution's ability to access previously banked human specimen resources, so this initial support is in both the investigator and the institution's best interest. Once established, there are regular costs associated with maintenance of the repository (e.g., staff to screen and consent participants and follow-up with participants, staff to manage freezers, maintenance of databases, freezer maintenance, etc.) that may not necessarily factor into a price-per-biospecimen. Planning for these needs is an important step in creating a sustainable biospecimen repository (Uzarski et al. 2015).

It is important to distinguish at the outset of a biospecimen repository whether or not the biospecimens collected under the biobanking protocol reside primarily with the investigator or with the institution. Investigators may not necessarily remain affiliated with the institution where the repository has been established, and it is important to plan for this contingency ahead of time. If the investigator has funding tied to the repository and is continuing to use these biospecimens in their research initiatives, they may want to take the biospecimens or aliquots of the biospecimens and a copy of any clinical annotations with them when they leave the institution. Owing to changes in institutional structure, biospecimen repositories may also change institutional affiliation after their establishment, and having a plan in place for these contingencies is essential.

There are many different positions, models and policies on biospecimen "ownership," and these will vary by individual and institution (Cadigan et al. 2011; Hakimian and Korn 2004; Hakimian 2004).[1] These positions or models can largely be divided into whether the relationship between the investigator or institution and biospecimens are seen as one of "ownership," versus one of "stewardship" (Dressler 2007; Ness and American College of Epidemiology Policy 2007).[2] These positions or models may also vary depending on whether these biospecimens were collected for the purposes of research or as clinical samples. Even though biospecimens were collected specifically for the purposes of research, if they arose as a result of a patient's clinical care (e.g., collected during a surgery), there may be an occasion when research biospecimens need to be used for all clinical work to be carried out. Hence, strong collaboration with clinicians is necessary in order to make sure that specimens intended for research are not released until the pathologist has completed all procedures needed for diagnosis. A clear structure of ownership and governance that balances the needs of researchers and institutions, while also taking into account the desires of participants, helps to ensure the successful and compliant use of the biospecimens for research.

## Conclusions

The age of personalized medicine has made the development and sustainability of biospecimen repositories containing high-quality and well-annotated biospecimens critical. Starting and managing a useful and sustainable biospecimen repository requires a significant amount of planning, organization, and management. Thoughtful planning and strict standard operating procedures for access and use allow for the greatest usability of the biospecimens, as well as allows for optimal opportunities for collaboration on research projects. Biospecimen repositories should enable maximal flexibility in use—and anticipate such when developing their policies—as research methodologies and scientific questions evolve rapidly and can be predicted to do so in the future. Institutional support for these biospecimen repositories increases the ability for these to lead to strong science and will increase the institutions competitiveness for future research funding and support.

## Notes

1. The legal issues surrounding biospecimen ownership are discussed elsewhere in this volume.

2. See chapters 1 and 2 in the present volume for additional discussion of these positions.

## References

Alsheikh-Ali, A. A., W. Qureshi, M. H. Al-Mallah, and J. P. Ioannidis. 2011. Public availability of published research data in high-impact journals. *PLoS One* 6 (9): e24357. doi:10.1371/journal.pone.0024357.

Biorepositories and Biospecimen Research Branch. 2014. Biospecimen Collection, Processing, Storage, Retrieval, and Dissemination. National Cancer Institute, Division of Cancer Treatment and Diagnosis, Cancer Diagnosis Program. http://biospecimens.cancer.gov/bestpractices/to/bcpsrd.asp.

Bowton, E., J. R. Field, S. Wang, J. S. Schildcrout, S. L. Van Driest, J. T. Delaney, J. Cowan, P. Weeke, J. D. Mosley, Q. S. Wells, J. H. Karnes, C. Shaffer, J. F. Peterson, J. C. Denny, D. M. Roden, and J. M. Pulley. 2014. Biobanks and electronic medical records: Enabling cost-effective research. *Science Translational Medicine* 234: 234cm3. doi:10.1126/scitranslmed.3008604.

Cadigan, R. J., M. M. Easter, A. W. Dobson, A. M. Davis, B. B. Rothschild, C. Zimmer, R. Sterling, and G. Henderson. 2011. "That's a good question": University researchers' views on ownership and retention of human genetic specimens. *Genetics in Medicine* 13 (6): 569–575. doi:10.1097/GIM.0b013e318211a9c2.

De Cecco, L., V. Musella, S. Veneroni, V. Cappelletti, I. Bongarzone, M. Callari, B. Valeri, M. A. Pierotti, and M. G. Daidone. 2009. Impact of biospecimens handling on biomarker research in breast cancer. *BMC Cancer* 409 (9). doi:10.1186/1471-2407-9-409.

Dressler, L. G. 2007. Biospecimen "ownership": Counterpoint. *Cancer Epidemiology, Biomarkers & Prevention* 16 (2):190–191. doi:10.1158/1055-9965.EPI-06-1004.

Genetic Information Nondiscrimination Act. 2008, Public Law 110–233, 110th Congress, May 21, 2008.

Giesbertz, N. A., A. L. Bredenoord, and J. J. van Delden. 2012. Inclusion of residual tissue in biobanks: Opt-in or opt-out? *PLoS Biology* 10 (8): e1001373. doi:10.1371/journal.pbio.1001373.

Goldenberg, A. J., K. J. Maschke, S. Joffe, J. R. Botkin, E. Rothwell, T. H. Murray, R. Anderson, N. Deming, B. F. Rosenthal, and S. M. Rivera. 2015. IRB practices and policies regarding the secondary research use of biospecimens. *BMC Medical Ethics* 32 (16). doi:10.1186/s12910-015-0020-1.

Grady, C., L. Eckstein, B. Berkman, D. Brock, R. Cook-Deegan, S. M. Fullerton, H. Greely, M. G. Hansson, S. Hull, S. Kim, B. Lo, R. Pentz, L. Rodriguez, C. Weil, B. S. Wilfond, and D. Wendler. 2015. Broad consent for research with biological samples: Workshop conclusions. *American Journal of Bioethics* 15 (9): 34–42. doi:10.1080/15265161.2015.1062162.

Hakimian, R. 2004. National Cancer Institute Cancer Diagnosis Program: 50-state Survey of Laws Regulating the Collection, Storage, and Use of Human Tissue Specimens and Associated Data for Research. US Department of Health and Human Services, National Institutes of Health.

Hakimian, R., and D. Korn. 2004. Ownership and use of tissue specimens for research. *JAMA* 292 (20): 2500–2505. doi:10.1001/jama.292.20.2500.

Hallmans, Göran, and Jimmie B. Vaught. 2011. Best Practices for Establishing a Biobank. In *Methods in Biobanking*, ed. Joakim Dillner. Humana Press.

Helgesson, G., J. Dillner, J. Carlson, C. R. Bartram, and M. G. Hansson. 2007. Ethical framework for previously collected biobank samples. *Nature Biotechnology* 25 (9): 973–976. doi:10.1038/nbt0907-973b.

Hong, S. H., H. A. Baek, K. Y. Jang, M. J. Chung, W. S. Moon, M. J. Kang, D. G. Lee, and H. S. Park. 2010. Effects of delay in the snap freezing of colorectal cancer tissues on the quality of DNA and RNA. *Journal of the Korean Society of Coloproctology* 26 (5): 316–323. doi:10.3393/jksc.2010.26.5.316.

International Society for Biological and Environmental Repositories. 2012. 2012 best practices for repositories collection, storage, retrieval, and distribution of biological materials for research international society for biological and environmental repositories. *Biopreservation and Biobanking* 10 (2):79–161. doi:10.1089/bio.2012.1022.

Mager, S. R., M. H. Oomen, M. M. Morente, C. Ratcliffe, K. Knox, D. J. Kerr, F. Pezzella, and P. H. Riegman. 2007. Standard operating procedure for the collection of fresh frozen tissue samples. *European Journal of Cancer* 43 (5): 828–834. doi:10.1016/j.ejca.2007.01.002.

Morente, M. M., R. Mager, S. Alonso, F. Pezzella, A. Spatz, K. Knox, D. Kerr, W. N. Dinjens, J. W. Oosterhuis, K. H. Lam, M. H. Oomen, B. van Damme, M. van de Vijver, H. van Boven, D. Kerjaschki, J. Pammer, J. A. Lopez-Guerrero, A. Llombart Bosch, A. Carbone, A. Gloghini, I. Teodorovic, M. Isabelle, A. Passioukov, S. Lejeune, P. Therasse, E. B. van Veen, C. Ratcliffe, and P. H. Riegman. 2006. TuBaFrost 2: Standardising tissue collection and quality control procedures for a European virtual frozen tissue bank network. *European Journal of Cancer* 42 (16): 2684–2691. doi:10.1016/j.ejca.2006.04.029.

National Institutes of Health. 2003. Final NIH Statement on Sharing Research Data. http://grants.nih.gov/grants/guide/notice-files/NOT-OD-03-032.html.

National Institutes of Health. 2014. NIH Genomic Data Sharing Policy. http://grants.nih.gov/grants/guide/notice-files/NOT-OD-14-124.html.

Ness, R. B., and American College of Epidemiology Policy. 2007. Biospecimen "ownership": Point. *Cancer Epidemiology, Biomarkers & Prevention* 16 (2): 188–189. doi:10.1158/1055-9965.EPI-06-1011.

Parodi, Barbara, Paola Visconti, Tiziana Ruzzon, and Mauro Truini. 2013. Governance of biobanks for cancer research: Proposal for a material transfer agreement. In *Comparative Issues in the Governance of Research Biobanks*, ed. Giovanni Pascuzzi, Umberto Izzo and Matteo Macilotti. Springer.

Peakman, T. C., and P. Elliott. 2008. The UK Biobank sample handling and storage validation studies. *International Journal of Epidemiology* 37 (Suppl. 1): i2–i6. doi: 10.1093/ije/dyn019.

Pulley, J., E. Clayton, G. R. Bernard, D. M. Roden, and D. R. Masys. 2010. Principles of human subjects protections applied in an opt-out, de-identified biobank. *Clinical and Translational Science* 3 (1): 42–48. doi:10.1111/j.1752-8062.2010.00175.x.

Riegman, P. H., and E. B. van Veen. 2011. Biobanking residual tissues. *Human Genetics* 130 (3): 357–368. doi:10.1007/s00439-011-1074-x.

Roden, D. M., J. M. Pulley, M. A. Basford, G. R. Bernard, E. W. Clayton, J. R. Balser, and D. R. Masys. 2008. Development of a large-scale de-identified DNA biobank to enable personalized medicine. *Clinical Pharmacology & Therapeutics* 84 (3): 362–369. doi:10.1038/clpt.2008.89.

Rothwell, E., K. J. Maschke, J. R. Botkin, A. Goldenberg, T. H. Murray, and S. M. Rivera. 2015. Biobanking research and human subjects protections: Perspectives of IRB leaders. *IRB* 37 (2): 8–13.

Shabihkhani, M., G. M. Lucey, B. Wei, S. Mareninov, J. J. Lou, H. V. Vinters, E. J. Singer, T. F. Cloughesy, and W. H. Yong. 2014. The procurement, storage, and quality assurance of frozen blood and tissue biospecimens in pathology, biorepository, and biobank settings. *Clinical Biochemistry* 47 (4–5): 258–266. doi:10.1016/j.clinbiochem.2014.01.002.

Thasler, W. E., R. M. Thasler, C. Schelcher, and K. W. Jauch. 2013. Biobanking for research in surgery: Are surgeons in charge for advancing translational research or mere assistants in biomaterial and data preservation? *Langenbeck's Archives of Surgery* 398 (4): 487–499. doi:10.1007/s00423-013-1060-y.

Thiel, D. B., J. Platt, T. Platt, S. B. King, N. Fisher, R. Shelton, and S. L. Kardia. 2015. Testing an online, dynamic consent portal for large population biobank research. *Public Health Genomics* 18 (1): 26–39. doi:10.1159/000366128.

Tsikitis, V. L., K. C. Lu, M. Douthit, and D. O. Herzig. 2013. Surgeon leadership enables development of a colorectal cancer biorepository. *American Journal of Surgery* 205 (5): 563–565. doi:10.1016/j.amjsurg.2013.01.020.

Turashvili, G., W. Yang, S. McKinney, S. Kalloger, N. Gale, Y. Ng, K. Chow, L. Bell, J. Lorette, M. Carrier, M. Luk, S. Aparicio, D. Huntsman, and S. Yip. 2012. Nucleic acid quantity and quality from paraffin blocks: Defining optimal fixation, processing and DNA/RNA extraction techniques. *Experimental and Molecular Pathology* 92 (1): 33–43. doi:10.1016/j.yexmp.2011.09.013.

Uzarski, D., J. Burke, B. Turner, J. Vroom, and N. Short. 2015. A plan for academic biobank solvency-leveraging resources and applying business processes to improve sustainability. *Clinical and Translational Science*. doi: 10.1111/cts.12287.

Wahlberg, K., J. Huggett, R. Sanders, A. S. Whale, C. Bushell, R. Elaswarapu, D. J. Scott, and C. A. Foy. 2012. Quality assessment of biobanked nucleic acid extracts for downstream molecular analysis. *Biopreservation and Biobanking* 10 (3): 266–275. doi:10.1089/bio.2012.0004.

Wendler, D. 2013. Broad versus blanket consent for research with human biological samples. *Hastings Center Report* 43 (5): 3–4. doi:10.1002/hast.200.

Yong, W. H., S. M. Dry, and M. Shabihkhani. 2014. A practical approach to clinical and research biobanking. *Methods in Molecular Biology* (1180): 137–162. doi: 10.1007/978-1-4939-1050-2_8.

## 20 Operationalizing Institutional Research Biospecimen Repositories: A Plan to Address Practical and Legal Considerations

Kate Gallin Heffernan, Emily Chi Fogler, Marylana Saadeh Helou, and Andrew P. Rusczek

Research biospecimen repositories have evolved from relatively decentralized and laboratory-specific or study-specific efforts (i.e., a collection of tissue samples stored in a laboratory freezer) to a recognized (and regulated) field of science that includes organized institutional, regional, national, international, population-based, and even virtual repositories through which investigators can locate and request research materials based on intricate computerized catalogs of annotated specimens (DeSouza and Greenspan 2013; Vaught et al. 2012). The benefits of centralized institutional research biorepositories have been discussed conceptually and through the description of real-world experience (Rogers et al. 2011; Marsolo et al. 2012; Pulley et al. 2010; Roden et al. 2008). This chapter explores some of the practical and legal issues institutions need to consider before operationalizing an institutional research biorepository, including the components of a biorepository protocol, approaches to informed consent, managing third-party interests in the biorepository, and educational efforts at the patient and clinician level to increase awareness and participation. Although the scope and nature of institutional biorepositories will necessarily vary, this chapter aims to highlight operational hurdles that create institutional risk or that prevent institutions from reaping the full rewards of the resource, and to recommend best practices for institutions contemplating or commencing such an endeavor.

**Planning for an Institutional Biospecimen Repository: Drafting the Protocol and Exploring the Mechanics of Informed Consent**

Although assembling a research biorepository precedes the specific studies in which such specimens and associated information will be used, it is well established that, as with specimen collection performed for research purposes, the creation and maintenance of research biospecimen repositories are "research" activities and, to the extent the biorepositories include individually identifiable information, are "human subjects research" under the federal human subjects protection regulations, known as the Common Rule and codified by the Department of Health and Human Services (HHS) at 45 C.F.R. Part 46.[1] As such, the requirements for institutional review board (IRB) oversight and informed consent will generally apply to the development of an institutional biorepository, unless a determination can be made that the biorepository is exempt from such requirements or consent is otherwise waived. For example, to the extent an institution plans to collect only existing specimens and data for inclusion, the biorepository might qualify for exemption from oversight under 45 C.F.R. § 46.101(b)(4).[2]

When investing significant resources into an institutional biospecimen repository, organizations may wish to consider whether narrowing the scope of the collection to achieve exemption or definitional exclusion from the regulations unacceptably decreases its value. Even where IRB oversight is not technically required, it is prudent for institutions to describe the protocol for a planned biorepository in writing, because the protocol serves as the operational framework and provides a reference for identifying future changes in scope or intent that could trigger additional requirements.

In addition to protections for human subjects, the biorepository will be required to comply with the requirements of the Health Insurance Portability and Accountability Act (HIPAA) and its privacy regulations (the "Privacy Rule") for the use and disclosure of protected health information, if it is maintained by an organization that is a "Covered Entity" and if the information captured by the biorepository is identifiable under HIPAA's standards.[3]

Although biorepositories will necessarily differ based on their intended scope and purpose, the Appendix contains a general checklist to help institutions focus their initial thinking and planning for the creation of biorepositories.

**The Protocol**
Development of an institutional research biospecimen repository protocol involves considerations of both substance and process. A research biospecimen repository protocol serves as a road map for the development and ongoing operation of the repository—the architectural plans or standard operating procedure (SOP) according to which it functions—and encompasses the scientific and technical procedures for how data and specimens are stored, as well as the overarching ethical and procedural principles

that will govern the receipt, maintenance, and release of specimens and information for specific research uses. The protocol itself should provide enough information to allow an IRB to assess compliance with applicable laws and ethical principles, as well as understand how the management of the repository and downstream uses will operate in practice. The amount of information and the level of detail required in a protocol will vary based on the size of the repository, the nature of the research, and the ability to identify the persons from whom the specimens are collected. As a practical matter, assessing these issues will require the involvement and input of many different departments within an organization before the protocol is submitted for review by the IRB.

There are certain topics that any institution seeking to develop an institutional biorepository should consider and address in the protocol, including the following:

- the purpose and management of the biorepository
- how specimens and associated clinical information will be accepted, initially and with respect to any updates
- how informed consent will be obtained (if applicable)
- storage and other operational considerations, including the mechanics of how information and specimens will be maintained and protected, and any quality-assurance procedures
- how specimens and information will be released from the biospecimen repository for secondary research uses.

Points to be considered under each topic are detailed below.

**About the Biospecimen Repository**  A full description of the repository should include the following:

- The scientific value of the biorepository. Describing relevant scientific literature regarding the benefits of large-scale, cross-sectional institutional biorepositories can educate the IRB or other internal review committees about the value of such an endeavor to the institution's goals.
- The purpose of the biorepository (i.e., the overall goals, objectives, and any known specific study aims for which the contents will be used and how decisions about specific secondary research uses will be made). Limiting the purpose by type of secondary research protocol (e.g., cardiovascular research, diabetes research, cancer research) may not be possible or necessary. However, examples of the types of work that will be supported by the biorepository, if known, may provide a more complete picture of how the

biorepository will serve the research mission of the institution and benefit the broader research community.
• Justification for the need for the biorepository. Responsible planning for a biorepository should include an assessment of the anticipated demand for this type of institutional resource, and how the planned enrollment pace and volume of collections will serve to meet the demand.
• The intended oversight and management for the biospecimen repository. For example, what are the guiding principles, governance structure (i.e., steering committee, director, and other supporting staff, as well as who will serve as the principal investigator), and what policies and procedures, if any beyond the protocol, will guide the biorepository's compliance with applicable laws, the terms of the protocol, and other institutional policies that might apply to the repository's operations.

**Materials In** The protocol should describe the following with as much specificity as is possible:

• What types of specimens will be included (e.g., blood, tissue, serum, DNA, RNA) and the sources (e.g., clinical excess, unused research specimens, prospective research collection).
• For any prospectively collected specimens, the approved collection methods (e.g., blood draw, lumbar puncture, buccal swab), along with any associated risks.
• The total number of donors expected for the initial IRB approval period (which number, once approved, would need to be increased by amendment or at continuing review if it is anticipated to increase).
• Any inclusion and exclusion criteria delineating who may participate in the biospecimen repository.
• How participants will be recruited. (For example, will potential participants be drawn from patients presenting at the institution for other health care? Will they be affirmatively recruited from broader pools?)
• What screening procedures will be required for those participants prospectively donating specimens.
• What individually identifying information will be collected, if any.
• If existing specimens previously collected for other purposes will be included, information regarding the circumstances of the original collections and the scope of any previously provided donor consent.

**Informed Consent** When outlining the process for obtaining informed consent, the protocol should cover the following:

- The timing for when consent will be obtained.
- Who will obtain consent (e.g., researchers, clinicians, or registration staff).
- What procedures will be used to obtain permission from individuals who are unable to consent for themselves.
- What consent forms will be used.
- How subjects who wish to withdraw their consent may do so.
- How subjects will be informed of any research results, if applicable, and whether such results will be individual or aggregated.
- What happens to a subject's specimens and information if consent is withdrawn (i.e., is there any ability to "claw back" specimens and information from downstream users or an affirmative statement that there is not).
- If consent will not have been obtained for all or some of the specimens that will be included in the biorepository, the conditions supporting a waiver of consent.
- A requirement that forms, Web portals, emails, or other visual materials to be used for participant recruitment, education or consent be submitted for IRB review.

**Maintaining the Materials**  The protocol should describe the specimen storage and security procedures, including:

- The physical location of the biorepository.
- How long it is anticipated that specimens and information will be stored.
- Scientific conditions and parameters for preserving the specimens.
- Technical privacy and security safeguards that will be imposed to protect any identifiable information.
- How, if at all, specimens will be linked to the subjects' identities and, if so, who will have access to identities or other identifying data (or, if a coding system is used, who will manage the key to the code).
- If specimens will be de-linked, how, when, and by whom this will be accomplished and how, if at all, specimens will be identified in the repository (e.g., what characteristics will be associated with each specimen, how and with what data points will they be annotated).
- If specimens will continue to be linked to subjects' identities and the biorepository is intended to reflect current annotated medical data, the general principles for how specimens and associated information will be updated (e.g., whether the biorepository will include a real-time connection to the institution's electronic medical record system to reflect up-to-date health information).

**Materials Out** The protocol should describe who will have access to the specimens and information in the biorepository, outlining any general principles applicable to access, including the following:

• Whether the biospecimen repository will be available for use only by researchers internal to the institution, by all academic and non-profit researchers, or also by commercial third parties.
• The process for how requests will be made to the biospecimen repository (what information will be required to be submitted, including, e.g., any template request form, conflict of interest forms, certification of compliance with applicable laws and the SOPs of the biorepository, and demonstration of IRB approval if required).
• Who has authority to consider and respond to such requests, and under what standards will the biospecimen repository make access determinations (a related SOP can be cross-referenced in lieu of including this in the protocol).
• What restrictions will apply to the use of specimens.
• What restrictions will be imposed on the format in which specimens and information may be released (i.e., identifiable vs. coded vs. fully de-identified).
• How will the biorepository ensure that release of specimens is in accordance with the terms of any consent, HIPAA authorization, or the scope of waivers of consent and authorization (if applicable).
• Whether participants will receive updates regarding downstream research involving their specimens.

### The Informed Consent Process

The successful implementation of an institutional research biospecimen repository protocol depends on a comprehensive and compliant approach to informed consent, to ensure that the repository will be sufficiently populated and that appropriate downstream research uses are not obstructed due to a lack of prospective preparation. However, the nature of meaningful consent in the biospecimen repository context and the risks that need to be articulated are necessarily different than in the context of informed consent to a specific research study. Other chapters in this book tackle in greater specificity some of the ethical considerations in biorepository consent and the content of such consent; here we focus more specifically on the process challenges.

Institutions contemplating the development of an institutional research biospecimen repository may want to consider creative approaches to seeking informed consent from potential donors. "Front door" consent is one mechanism that has been successfully used by institutions to reach a broad base of patients registered with or seeking care at the institution. This approach presents patients with participation in a repository through existing communication portals or during other planned moments of patient interaction. For example, institutions with Web-based appointment systems or electronic medical record (EMR) access may use such systems to embed requests for participation. Other approaches include the use of tablets or other electronic kiosks at walk-in clinics, within hospital waiting rooms, or near the registration desk upon entry. Logistically, implementing a successful front-door consent process can depend on the reviewing IRB approving alternative mechanisms to consent, many of which may rely on electronic signatures. If legally effective under applicable state law (which may not be a straightforward analysis in research involving on-line cross-jurisdictional enrollment), electronic signatures are recognized as effective for Common Rule and HIPAA purposes and formal IRB waiver or alteration of the standard elements of consent and authorization would not be required (OHRP 2011; OCR 2008).[4]

It is important to note that front-door consent is a *mechanism* for obtaining informed consent to a biorepository; it does not dictate the nature or scope of the consent obtained. As reflected in other chapters, informed consent itself can be broad, protocol-specific, tiered, or dynamic, and can be based on an opt-in or opt-out premise. Which approach is suited to a given institutional biospecimen repository will depend on the purpose and scope of the repository and the policies of the institution. If a tiered consent approach is selected (i.e., allowing participants the ability to direct future uses in large or small ways) (Mello and Wolf 2010), the institution must ensure it has the infrastructure to track and manage these selections. Recently, "dynamic" consent, a technological method through which donors could update, in real time, their preferences for future uses via smart phone and other Web-based technology, has been proffered as a method to more accurately reflect the autonomy and preferences of specimen donors (Kaye et al. 2015). However, the challenges of managing the downstream implications for evolving preferences, particularly as more specimens become subject to the requirements of consent, are many.

The approach to consent may differ for various sources of specimens and information. For example, specimens prospectively obtained for research purposes will not be eligible for a waiver of consent. Specimens that exist at the time the biorepository is created, for which consent to future research was not obtained, on the other hand, may qualify for a waiver. What level of consent to require from patients whose care at the institution may result in clinical excess requires particular attention. Historically, general statements in standard surgical and procedural consent forms notified patients that clinical excess may be discarded or put to education and research uses. With an organized biospecimen repository, institutions should think proactively whether clinical excess will undergo the same process as research specimen collection (i.e., informed consent), or whether an alternative process is permissible (for example, special notification regarding the biospecimen repository with an opt-in or opt-out process, or continued reliance on existing procedure consent forms, each in combination with an IRB waiver of consent if the biospecimens are identifiable). If clinical excess and the collection of extra specimens are treated similarly, a decision not to participate or a withdrawal from the research biorepository may have implications for both types of specimens. The potential for inconsistencies with information provided in standard procedure forms will need to be avoided.

### Planning for Third-Party Interests in Specimens

Institutions should consider the ways in which third parties will interact with the biorepository in order to delineate any competing rights and obligations. The rights of the institution, those of the specimen donors, those of the investigators, and those of commercial or other sponsors of primary research from which specimens will be collected or donated to the biospecimen repository will intersect in various ways, through various points of contact and legal documentation of rights and responsibilities. Addressing priorities among various parties' interests in advance may avoid unnecessary conflict once biorepository operations commence.

### Rights of Specimen Donors

A research biospecimen repository depends, necessarily, on the willingness of patients and research subjects to donate of themselves—literally—to science. A patient who provides a specimen has certain expectations—and potentially legal rights—regarding how that specimen is obtained and used. The current relevance of these issues is readily apparent in certain high-profile controversies surrounding the use of stored specimens, which are discussed in other chapters of this book in more detail (Callaway 2013). Although an inalienable right, property or otherwise, to excised specimens has not been legally recognized, the issue continues to animate news stories, court cases, and state legislative efforts.[5] Some states go so far as to recognize explicitly a property interest in DNA, while others have unsuccessfully introduced bills to that effect.[6] The available case law on the issue of specimen ownership is fact-specific and generally decided under state law, including principles of gift law. However, it is clear that the content of the informed consent form will be critical in the event that a patient later disputes the institution's rights with respect to donated specimens or commercial profits that arise from their use.[7] Institutions must walk the line between explaining the limits of specimen donors' rights and offending the Common Rule's prohibition on exculpatory language.[8] Nonetheless, it is important to explain the limits of any rights to commercial proceeds and whether specimens can be re-directed to other uses following donation to the institutional biospecimen repository, including whether or not a withdrawal from participation in the biorepository equates to a right to have the specimen physically returned to the donor.[9]

**Rights of Investigators**
Internal researchers responsible for contributing certain specimens or subsets of the biospecimen repository (e.g., a breast cancer specialist helping to facilitate the collection and deposit of breast cancer tumor specimens into the repository) may have expectations of credit or ownership that flow from their contributions. Institutions have an interest in preserving control over a biospecimen repository resource and maintaining the stability of its contents. Clear policies articulating intellectual property rights and the institution's authority to maintain and control institutional databases or biospecimen repositories to which its employees contribute set appropriate expectations for investigators and avert disputes. For the same reasons, policies should address the rights and obligations of departing investigators with respect to resources shared through repositories.[10]

Institutions with specialized or comprehensive biospecimen repositories also should proactively manage claims by third-party researchers using the resource. If a third-party researcher receives access under the biospecimen repository protocol's approved criteria, the primary operational consideration is the contract delineating the terms under which the specimens and associated information will be provided. It is best practice for the institution to enter into a Materials Transfer Agreement (MTA) with the third-party researcher that addresses, among other topics, the following:

- permitted uses of the specimens and associated data
- ownership of the specimens and data and any derivatives (e.g., cell lines)

- if de-identified or limited data set information is provided, a commitment by the researcher not to attempt to re-identify the information
- whether review and approval of the research by an IRB or ethics committee external to the institution is a condition of receipt
- the institution's rights, if any, in any intellectual property developed from the use of the specimens or data
- possible indemnification by the researcher for the secondary research use of the specimens or data or violation of the terms of the MTA (or applicable laws)
- the limits on the source institution's liability for specimens' fitness for a particular purpose
- privacy and security requirements concerning the maintenance of the specimens and data, including compliance with applicable privacy laws (e.g., HIPAA and state privacy laws or, if the recipient is in a foreign jurisdiction, foreign privacy laws) and the parties' obligations in the event of a breach
- costs for the time and effort of providing the specimens and data
- publication of research results and acknowledgment of the source institution
- requirements for returning or destroying excess specimens and data after the research has concluded or following participant withdrawal, if applicable
- obligations on the recipient to pass through any restrictions to further downstream users, assuming further disclosure and sharing is permissible
- reporting research results back to the source institution or otherwise making them available to the donors of the specimens and data.

The terms of any MTA should be consistent with any related requirements under the protocol or SOPs governing the operations of the biospecimen repository, as well as the terms of any informed consent and authorization.

**Rights of Commercial Sponsors and Other Similarly Situated Third Parties**
In conducting a research study sponsored by a pharmaceutical company or other third-party, an institution may wish to incorporate into its biospecimen repository *excess* specimens collected during the research study or collect *extra* specimens for the repository in tandem with the research study's procedures. The institution should not incorporate excess research study specimens into its repository without the written approval of the third-party sponsor and without ensuring that the specimen donor has

provided his or her informed consent *in addition* to the consent to participate in the underlying trial. Failure to do so probably would constitute a significant violation of the terms of the applicable research agreement. The sponsor might also assert ownership rights in the specimens or any intellectual property derived from their use. If the sponsor is willing to give its approval for the retention of excess specimens, this permission, and any rights retained by the sponsor in such materials, should be documented in the existing research agreement or a separate written agreement. Even the simultaneous collection of *extra* specimens for the repository in the course of an on-going sponsored research study may be something to which a sponsor objects, counseling in favor of frank discussions with the sponsor ahead of time. For example, some research agreements contain terms prohibiting the institution from enrolling trial participants in other simultaneous research, which, if broadly construed, may prevent trial participants from providing separate consent to donate specimens to the repository.

**Education Efforts**

Education and engagement of the institution's patient and clinician communities will help to ensure the success of a centralized institutional biospecimen repository, and in particular to the success of front-door consent. If patients or clinicians lack understanding of the biospecimen repository's goals and operations, or have unaddressed concerns regarding the perceived risks and burdens of their participation and involvement, then enrollment may be slow or low, and the repository may not acquire a broad or representative spectrum of specimens.

With respect to the patient community, general biospecimen repositories may face greater challenges in promoting knowledge and investment than disease-specific ones (Simon et al. 2011), the latter of which may benefit from already sophisticated and motivated patient populations who have more experience with the medical community in general and with research in particular. Studies have indicated that with even modest educational efforts, the general public readily recognizes the medical and scientific value of biospecimen repositories (L'Heureux et al. 2013; Simon et al. 2011). However, such appreciation does not necessarily translate into high rates of participation. Rather, individuals express numerous questions and worries regarding biospecimen repository operations, ranging from ownership of the contributed specimens, to the scope of third-party access to and use of the resource (particularly by industry) and associated privacy concerns, to the potential for misuse by insurers or others, to the extra time or potential discomfort required to participate (such as additional visits to the institution or anxiety about an additional needlestick) (L'Heureux et al. 2013; Simon et al. 2011; Kaufman et al. 2009).

Understandably, much of the focus of institutional efforts in addressing these patient concerns, as well as of academic, legal, and ethical discussion of patient engagement with biospecimen repositories, is on individual informed consent (Murphy et al. 2009). As was noted earlier, in a front-door enrollment model, patients often hear about the repository through an individualized recruitment and consent process that occurs at the time of an encounter with their health provider or at patient registration. The informed consent process alone, however, may not optimize patient education or engagement with the repository. Many patients do not fully understand the information presented during the consent process (L'Heureux et al. 2013), the risk of which increases if consent is being obtained by clinical or patient registration staff rather than by repository or other research representatives. Depending on the purpose of their clinical visit, patients may also be distracted at the time they are approached for consent and unable or unwilling to give consideration to biospecimen repository participation. In addition, the consent process may not address the full range of issues about which patients may have questions (most consent processes do not proactively discuss ownership of specimens, for example). Finally, regardless how robust, if the consent process is the first or only time that patients hear about the repository, they may be insufficiently prepared or interested to engage.

To enhance and sustain patient understanding and support of a centralized biospecimen repository, institutions should consider employing a broader set of educational tools that reach patients both before and after the consent encounter and that incorporate interactive group or community learning opportunities. For example, a website dedicated to the repository can post the informed consent form for patients' review in advance of their next clinical visit, but can also provide and link fuller descriptive information about the repository, its goals, and its operations, and in multiple user-friendly formats. These may include videos (Partners Health-Care Biobank 2015b) or frequently asked questions (Partners Health Care Biobank 2015a; Vanderbilt University BioVU 2015c). Newsletters or press releases from the institution can provide updates about the progress of the repository in reaching specimen collection goals and information about specific institutional and third-party research projects using the specimens and the general results arising out of such research (Vanderbilt University

BioVU 2015a). These types of tools allow patients and prospective donors to review repository participation at their leisure and before the recruitment and consent process, which allows greater opportunity to formulate questions and identify specific concerns. Such tools also provide for dynamic, ongoing communications from the institution, which may better maintain community interest and awareness regarding the repository (Yamada et al. 2013). Other methods of increasing and sustaining patient knowledge and support may include placing posters, pamphlets, or displays throughout the institution (not just in provider offices), on-line resources including social media and patient portals, and hosting periodic educational sessions for patients and the general public about the repository. The latter offer individuals the opportunity to benefit from hearing others' questions regarding the repository and learning about the institution's research programs in general. By bringing patients together, these sessions also may help create a sense of community around the institution and better inform the institution of specific local issues. These effects may, in turn, assist the institution to do the necessary work to tailor its repository consent and consent procedures to the local context (Simon et al. 2011).

Institutions may also wish to consider undertaking a broader educational effort that places the biospecimen repository in the context of the institution's overall research mission. In the authors' experience, some of the patient questions, concerns, and complaints that arise during repository recruitment via front-door consent, such as what privacy protections apply to information in the repository and what permissions are necessary from them to store and use their specimens and information for unspecified future research, are equally applicable to, or stem from, questions and concerns about the research enterprise in general. Many, perhaps most, patients do not understand what types of research can be done under the existing regulatory structure without their consent, or what other legal, ethical, and institutional norms and protections govern the conduct of research. Without such understanding, they cannot place the institution's approach with respect to the biospecimen repository within any general framework. As previously noted, if front-door consent is used as a mechanism to obtain broad consent to the use of clinical excess in addition to the prospective collection of research specimens, the biospecimen repository consent may be inconsistent with other clinical procedural consent forms, and such inconsistencies (in addition to needing to be resolved for their own sake) may exacerbate patient misunderstandings. Studies confirm that the public favors reducing or eliminating gaps in awareness of the institution's research activities, and in particular of research uses of their specimens and information (L'Heureux et al. 2013).

One way to implement such a broad educational effort would be through an institution-wide informational campaign focused on research, which could take the form of a written document and/or electronic module available to the public on the institution's website and provided to all patients upon their initial interaction with the institution for clinical care. The notification would provide information, in simple and comprehensible terms, about the conduct and oversight of research at the institution, including the collection, consent, use, and distribution of specimens, before patients are specifically approached about biospecimen repository participation. The repository, as well as any other major or centralized institutional research initiatives, could be referenced in the notification, which would provide the background context important to patients' understanding of the biospecimen repository specifically.

Educational tools specific to the centralized biospecimen repository and broader educational efforts like a research notification would have value for the repository beyond the patient community, specifically in garnering support of clinicians. Distinct from but potentially causally linked to low patient understanding or support of central repositories may be a low level of understanding and support for the biospecimen repository among the institution's clinicians. Among the more immediate challenges for a centralized repository with respect to clinicians are concerns about the potential drain on clinical and administrative staff's time to support biospecimen repository recruitment and enrollment, and reservations about such staff's ability to respond adequately to patient questions and complaints about the repository activities (during and after enrollment). When an institution has the resources to deploy biospecimen repository or other research staff in patient clinics or otherwise alongside providers or patient registration specialists, it can be very beneficial to the involvement and experience of both patients and clinicians in regard to the biospecimen repository (Dalton-Salahuddin and Sieffert 2014). However, many institutions may not have this option. Moreover, as with patients, lack of understanding or interest in the institution's overall research mission by clinicians who do not routinely conduct research may be a less obvious but key impediment to the success of the repository.

Education of clinicians about the biospecimen repository and about the institution's research mission more generally can be accomplished through many of the same mechanisms as may be utilized for patients, including Web-based and other clinician-focused communications and materials

(Vanderbilt University BioVU 2015b), in-person informational sessions, regular updates about research conducted using repository resources, and distribution of a "research notification." As with the patient community, efforts to solicit and consider the questions, concerns, and perspectives of clinician before requesting clinician involvement in the biospecimen repository may be more likely to engender and sustain clinician knowledge, support, and engagement than a one-time or static interaction at the time of biospecimen repository launch.

## Conclusion

Much is required in order to establish a compliant, smoothly functioning institutional biospecimen repository. The enterprise demands the forethought and expertise of many disciplines, and it benefits from an early commitment to designing a comprehensive protocol, implementing ethical, carefully chosen informed consent procedures, and crafting a superstructure of policies and contracts to anticipate and avoid misaligned expectations. Yet a biospecimen repository, however well it is conceived, is only a vessel. Its robustness, quality, and true scientific potential require ongoing education of at least two major stakeholder groups: patients, who are too often under-informed of the existence and role of repositories, and clinicians, whose misconceptions about biospecimen repositories may needlessly inhibit their support and participation.

## Appendix: Biospecimen Repository Planning Checklist for Kickoff Meeting

As is reviewed in this chapter, institutional biospecimen repositories will necessarily differ based on their intended scope and purpose. However, regardless of size, scope, and focus, it is critical that an institution bring together appropriate institutional representatives in designing, developing, and managing the repository. The privacy office, IT department, research administration, medical records, building and facilities, clinical heads, investigators, the IRB director, and institutional legal counsel, among others, should all have input into the project protocol and accompanying policies and procedures. At a hypothetical kickoff meeting, the convened discussants should do the following:

- Define the parameters of the repository: what specimen sources will be utilized (clinical excess, unused research specimens, prospective research collection); what level of identifiability will be observed; will it be a static or

"living" repository with real-time informational updates; for whom is the resource being developed and what limitations will be imposed on users; will donors have the ability to provide tiered or dynamic consent to secondary uses; will broad consent be sought for clinical excess.
- Explore the anticipated demand for the repository, and how do the planned enrollment pace and volume of collections serve to meet the demand (and avoid wasted resources).
- Discuss with the clinicians and the building and facilities management group how best to preserve the specimens once stored in the repository (e.g., the types of freezers, monitoring systems, refrigerators, centrifuges, computers the repository may need, and required cooling capacity at potential location of biospecimen repository). If on-site, discuss the potential spaces the repository may occupy; if off-site, consider the required terms for any agreement with the off-site location that would govern the arrangement.
- If identifiable health information associated with the stored specimens will be collected, discuss with the IT department and the privacy office (and potentially the medical records department), as applicable, the institution's ability to de-identify the data, implement a coding mechanism for the data, link the repository with real-time EMR updates, and apply administrative, technical, and physical safeguards needed to protect the privacy and confidentiality of data.
- Identify the appropriate clinicians, staff, and administrative personnel who will be responsible for managing the day-to-day operations of the biospecimen repository (and any new FTEs for which approval must be requested), develop a plan to ensure they will receive proper training, and decide what type of governance structure will be in place and how it will be documented.
- Flag specific issues of concern for the IRB Director to raise with the full board or IRB executive committee as appropriate.
- Establish a process for providing researchers (internal and external) access to the repository and a method for tracking who accesses what materials.
- Establish a method for prioritizing specimen requests in situations where the available quantity of biospecimen is limited.
- Discuss the financial sustainability of an institutional biospecimen repository; what are the start-up costs and projected operating costs, what revenue centers will finance it, and is there a plan to make it financially sustainable/cost neutral. Will there be a service fee for researchers to obtain the stored biospecimens?

## Notes

1. 45 C.F.R. § 46.102 (2001). A final rule revising the Common Rule was published on January 19, 2017; most of its changes are effective as of January 19, 2018. Under the revised Common Rule, identifiable biospecimens remain within the definition of "human subject," and the creation and maintenance of biorepositories housing such biospecimens are "human subject research" activities.

2. The exemption in 45 C.F.R. § 46.101(b)(4) remains, with slight modification, in the revised Common Rule at 45 C.F.R. § 46.101(d)(4)(ii). Furthermore, the revised Common Rule at 45 C.F.R. § 46.103(d)(7) exempts the creation and maintenance of biorepositories for which broad consent, as defined in the revised regulations, is utilized. This exemption is conditioned on an IRB determination that broad consent is obtained and documented, unless documentation may be waived, and that there are adequate provisions to protect subject privacy and the confidentiality of data, assuming the creation of the biorepository represents a change in how the biospecimens are stored that has been implemented for research purposes.

3. 45 C.F.R. § 160.103.

4. This long-standing interpretation by the Office for Human Research Protections was codified explicitly in the revised Common Rule, effective January 19, 2018, at 45 C.F.R. §§ 46.102(m), 46.117(a) (2018)."

5. Russell Korobkin's discussion in chapter 2 provides a more detailed exploration of property rights as they relate to human specimens.

6. See, e.g., Alaska Stat. Ann. §18.13.010(a)(2)(2004), Fla. Stat. Ann. § 760.40(2)(a) (2009), Colo. Rev. Stat. Ann. § 10-3-1104.7(1)(a)(2009), Ga. Code Ann., § 33-54-1(1) (2015), La. Rev. Stat. Ann. § 22:213.7(E)(2011); see also S.B. 1080, 187th Gen. Ct. (Mass. 2011).

7. See, e.g., *Washington Univ. v. Catalona*, 437 F. Supp. 2d 985 (2006) (finding that, based in part on the language in the informed consent form, Washington University owned all biological materials, including blood, tissue, and DNA, in its biospecimen repository), aff'd, 490 F.3d 667 (8th Cir. 2007); *Moore v. Regents of the University of California*, 793 P.2d 479 (Cal. 1990) (finding in part that "a physician who is seeking a patient's consent for a medical procedure must, in order to satisfy his fiduciary duty and to obtain the patient's informed consent, disclose personal interests unrelated to the patient's health, whether research or economic, that may affect his medical judgment"); but see *Greenberg v. Miami Children's Hosp. Research Inst., Inc.*, 264 F. Supp. 2d 1064 (S.D. Fla. 2003) (finding in part that a researcher had no obligation to include economic interests in the informed consent form for the research use of specimens).

8. 45 C.F.R. § 46.116.

9. *See Washington Univ. v. Catalona*, 490 F.3d 667, 671, 676–77 (8th Cir. 2007).

10. See, e.g., *Washington Univ. v. Catalona*, 490 F.3d 667, 676 n.8 (8th Cir. 2007).

## References

Callaway, Ewen. 2013. Most popular human cell in science gets sequenced. *Nature News*. doi:10.1038/nature.2013.12609.

Dalton-Salahuddin, Sonya L., and Nicole Sieffert. 2014. Obtaining Biorepository "Front Door" Consent in the Pre-Op Anesthesia Clinic to Benefit Patients, Protocols, and Pre-Op Staff. Poster, PRIM&R's Advancing Ethical Research Conference, Baltimore. http://www.primr.org/aer14/posters/pdfs/.

De Souza, Yvonne G., and John S. Greenspan. 2013. Biobanking past, present and future: Responsibilities and benefits. *AIDS* 27 (3): 303–312. doi:10.1097/QAD.0b013e32835c1244.

Kaufman, David J., Juli Murphy-Bollinger, Joan Scott, and Kathy L. Hudson. 2009. Public opinion about the importance of privacy in biobank research. *American Journal of Human Genetics* 85 (5): 643–654. doi:10.1016/j.ajhg.2009.10.002.

Kaye, Jane, David Lund, Edgar A Whitley, Michael Morrison, Harriet Teare, and Karen Melham. 2015. Dynamic consent: A patient interface for twenty-first century research networks. *European Journal of Human Genetics* 23 (2): 141–146. doi:10.1038/ejhg.2014.71.

L'Heureux, Jamie, Jeffrey C. Murray, Elizabeth Newbury, Laura Shinkunas, and Christian M. Simon. 2013. Public perspectives on biospecimen procurement: What biorepositories should consider. *Biopreservation and Biobanking* 11 (3): 137–143. doi:10.1089/bio.2013.0001.

Marsolo, Keith, Jeremy Corsmo, Michael G. Barnes, Carrie Pollick, Jamie Chalfin, Jeremy Nix, Christopher Smith, and Rajesh Ganta. 2012. Challenges in creating an opt-in biobank with a registrar-based consent process and a commercial EHR. *Journal of the American Medical Informatics Association* 19: 1115–1118. doi:10.1136/amiajnl-2012-000960.

Mello, Michelle M., and Leslie E. Wolf. 2010. The Havasupai Indian Tribe case—Lessons for research involving stored biologic samples. *New England Journal of Medicine* 363 (3): 204–207. doi:10.1056/NEJMp1005203.

Murphy, Juli, Joan Scott, David Kaufman, Gail Geller, Lisa LeRoy, and Kathy Hudson. 2009. Public perspectives on informed consent for biobanking. *American Journal of Public Health* 99 (12): 2128–2134. doi:10.2105/AJPH.2008.157099.

OCR (Office for Civil Rights). 2008. "Frequently Asked Questions: How do HIPAA Authorizations Apply to an Electronic Health Information Exchange Environment?"

US Department of Health and Human Services. http://www.hhs.gov/ocr/privacy/ hipaa/faq/health_information_technology/554.html.

OHRP (Office for Human Research Protections). 2011. "Frequently Asked Questions." US Department of Health and Human Services. Last modified January 20. http://www.hhs.gov/ohrp/policy/faq/informed-consent/can-electronic-signature -be-used-to-document-consent.html.

Partners HealthCare Biobank. 2015a. "Frequently Asked Questions." https://biobank .partners.org/faqs.

Partners HealthCare Biobank. 2015b. "Home Page." https://biobank.partners.org.

Pulley, Jill, Ellen Clayton, Gordon R. Bernard, Dan M. Roden, and Daniel R. Masys. 2010. Principles of human subjects protections applied in an opt-out, de-identified biobank. *Clinical and Translational Science* 3 (1): 42–48. doi:10.1111/j.1752-8062.2010 .00175.

Roden, D. M., J. M. Pulley, M. A. Basford, G. R. Bernard, E. W. Clayton, J. R. Balser, and D. R. Masys. 2008. Development of a large-scale de-identified DNA biobank to enable personalized medicine. *Clinical Pharmacology and Therapeutics* 84 (3): 362–369. doi:10.1038/clpt.2008.89.

Rogers, Joyce, Todd Carolin, Jimmie Vaught, and Carolyn Compton. 2011. Biobankonomics: A taxonomy for evaluating the economic benefits of standardized centralized human biobanking for translational research. *Journal of the National Cancer Institute Monographs* 42: 32–38. doi:10.1093/jncimonographs/lgr010.

Rothstein, Mark A. 2010. Is deidentification sufficient to protect health privacy in research? *American Journal of Bioethics* 10 (9): 3–11. doi:10.1080/15265161.2010.494 215.

Simon, Christian M., Jamie L'Heureux, Jeffrey C. Murray, Patricia Winokur, George Weiner, Elizabeth Newbury, Laura Shinkunas, and Bridget Zimmerman. 2011. Active choice but not too active: Public perspectives on biobank consent models. *Genetics in Medicine* 13 (9): 821–831. doi:10.1097/GIM.0b013e31821d2f88.

Steinsbekk, Kristin Solum, Bjørn Kåre Myskja, and Berge Solberg. 2013. Broad consent versus dynamic consent in biobank research: Is passive participation an ethical problem? *European Journal of Human Genetics* 21: 897–902. doi:10.1038/ ejhg.2012.282.

Vanderbilt University BioVU. 2015a. "BioVU Newsletter." https://victr.vanderbilt .edu/pub/biovu/?sid=221.

Vanderbilt University BioVU. 2015b. https://victr.vanderbilt.edu/pub/biovu/?sid =226.

Vanderbilt University BioVU. 2015c. https://victr.vanderbilt.edu/pub/biovu/?sid=218.

Vaught, Jimmie B., Marianne K. Henderson, and Carolyn C. Compton. 2012. Biospecimens and biorepositories: From afterthought to science. *Cancer Epidemiology, Biomarkers & Prevention* 21 (2): 253–255. doi:10.1158/1055-9965.EPI-11-1179.

Yamada, Kathryn A., Akshar Y. Patel, Gregory A. Ewald, Donna S. Whitehead, Michael K. Pasque, Scott C. Silvestry, Deborah L. Janks, Douglas L. Mann, and Jeanne M. Nerbonne. 2013. How to build an integrated biobank: The Washington University Translational Cardiovascular Biobank & Repository experience. *Clinical and Translational Science* 6 (3): 226–231. doi:10.1111/cts.12032.

# Contributors

**Rebecca A. Anderson, PhD, RN** Director, Utah Center of Excellence in Ethical, Legal, and Social Implications of Genetics Research (UCEER); Assistant Director, Genetic Science in Society (GeneSIS) Center, University of Utah

**Heide Aungst, MA** Senior Outreach Specialist, Center for Pediatric Genomics, Cincinnati Children's Hospital Medical Center; previously Narrative Medical Writer with the Departments of Bioethics and Family & Community Health Research at Case Western Reserve University

**Avery Avrakotos** MPH/MPP candidate, University of Michigan School of Public Health and Gerald R. Ford School of Public Policy

**Mark Barnes, JD, LLM** Partner, Ropes & Gray LLP; Faculty co-Director, Multi-Regional Clinical Trials Center (MRCT) of Harvard and Brigham and Women's Hospital; Lecturer, Yale Law School and Yale School of Medicine

**Jill S. Barnholtz-Sloan, PhD** Professor, Associate Director for Bioinformatics/Translational Informatics, Case Comprehensive Cancer Center and Case Western Reserve University School of Medicine

**Benjamin Berkman, JD, MPH** Faculty, Department of Bioethics, National Institutes of Health; Deputy Director, Bioethics Core, National Human Genome Research Institute

**Barbara E. Bierer, MD** Professor of Medicine, Harvard Medical School; Faculty co-Director, Multi-Regional Clinical Trials (MRCT) Center of Harvard and Brigham and Women's Hospital; Director, Regulatory Foundations, Ethics and the Law Program, Harvard Catalyst, Harvard Medical School;

Senior Physician, Division of Global Health Equity, Department of Medicine, Brigham and Women's Hospital

**Jeffrey R. Botkin, MD, MPH** Professor of Pediatrics, Chief, Division of Medical Ethics and Humanities, Associate Vice President for Research, University of Utah School of Medicine

**Dan Brock, PhD** Frances Glessner Lee Professor of Medical Ethics, Department of Social Medicine, Harvard Medical School; Director, Harvard University Program in Ethics and Health

**Ellen Wright Clayton, MD, JD** Craig-Weaver Professor of Pediatrics, Professor of Law, Professor of Health Policy, and Member, Center for Biomedical Ethics and Society, Vanderbilt University Medical Center and Vanderbilt University

**I. Glenn Cohen, JD** Professor of Law; Faculty Director, Petrie-Flom Center for Health Law Policy, Biotechnology, and Bioethics, Harvard Law School

**Lisa Eckstein, SJD** University of Tasmania Faculty of Law

**Barbara J. Evans, PhD, JD, LLM** Professor of Law and George Butler Research Professor; Director, Center on Biotechnology & Law, University of Houston Law Center

**Emily Chi Fogler, JD** Counsel, Verrill Dana LLP

**Nanibaa' A. Garrison, PhD** Assistant Professor, Division of Bioethics, Department of Pediatrics, University of Washington School of Medicine; Faculty, Treuman Katz Center for Pediatric Bioethics, Seattle Children's Hospital and Research Institute; Adjunct Assistant Professor, Department of Bioethics and Humanities, University of Washington School of Medicine

**Pamela Gavin, MBA** Chief Operating Officer, National Organization for Rare Disorders

**Aaron J. Goldenberg, PhD, MPH** Associate Professor and Director of Research, Department of Bioethics, Case Western Reserve University School of Medicine; Associate Director, Center for Genetic Research Ethics and Law

**Christine Grady, MSN, PhD** Chief, Department of Bioethics, National Institutes of Health Clinical Center

Kate Gallin Heffernan, JD  Partner, Verrill Dana LLP

Marylana Saadeh Helou, JD  Associate, Verrill Dana LLP

Sara Chandros Hull, PhD  Faculty, Department of Bioethics, National Institutes of Health; Director, Bioethics Core, National Human Genome Research Institute.

Elisa A. Hurley, PhD  Executive Director, Public Responsibility in Medicine and Research (PRIM&R)

Steven Joffe, MD, MPH  Vice Chair of Medical Ethics, Emanuel and Robert Hart Associate Professor of Medical Ethics and Health Policy, Associate Professor of Pediatrics, Department of Medical Ethics and Health Policy, University of Pennsylvania Perelman School of Medicine

Erin Johnson, PhD  Research Associate, College of Nursing, University of Utah; Associate Director of the Utah Center of Excellence in Ethical, Legal, and Social Implications of Genetics Research

Julie Kaneshiro, MA  Deputy Director, Office for Human Research Protections, United States Department of Health and Human Services

Aaron S. Kesselheim, MD, JD, MPH  Associate Professor of Medicine, Harvard Medical School; Director, Program On Regulation, Therapeutics, And Law (PORTAL), Division of Pharmacoepidemiology and Pharmacoeconomics, Brigham and Women's Hospital

Isaac S. Kohane, MD, PhD  Marion V. Nelson Professor and Chair, Department of Biomedical Informatics, Harvard Medical School

David Korn, MD  Consultant in Pathology, Massachusetts General Hospital and Professor of Pathology, Harvard Medical School. Previously, inaugural Vice-Provost for Research of Harvard University; Vice-President, Dean of Medicine, and Professor and Founding Chair of Pathology, emeritus, Stanford University.

Russell Korobkin, JD  Richard C. Maxwell Professor of Law, UCLA School of Law

Bernard Lo, MD  Professor of Medicine Emeritus, Director of the Program in Medical Ethics Emeritus, University of California San Francisco; President, Greenwall Foundation

**Geoffrey Lomax, DrPH** Senior Officer for CIRM Strategic Infrastructure, California Institute for Regenerative Medicine

**Kimberly Hensle Lowrance, EdM** Former Managing Director, Public Responsibility in Medicine and Research (PRIM&R)

**Holly Fernandez Lynch, JD, M.Bioethics** Executive Director, Petrie-Flom Center for Health Law Policy, Biotechnology, and Bioethics, Harvard Law School; Faculty, Center for Bioethics, Harvard Medical School; Member, US Department of Health and Human Services Secretary's Advisory Committee on Human Research Protections (SACHRP)

**Bradley A. Malin, PhD** Professor of Biomedical Informatics, School of Medicine; Professor of Biostatistics, School of Medicine; Associate Professor of Computer Science, School of Engineering; Director, Co-Health Data Science Center, Vanderbilt University

**Karen J. Maschke, PhD** Research Scholar; Editor, IRB: Ethics & Human Research, The Hastings Center

**Eric M. Meslin, PhD, FCAHS** President, Council of Canadian Academies

**P. Pearl O'Rourke, MD** Director, Human Research Affairs, Partners HealthCare, Boston; Associate Professor Pediatrics, Harvard Medical School

**Quinn T. Ostrom, MA, MPH** Research Coordinator, Case Comprehensive Cancer Center and Case Western Reserve University School of Medicine

**David Peloquin, JD** Associate, Ropes & Gray LLP

**Rebecca Pentz, PhD** Professor of Research Ethics, Emory School of Medicine

**Jane Perlmutter, PhD, MBA** Founder and President, Gemini Group

**Ivor Pritchard, PhD** Senior Advisor to the Director of the Office for Human Research Protections, United States Department of Health and Human Services

**Suzanne M. Rivera, PhD, MSW** Vice President for Research and Technology Management, Case Western Reserve University; Assistant Professor, Departments of Bioethics and Pediatrics at CWRU School of Medicine; Faculty Member, Rainbow Babies and Children's Hospital, Center for Child Health and Policy

**Erin Rothwell, PhD** Associate Professor, College of Nursing and Division of Medical Ethics and Humanities at University of Utah. Co-Director of Research at the Utah Center of Excellence in Ethical, Legal, and Social Implications of Genetics Research and Women and Children's Healthcare

**Andrew P. Rusczek, JD, MBE** Partner, Verrill Dana LLP

**Rachel E. Sachs, JD, MPH** Associate Professor of Law, Washington University in St. Louis School of Law; previously, Academic Fellow, Petrie-Flom Center for Health Law Policy, Biotechnology, and Bioethics, Harvard Law School

**Carol Weil, JD** Program Director, Ethical and Regulatory Affairs, Cancer Diagnosis Program, National Cancer Institute

**David Wendler, PhD** Senior Investigator, Head, Section on Research Ethics, Department of Bioethics, National Institutes of Health Clinical Center

**Benjamin S. Wilfond, MD** Professor and Head, Division of Bioethics, Department of Pediatrics, University of Washington School of Medicine; Director, Treuman Katz Center for Pediatric Bioethics, Seattle Children's Hospital and Research Institute; Adjunct Professor, Department of Bioethics and Humanities, University of Washington School of Medicine

**Susan M. Wolf, JD** McKnight Presidential Professor of Law, Medicine & Public Policy; Faegre Baker Daniels Professor of Law; Professor of Medicine; Chair, Consortium on Law and Values in Health, Environment & the Life Sciences, University of Minnesota

# Index

Abandonment, law of, 54
Accountability of investigators, 363
 awareness and, 358–359
Advance Notice of Proposed
 Rulemaking (ANPRM), 7, 8, 27,
 322, 323, 327
Advocacy, patient. *See* Patient advocacy;
 Patient advocates
Advocacy organizations, health, 335
Agency. *See also* Autonomy
 of research subjects, 100 (*see also*
 *Greenberg v. Miami Children's*
 *Hospital Research Institute*)
AIDS Specimen Biobank (ASB) at UCSF,
 336
Albright, Madeleine, 58
*All of Us* Research Program, 225. *See*
 *also* Precision Medicine Initiative
 Cohort Program
Allyse, M., 177
Altruism, donating specimens out of,
 126
American Academy of Pediatrics (AAP),
 246, 249
American Society of Human Genetics
 (ASHG), 249
Analytical validity, 341
Anonymization, 111, 235
 in Brazil, 40
 Data Protection Directive and, 38
Anonymized/anonymous data, 3, 195

Council of Europe (COE) and, 37
 vs. de-identified data, 345
 public support for secondary use of,
 110, 112, 248
Aristotle, 205
Arizona State University (ASU). *See*
 Havasupai people, 2003–2010
 blood sample controversy;
 *Havasupai Tribe of the Havasupai*
 *Reservation v. Arizona Board of*
 *Regents and Therese Ann Markow*
Autonomy, 7–9, 132. *See also specific*
 *topics*
 justice and, 91, 94–99, 101, 104, 133,
 134
 vs. privacy, 60–61
 secondary research and, 163

Ballantyne, A., 173
Bardill, J., 174
*Bearder v. Minnesota*, 321
Beauchamp, Tom L., 97
Beecher, Henry, 95
Belmont Principles, 91, 132–134
 justifiable persuasion and, 211
 undue influence and, 207, 211
Belmont Report: Ethical Principles
 and Guidelines for the Protection
 of Human Subjects of Research,
 69, 87, 91, 95, 97, 104, 206,
 293

Belmont Report (cont.)
    Common Rule and, 70, 101
    informed consent and, 69
    National Research Act and, 70
    overview, 69
    privacy and, 73
    regulations, OHRP guidance, and, 202–203
    risk of harms and, 144
Beneficence principle, 69, 89–91, 97, 133–134, 204, 213, 214, 252
Bequests and gifts, *inter vivos*, 51–54
Bergner, Amanda L., 263t, 266
Biden, Joseph, 346
Biobanking, basics of, 368–370
Biobanks, 338–339
    defined, 40, 319
    types of, 373
Bioethicists, 343
Biorepositories (biospecimen repositories), 385–386. *See also* Biospecimen repository governance; *specific topics*
    defined, 40
    defining the purpose of, 228–229
    terminology, 318–319, 338–339
    types of, 373
Biorepositories and Biospecimen Research Branch (BBRB), 319
Biorepository/cohort design and participant recruitment, 228–230
Biorepository maintenance and management, 231
    issues relating to combining cohorts, 232–233
    managing access to samples and data, 231–232
    protecting participants' rights, 231
    research oversight and control, 233–234
    return of results, 234–236
    translation of benefits back to communities, 236

Biospecimen and data collection, assuring quality in, 370–372
Biospecimen annotation and longitudinal collection data of outcomes data, 374–376
Biospecimen exceptionalism, 111, 143, 153
Biospecimen ownership, 376–377. *See also* Property
Biospecimen repositories. *See* Biorepositories
Biospecimen repository governance, 300
    "consent to governance" approach, 310
    factors with implications for designing, 310–311
    raising the bar for, 307–311
    UK Biobank and the road to, 300–302
    what it should look like, 303–306
Biospecimen repository landscape in United States, 306–307
Biospecimen research
    inconsistent nature of, 359–360 (*see also specific topics*)
    similarities and differences with clinical care, 360–361
Biospecimen resources, evolving new models for, 326–328
Biospecimens. *See also specific topics*
    defined, 253n1
    use of the term, 253n1
Blanket consent, 60, 161–163, 171, 174
    vs. broad consent, 63n43, 161, 167–168, 170f, 172, 369
    definition and nature of, 63n43, 167–168, 170f, 369
Botkin, J., 172
Brazil, 39–40
Broad consent, 72, 129, 133, 161, 178, 233, 327, 358, 369–370. *See also specific topics*
    advantages, 162

# Index

vs. blanket consent, 63n43, 161, 167–168, 170f, 172, 369–370
Common Rule and, 133, 185, 192, 327
CR Final Rule and, 7, 8, 33–35, 37, 42, 348, 355, 369
criticisms of, 72, 234
data sharing, trust, and, 193–195
defined, 63n43, 72, 161, 167, 170f
objections to, 173–174
participants' perspectives on, 195–196
proposal for, 174–177
in rare disease populations, empirical studies on, 262, 264t
vs. specific consent, 9, 172, 196, 258–260 (*see also* Study-specific consent)
systematic reviews of public attitudes about, 260–261
*Brotherton v. Cleveland*, 49
Broussard, Allen, 48
Bush, George W., 274, 277

California Institute for Regenerative Medicine (CIRM), 274–285
iPSC initiative, 279–280
Medical and Ethical Standards Working Group, 276, 277, 281, 282
California Supreme Court. *See Moore v. Regents of the University of California*
Cancer. *See* HeLa cells; Lacks, Henrietta; National Cancer Institute
Capron, Alexander, 329
Cardozo, Benjamin, 327–329
*Catalona, Washington Univ. v.*, 26–27, 52–53, 58, 99–100, 131, 399n7
Cellular Dynamics International (CDI), 275
Checklist consent, 170f, 172
Childress, James F., 97
China, 38–39
Claeys, Eric R., 119

Clinical Laboratory Improvement Amendment (CLIA) certified labs, 341
Clinical Trials Transformation Initiative (CTTI), 344
Clinical utility, 341
Clinical validity, 341
Code of Best Practices for Human Biospecimen Research, 102
Coercion, 202, 203. *See also* Undue influence
package deals as coercive, 205–207
Commercial research, rationality of the bias against, 108–111
Commercial sponsors and other similarly situated third parties, rights of, 392–393
Commercial use of the induced pluripotent stem (iPS) cell repository, 278–279
Common law, 54
of informed consent, 59–61
of property, 26–27
Common Rule (Federal Policy for the Protection of Human Subjects), 22, 30, 61, 144, 391
ANPRM and, 7, 322, 323, 327
biospecimens and, 22, 32–33, 44n18, 61, 70, 153
broad consent and, 133, 185, 192, 327
CIRM and, 276, 282
CR Final Rule and, 7, 27, 30–32
DBS and, 250, 252
electronic signatures as effective for, 389
federally funded research and, 118
final revisions to (*see* CR Final Rule)
HIPAA and, 35, 36, 115, 116
and identifiable, nonidentifiable, and de-identified information, 32–33, 35, 74, 347 (*see also under* De-identified biospecimens)
independent oversight and, 176

Common Rule (cont.)
informed consent and, 29, 60, 61, 70, 76, 113, 389
IRBs and, 9, 29, 31, 56–57, 60, 73–74, 76, 102, 114–117, 193, 389, 399n2
National Bioethics Advisory Commission (NBAC) and, 319–320
National Commission and, 112–113
Newborn Screening Saves Lives Act and, 252
Newborn Screening Saves Lives Reauthorization Act and, 169
NPRM and, 7, 31, 61, 68, 101, 102, 143, 252, 323, 324, 327
origins and history of, 5–6, 70, 112, 113, 324
overview and nature of, 5, 27, 28, 56, 70, 101, 250, 384
PPSC and, 112
privacy and, 73, 115, 116
Privacy Rule and, 35, 115, 116
prohibition on exculpatory language, 391
revised, 9, 12, 27, 68, 117, 169, 185, 250, 252, 323, 324, 347, 353, 399n2, 399n4 (*see also* CR Final Rule)
revisions to, 22, 31–35, 44n11, 61, 114, 129, 133, 153, 192–193, 196, 322, 399n1 (*see also* Notice of Proposed Rulemaking)
waiver provision, 114–118
withdrawal from research and, 282
Communication with donors, ongoing, 177
Community Advisory Boards (CABs), 237
Community-based participatory research (CPBR), 128–129, 321
Community engagement, 192, 311
Companion studies, 201–215
defined, 201

Compensation, 49, 55–58
Conditional vs. unconditional gifts and donations, 51–52. See also *Catalona, Washington Univ. v.*
Consent, 327, 367. See also Informed consent; *specific topics*
considering costs and burdens of obtaining, 172–173
empirical support for, 170–172
ethical reasons in favor of, 169–170
ever-changing purpose of, 358
initial, 175–176
types of, 70–72, 170f, 170t, 172 (*see also* Blanket consent; Broad consent; Front-door consent; Study-specific consent; Tiered consent)
Consent process(es), 368–370, 388–390
designing, 230
institutional biospecimen repositories and, 388–390
investigator's commitment during, 353–364
length of, 267
"Consent to governance" approach, 310
Consistency, 22–23
Council for International Organizations of Medical Sciences, 56
Council of Europe (COE), 36, 37, 42
Covered entities, 30
defined, 44n14
CR Final Rule, 30, 43n7, 168–169, 205, 213, 234, 252, 327, 348. See also Common Rule, revised
ANPRM and, 27, 323
broad consent and, 7, 8, 33–35, 37, 42, 348, 355, 369
Common Rule and, 7, 27, 30–32, 114, 117, 399n1
identifiability and, 7, 30
IRBs and, 33, 34
NPRM and, 7, 9, 33–35, 101, 151, 168, 252, 323, 324, 355

Index                                                                413

preamble, 8
terminology and, 32–33
Cuttler, Leona, 4–5

Dang, Julie H. T., 127, 128
Darquy, Silviane, 263t, 266
Database of Genotypes and Phenotypes
    (dbGaP), 151, 191, 195
Data Protection Directive, 37–38,
    44n19
Data sharing, 368–369
    broad consent, trust, and, 193–195
    tribal sovereignty and, 190–192
Data use agreement (DUA), 31
Dawson, Angus, 90
De-identification. *See also under*
        Common Rule
    continuum of, 345–346
    of dried bloodspots (DBS), 246, 250,
        323
    and genetic/genomic research,
        345–346
    and return of results, 235
De-identified biospecimens, 61, 102,
    125, 320, 321, 346. *See also*
    Re-identification
    benefits to having, 369
    broad consent and, 173
    Common Rule and, 30, 35, 61, 101,
        102, 153, 176, 347
    CR Final Rule and, 30, 323, 324
    importance of access to, 320
    IRBs and, 102, 294
    NPRM and, 7, 101, 148, 323
    Office for Protection from Research
        Risks (OPRR) and, 319
De-identified data, 147, 148
    vs. anonymized data, 345
    HIPAA and, 30, 35, 146, 149
De-identified iPS cell lines, 283
Department of Health and Human
    Services (DHHS/HHS), 319. *See
    also* Office for Human Research

Protections; Secretary's Advisory
    Committee for Heritable Diseases
    in Newborns and Children
Diabetes Project. *See* Havasupai
    people, 2003–2010 blood sample
    controversy
Discarded tissue and specimens, 3, 125,
    294, 368, 371, 390. *See also* Excess
    specimens
Disease-team collaboration for
    prioritization of biospecimen
    repository resources, 373–374
Diversity, 229, 338
    defining, 229
Donation utility, 340–342
Donor education, 338–339, 348
Donors, 135n1. *See also specific topics*
    reasons for giving specimens, 125–127
    respect for, 348–349
    rights of, 390–391
Donors' wishes
    as binding, 132
    honoring, 130, 132
    challenges in, 130–131
Dove, E. S., 305–306
Dresser, Rebecca, 343
Dried bloodspots (DBS) from newborn
    screening programs, 243–247,
    252–253
    changes in federal policy regarding,
        249–252
    professional organizations' stands
        regarding, 248–249
    public attitudes regarding retention
        and use of, 247–248
    retention time for, 244, 245f
Dukepoo, Frank, 188
Dynamic consent, 259, 297, 303, 328,
    329, 389, 398
    ongoing communication with donors
        and, 177
    overview and nature of, 177, 389
    stem cells and, 284

Education efforts, 393–397
Electronic medical records (EMRs), 375, 389
Emanuel, Ezekiel, 88
Eminent domain, law of, 88
Eminent domain jurisprudence, 107. *See also* Taking of property for public purposes
Empowered patient involvement, 342–343
Empowerment of patient advocates, 342–343
Ethical considerations, 133–135
  principles guiding human research ethics, 133–134
Etiologic complexity of genetic conditions, 280
Europe, 36–38
European Union (EU) privacy framework, 37–38
Excess, clinical, 390, 395, 397–398
Excess specimens, 88, 94, 293, 294, 368, 392, 393. *See also* Discarded tissue and specimens

Families. *See* Rare disease patients and their family members
Family and patient revolution, 324–326
Family organizations, 318
Federal Policy for the Protection of Human Subjects. *See* Common Rule
Final Rule. *See* CR Final Rule
Food and Drug Administration (FDA), 28, 29, 39, 60, 221, 341
Front-door consent, 389, 393, 395
Future research, difficulty describing, 44n16

Gaskell, George, 127
Gelsinger, Jesse, 21
General Data Protection Regulation, 38

Genetic resources and materials, 39
Genomics, 339. *See also specific topics*
  translational, 318, 328
Genomics revolution, 319
Gey, George, 3
Gifts
  conditional vs. unconditional, 51–52
  of human tissues, 51–52
  *inter vivos*, 51–54
Golde, David, 54–55, 57–60. *See also Moore v. Regents of the University of California*
Goldenberg, Aaron J., 128
Governance. *See* Biospecimen repository governance; Participatory governance model
Grady, Christine, 72
Graham, John, 300
*Greenberg v. Miami Children's Hospital Research Institute*, 43n5, 50, 53, 58, 100, 131
Group harms, 78
Gymrek, M., 147

Haldeman, Kaaren M., 311
Harris, John, 88
Havasupai people
  2003 Banishment Order issued by, 187
  2003–2010 blood sample controversy, 23, 77–78, 194–196 (*see also Havasupai Tribe of the Havasupai Reservation v. Arizona Board of Regents and Therese Ann Markow*)
  Autonomy principle and, 98–99
  broad consent and, 133, 186, 188, 190, 192–194, 196
  data sharing, tribal sovereignty, and, 190–192
  Justice principle and, 98
  overview, 3, 72, 98–99, 133, 186–188
  study-specific consent and, 190, 192

*Havasupai Tribe of the Havasupai Reservation v. Arizona Board of Regents and Therese Ann Markow*, 3, 72, 98, 186, 187, 321
  court ruling in, 3, 321
  impact on practices for obtaining consent, 193–194
Hawkins, Alice K., 304
Health advocacy organizations, 335
Health and Human Services (HHS). *See* Department of Health and Human Services
Healthcare provider organizations (HPOs), 307
Health equity, 228, 229
Health Insurance and Portability Accountability Act of 1996. *See* HIPAA
HeLa cells, 3, 4, 67–68, 75–76, 79n1, 320. *See also* Lacks, Henrietta
Helsinki, Declaration of, 87, 95–96, 101
Henderson, Gail E., 306, 319
HIPAA (Health Insurance and Portability Accountability Act of 1996), 31, 36, 37, 93, 146, 152, 363. *See also* Privacy Rule
  authorization, 31, 44n16, 363
  waiver of, 31, 35, 115, 116, 118
  biorepositories and, 384
  breaches of, 148, 149
  Clinical Laboratory Improvement Amendment (CLIA) policy and, 341
  Common Rule and, 35, 36, 115, 116
  de-identified data and, 30, 35, 146, 149
  IRBs and, 31, 116, 118
  protected health information (PHI) and, 30, 31
  Safe Harbor model of de-identification, 146, 149
  Security Rule, 149
  21st Century Cures Act and, 44n16
  waiver provisions, 31, 35, 115, 116, 118
HIV/AIDS movement, 336
Homer, N., 145
Human Genetic Resources Administration of China (HGRAC), 39
Human Research Protections, Office for. *See* Office for Human Research Protections
"Human subject." *See also* Participants
  definitions and scope of the term, 7, 8, 28, 29, 32, 133, 153, 319, 399n1
"Human subjects research"
  definitions and scope of the term, 28, 61, 70, 77
  informed consent and, 60 (*see also* Informed consent)

Identifiability (of biospecimens), 73–78, 204–205. *See also under* Common Rule; De-identification; Re-identification; *specific topics*
  concerns about, 74–75
  defined, 74
  vs. non-identifiability, 75
Identifiable biological materials vs. non-identifiable biological materials, 36
Identifiable biospecimens
  definitions and meanings of the term, 31–33
  ethics of access to, 111–114
Identifiable private information. *See also under* Common Rule
  definitions and meanings of the term, 32–33
*Immortal Life of Henrietta Lacks, The* (Skloot), 3, 67, 68. *See also* HeLa cells
Independent advisory board, 308–309
Independent oversight, 176–177
India, 40–41

Indian Council of Medical Research (ICMR), 40–41, 44n22
Induced pluripotent stem (iPS) cell research, use of biospecimens in, 274–275, 285
  commercial use, 278–279
  donor consent, 277–284
  ethics policy considerations, 275–277
  previously banked research specimens, 284–285
  return of results, 279–282
  sensitive use of specimens, 283–284
  withdrawal from research, 282–283
Induced pluripotent stem (iPS) cells, 273–274
  genetic instability, 280
Informational studies, 107
Informed consent, 69–73, 343–345. *See also* Consent; *specific topics*
  common law principle of, 59–61
  components, 202–203
  definitions, 96
  important matters in, 344
  privacy and protection of donors' information, 345–347
  protocols and, 386–387
  providing aggregate and/or individual results to donors, 347–348
  relevance to biospecimen research, 59
Informed consent process. *See* Consent process(es)
Institutional biospecimen repository(ies), 373
  planning for an, 383–384
  the informed consent process, 388–390
  the protocol, 384–388
Institutional review boards (IRBs), 4, 9, 99, 203, 281, 282, 309, 325, 368, 369, 385
  biobanking governance and, 370
  broad consent and, 196, 279, 399n2

chairpersons, 188, 193, 194
Common Rule and, 9, 28–29, 31, 56–57, 60, 73–74, 76, 102, 114–117, 193, 389, 399n2
CR Final Rule and, 33, 34
criteria for IRB review, 73–74
dried bloodspots (DBS) and, 246
FDA and, 28, 29
genetic research and, 358
HIPAA and, 31, 116, 118
institutions and, 374
National Commission and, 112–113
NPRM and, 102
oversight, 29–30, 102, 384
patient advocates and, 296
Privacy Rule and, 115, 116
protected health information (PHI) and, 30
public benefit standard and, 114–118
specific consent and, 194, 196
waivers of consent and, 29, 114, 205, 246, 294, 320, 362, 389, 390
Institutions, collaboration within and between, 374
International biobanking, challenges in, 41–42
International Society for Biological and Environmental Repositories (ISBER), 319
International Society for Stem Cell Research (ISSCR), 276
*Inter vivos* gifts and bequests, 51–54
Investigator oath, 358–359, 361–363
  components of, 361
  statements in, 357t
Investigators, rights of, 391–392
Ischemic time associated with each specimen, tracking the, 371

Jacobson, Peter D., 113
Jamal, Leila, 263t, 266, 267, 269
Jennings, Bruce, 90
Joly, Y., 305

Justice
  and building trust, 79n2
  principle of, 91, 97, 98
  autonomy and the, 91, 94–99, 101, 104, 133, 134

Katz, Jay, 327–329
Kickoff meeting, biospecimen repository planning checklist for, 397–398
Knoppers, B. M., 305
Kvale, S., 356

Lacks, Henrietta, 3, 22, 67–68, 70, 75–76, 79n1, 320, 338. *See also* HeLa cells
Legal landscape, 99–101
Leukodystrophy (LD), 262, 266–268
Loans in trust, biospecimens as, 174
Long, Millie D., 126–127

Markow, Therese, 186
Martin, John, 186
Matalon, Reuben, 50
Materials out (institutional biospecimen repositories), 388
Materials transfer agreement (MTA), 391–392
Materials used (institutional biospecimen repositories), 386
  maintaining the, 387
*McFall v. Shimp*, 48, 51
Medical advances, 221
Medical and Ethical Standards Working Group of CIRM, 276, 277, 281, 282
Medical waste, 55, 125
Might, Matthew, 325
Moore, John, 48, 54–55, 57–60, 99–100
*Moore v. Regents of the University of California*
  compensation and, 57–58
  court ruling in, 2–3, 54, 57, 59, 99–100, 131, 320

informed consent law, privacy concerns, and, 59–60
  law of abandonment and, 54
  overview, 2–3, 54–55, 99
  property law and, 26–27
  tissues as chattels and, 47–48
  transfer of property rights and, 54–55
Morgagni, Giovanni, 91
Murphy, Juli, 127
*Myriad, Association for Molecular Pathology v.*, 50

National Bioethics Advisory Commission (NBAC), 113, 319–320
National Cancer Institute (NCI), 319, 322, 323
National Commission for the Protection of Human Subjects of Biomedical and Behavioral Research, 112–114, 117, 118, 202–203. *See also* Belmont Report
National Institutes of Health (NIH), 145, 195, 367. *See also* National Cancer Institute; Precision Medicine Initiative
  Data Access Committee (DAC), 191
National Organ Transplantation Act (NOTA), 49, 56
National Research Act of 1974, 70, 112
*NCI Best Practices for Biospecimen Resources*, 319, 322, 323
Neumann, Larissa B., 303–304
Newborn screening bloodspots, 244–245. *See also* Dried bloodspots (DBS) from newborn screening programs
Newborn Screening Saves Lives Act, 249–250
Newborn Screening Saves Lives Reauthorization Act of 2014, 76, 169, 323
NIH Director, 308

Notice of Proposed Rulemaking (NPRM), 7, 61, 102, 148, 252, 322, 355
Common Rule and, 7, 31, 61, 68, 101, 102, 143, 252, 323, 324, 327
CR Final Rule and, 7, 9, 33–35, 101, 151, 168, 252, 323, 324, 355
Nuremberg Code, 2, 69–70, 87, 95
Nuu-Chah-Nulth people, genetic research with the, 188–190

Obama, Barack, 225, 277, 299
O'Doherty, Kieran C., 304–305, 311
Office for Human Research Protections (OHRP), 28, 168, 169, 203, 215. *See also* Notice of Proposed Rulemaking
 biospecimen repositories and, 35
 Code of Best Practices for Human Biospecimen Research and, 102
 Common Rule and, 28, 320
 compliance actions, 203–204
 frequently asked questions (FAQs), 203
 penalties imposed by, 146
 undue influence and, 204, 207, 211–213
Office for Protection from Research Risks (OPRR), 319
Office of Biorepositories and Biospecimen Research (OBBR), 319
Organs. *See also* National Organ Transplantation Act
 sale of, 49
Ownership, 376
 vs. stewardship, 376–377

Package consent, 161–162, 165. *See also* Package deals
Package deals, 201–202. *See also* Package consent
 as coercive, 205–207
 as creating undue influence, 207–213
 as inhibiting voluntary informed consent, 213–214
Participant information, 368–369
Participant-led research, 325–326
Participant-researcher relationship. *See also under* Partnership(s)
 shift in attitudes toward the, 321
Participants. *See also* "Human subject"; *specific topics*
 protecting the rights of, 231
 recruitment of, 228–230
Participatory governance model, 296, 299, 303, 305–306, 309–311
Participatory research model, 324. *See also* Community-based participatory research
Partnership governance, 304
Partnership(s). *See also* Patient engagement
 between donor and investigator, 353, 363, 364
 between participant and researcher, 296–297, 299, 321, 325, 326, 329
 from paternalism to, 296, 318, 329
 with the public, 164, 321 (*see also* Community-based participatory research)
Partridge, E. E., 128–129
Patent and Technology Office, U.S., 50
Paternalism, 94, 296, 318, 329
Patient advocacy, 335–337
Patient advocates
 empowerment of, 342–343
 giving a voice to, 296, 335, 336, 343
Patient and family organizations, 318
Patient and family revolution, 324–326
Patient autonomy. *See* Autonomy
Patient-centered drug development, movement toward, 338
Patient-centered model of collaboration, new, 318
Patient-Centered Outcomes Research Institute (PCORI), 336, 342

Patient education, 338–339, 348
Patient engagement, 23, 297, 336. *See also* Partnership(s)
shift toward, 325
Patients
history of exclusion from biospecimen research, 318–324
understanding of biomedical research, 337–338
Perlmutter, Jane, 335
Personal health data (PHD), 110
Pharmacoepidemiological research, 113
Ponder, M., 263t, 266
Population representativeness, 227, 228
Precision Medicine Initiative (PMI), 152, 229–231, 236, 237, 319, 328
*All of Us* Research Program, 225 (*see also* Precision Medicine Initiative Cohort Program)
fact sheet about, 299
as patient-centered, 317
reality vs. donors' perceptions of, 339–340
Working Group, 152, 226, 299, 307–310, 339
Precision Medicine Initiative Cohort Program (PMI-CP), 226, 296, 299, 307, 311
Director, 308
governance model, 299, 307–311
Steering Committee, 308
Primary studies, 205–215
definition and scope of the term, 201, 202
Privacy, 73–78. *See also specific topics*
vs. autonomy, 60–61
concerns about, 74–75
Privacy Act of 1974, 111
Privacy concerns and informed consent law, the complicating interaction of, 59–61
Privacy Protection Study Commission (PPSC), 111–113, 115–118

Privacy Rule (HIPAA). *See also* HIPAA
biorepositories and, 384
Common Rule and, 35, 115, 116
2013 Final Omnibus Rule Update, 169
future research and, 44n16
IRBs and, 116
public benefit standard and, 115, 116
waiver provisions, 35, 115, 116
Profit seeking and profit sharing. *See* Commercial research
Prohibitionist statutes concerning compensation for human biospecimens, 49, 55–56. *See also* Compensation
Property
defined, 48
tissues as, 47, 48
Property rights, 59
transfer of, 51–55
Prospective collection of biospecimens for research, 70–71
Protected health information (PHI), 30, 31
defined, 44n13
Protocol, institutional biospecimen, 384–388
Public benefit standard, 107–108
clarifying the public benefit standard for research, 116–120
as missing in current regulations, 114–116
Public Health Service (PHS) Syphilis Study, 70, 97

Quasi-property, tissues as, 47

Rare disease patients and their family members, attitudes of, 262
on access to data by secondary researchers, 264t, 266
on consent, 262, 263–265t, 267
on non-welfare interests, 265t, 267

Rare disease patients and their family members (cont.)
  on privacy and confidentiality, 265t, 266–267
  willingness and motivations to participate, 262, 264t, 266
Rare diseases, 257
  informed consent for genetic research on, 257–258, 268–270 (*see also* Rare disease patients and their family members)
Reciprocity of advantage, 120
Re-contacting donors, 71, 281, 282, 295
  attitudes and preferences regarding, 192
  coding and, 279
  consent for, 162, 281–282, 295
Recruitment of participants, 228–230
Recruitment processes, designing, 230
Re-identification (re-ID), 6, 75, 89, 102
  concerns about, 6, 102, 103, 108, 111, 112, 143, 283, 321
  CR Final Rule and, 151–152
  difficulties in, 319, 346
  efforts at, 144–150, 153n2
  harms from, 147
  NPRM and, 143, 151–152
  Precision Medicine Initiative (PMI) and, 152
  preventing, 77, 89, 102, 148–153
  public benefit standard and, 114, 117–118
  re-identification codes, 28
  re-identification key, 28
  risk of, 6, 34, 103, 111, 143–146, 149–150, 153, 211, 295, 324
  understanding the, 146–149
  what is involved in re-identifying de-identified biospecimens, 144–146
Relational-trust model, 325
Relational vs. transactional model of biobanks, 325

Religion and specimen donation, 129
Representative samples, 228. *See also* Underrepresentation in research
Research
  defined, 28
  evolving new models for, 326–328
  need for future debate and, 178
Researcher-participant relationship, shift in attitudes toward the, 321
Research initiatives, perception vs. reality of, 339–340
Research notification, distribution of a, 397
Research regulations and guidance, changing, 168–169
Residual newborn screening bloodspots. *See* Dried bloodspots (DBS) from newborn screening programs
Respect for persons, 70–71, 133, 152, 215. *See also* Autonomy
  Belmont Report and, 69, 73, 95, 132, 202
  beneficence and, 214, 252
Rhodes, R., 173
Risk-benefit analysis, 94–95
Risk-benefit balance for research using biospecimens and associated data, how to achieve, 151–153

Sale of organs and body parts, 49
Schaefer, G. Owen, 88
Schizophrenia, 3, 172, 186, 187
*Schloendorff v. Society of New York Hospital*, 327–328
Secondary research, 60, 163, 169, 216n3, 360, 362, 386
  broad consent and, 7, 8, 33, 35, 175, 192, 196
  Common Rule and, 70, 169, 192
  Council of Europe (COE) and, 37
  CR Final Rule and, 35, 324
  defined, 44n17, 60, 70

HIPAA and, 169, 363
identifiability of biospecimens and, 204
IRBs and, 33, 196
Notice of Proposed Rulemaking (NPRM) and, 252
package deals and, 208, 211, 213, 214
vs. prospective research, 70–71
Secondary researchers, access to data by, 264t, 266
Secretary's Advisory Committee for Heritable Diseases in Newborns and Children (SACHDNC), 248, 249
Secretary's Advisory Committee for Human Research Protection, 281
Self-interest, donating specimens out of, 126–127
Selling of organs and body parts, 49
Sensitivity, 341
*Shimp, McFall v.*, 48, 51
Short tandem repeats (STRs), 145
Skloot, Rebecca, 3, 67–68, 320
Slavin, Ted, 49
Social benefit, goal of, 9
Social contract, specimen donation as, 132
Specific consent, 71, 161, 175, 192–193. *See also* Consent; Study-specific consent; *specific topics*
vs. broad consent, 9, 172, 196, 258–260
definition and nature of, 71
IRBs and, 194, 196
Specificity, 341
Specimen donors. *See* Donors
Specimens. *See* Biospecimens
Standards, sustaining public support through rigorous oversight and enforcement of, 102–104
Standards Working Group. *See* Medical and Ethical Standards Working Group of CIRM

Stem cells. *See* Induced pluripotent stem (iPS) cells
Study-specific consent, 170f, 185, 190, 192, 193, 353. *See also* Specific consent
Subjects. *See* "Human subject"; Participants
Sustainability of biobanking, working toward long-term, 376–377
Syphilis experiment, Tuskegee, 70, 97

Tabor, Holly K., 263t, 266
Taking of property for public purposes, 88, 107, 116, 118–120
Taking theory, 119–120
Targeted attacks, 147
Tautenberger, Jeffrey, 92
Testing, 340–341
terminology in, 341
Third-party interests in specimens, planning for, 390–393
Tiered consent, 40, 73, 163, 389, 398
costs of obtaining, 172
overview and nature of, 40, 71–72, 389
Tilousi, Carletta, 186, 189
Tissue bank, 98, 125. *See also* Biorepositories
Tissues
as chattels, 47–49
information within, 49–50
Transfer of property rights, 51–55
Translational genomics, 318, 328
Translational research, 109, 226, 328
Transplants. *See* National Organ Transplantation Act
Tribal sovereignty and data sharing, 190–192
Tribes. *See also specific tribes*
genetic research with, 185–190
Trinidad, Susan Brown, 129

Trust, as factor in donating specimens, 126–129
Tuskegee syphilis experiment, 70, 97
21st Century Cures Act, 44n12, 44n16

UK Biobank, 301
　and the road to governance, 300–302
Ulrich, Michael, 281
Underrepresentation in research, problem of, 227
Underserved populations
　defining, 229–230
　scholarship on research and, 226–228
Undue inducement, 56
Undue influence, 203, 211–212
　factors that determine, 209–211
　interpretations of, 211–212
　Office for Human Research Protections (OHRP) and, 204, 207, 211–213
　package deals as creating, 207–213
Uniform Anatomical Gift Act (UAGA), 49, 56
United States, 25–26
　legal framework of biospecimen repositories, 35–36
　privacy law, 30–31
　property law, 26–27
　recent changes to the Common Rule, 31–35 (*see also* Common Rule, revisions to)
　regulations on research involving human subjects, 27–30
United States Patent and Technology Office (USPTO), 50

Validity
　analytical, 341
　clinical, 341
Vayena, Effy, 325
*Venner v. State of Maryland*, 55
Venter, Craig, 148
Virchow, Rudolf, 92
Voluntariness, 202–203

Waiver of HIPAA authorization, 31, 35, 115, 116, 118
Waiver provisions. *See under* Common Rule; HIPAA; Privacy Rule
Waivers of consent. *See under* Institutional review boards
Ward, Nyk, 189
Waste, medical, 55, 125
Watson, James, 148
Wendler, David, 260
Wertheimer, Alan, 88, 205–207, 210–211
Wilsey, Matt, 325
Winickoff, David E., 303–305, 310

Yamanaka, Shinya, 273, 285

# Basic Bioethics

Arthur Caplan, editor

## Books Acquired under the Editorship of Glenn McGee and Arthur Caplan

Peter A. Ubel, *Pricing Life: Why It's Time for Health Care Rationing*

Mark G. Kuczewski and Ronald Polansky, eds., *Bioethics: Ancient Themes in Contemporary Issues*

Suzanne Holland, Karen Lebacqz, and Laurie Zoloth, eds., *The Human Embryonic Stem Cell Debate: Science, Ethics, and Public Policy*

Gita Sen, Asha George, and Piroska Östlin, eds., *Engendering International Health: The Challenge of Equity*

Carolyn McLeod, *Self-Trust and Reproductive Autonomy*

Lenny Moss, *What Genes Can't Do*

Jonathan D. Moreno, ed., *In the Wake of Terror: Medicine and Morality in a Time of Crisis*

Glenn McGee, ed., *Pragmatic Bioethics, 2nd edition*

Timothy F. Murphy, *Case Studies in Biomedical Research Ethics*

Mark A. Rothstein, ed., *Genetics and Life Insurance: Medical Underwriting and Social Policy*

Kenneth A. Richman, *Ethics and the Metaphysics of Medicine: Reflections on Health and Beneficence*

David Lazer, ed., *DNA and the Criminal Justice System: The Technology of Justice*

Harold W. Baillie and Timothy K. Casey, eds., *Is Human Nature Obsolete? Genetics, Bioengineering, and the Future of the Human Condition*

Robert H. Blank and Janna C. Merrick, eds., *End-of-Life Decision Making: A Cross-National Study*

Norman L. Cantor, *Making Medical Decisions for the Profoundly Mentally Disabled*

Margrit Shildrick and Roxanne Mykitiuk, eds., *Ethics of the Body: Post-Conventional Challenges*

Alfred I. Tauber, *Patient Autonomy and the Ethics of Responsibility*

David H. Brendel, *Healing Psychiatry:Bridging the Science/Humanism Divide*

Jonathan Baron, *Against Bioethics*

Michael L. Gross, *Bioethics and Armed Conflict: Moral Dilemmas of Medicine and War*

Karen F. Greif and Jon F. Merz, *Current Controversies in the Biological Sciences: Case Studies of Policy Challenges from New Technologies*

Deborah Blizzard, *Looking Within: A Sociocultural Examination of Fetoscopy*

Ronald Cole-Turner, ed., *Design and Destiny: Jewish and Christian Perspectives on Human Germline Modification*

Holly Fernandez Lynch, *Conflicts of Conscience in Health Care: An Institutional Compromise*

Mark A. Bedau and Emily C. Parke, eds., *The Ethics of Protocells: Moral and Social Implications of Creating Life in the Laboratory*

Jonathan D. Moreno and Sam Berger, eds., *Progress in Bioethics: Science, Policy, and Politics*

Eric Racine, *Pragmatic Neuroethics: Improving Understanding and Treatment of the Mind-Brain*

Martha J. Farah, ed., *Neuroethics: An Introduction with Readings*

Jeremy R. Garrett, ed., *The Ethics of Animal Research: Exploring the Controversy*

**Books Acquired under the Editorship of Arthur Caplan**

Nicholas Agar, *Truly Human Enhancement: An Argument for Moderate Human Enhancement*

Sheila Jasanoff, ed., *Reframing Rights: Bioconstitutionalism in the Genetic Age*

Christine Overall, *Why Have Children? The Ethical Debate*

Yechiel Michael Barilan, *Human Dignity, Human Rights, and Responsibility: The New Language of Global Bioethics and Bio-Law*

Tom Koch, *Thieves of Virtue: When Bioethics Stole Medicine*

Timothy F. Murphy, *Ethics, Sexual Orientation, and Choices about Children*

Daniel Callahan, *In Search of the Good: A Life in Bioethics*

Robert Blank, *Intervention in the Brain: Politics, Policy, and Ethics*

Gregory E. Kaebnick and Thomas H. Murray, eds., *Synthetic Biology and Morality: Artificial Life and the Bounds of Nature*

Dominic A. Sisti, Arthur L. Caplan, and Hila Rimon-Greenspan, eds., *Applied Ethics in Mental Healthcare: An Interdisciplinary Reader*

Barbara K. Redman, *Research Misconduct Policy in Biomedicine: Beyond the Bad-Apple Approach*

Russell Blackford, *Humanity Enhanced: Genetic Genetic Choice and the Challenge for Liberal Democracies*

Nicholas Agar, Truly *Human Enhancement: A Philosophical Defense of Limits*

Bruno Perreau, *The Politics of Adoption: Gender and the Making of French Citizenship*

Carl Schneider, *The Censor's Hand: The Misregulation of Human-Subject Research*

Lydia S. Dugdale, ed., *Dying in the Twenty-First Century: Towards a New Ethical Framework for the Art of Dying Well*

John D. Lantos and Diane S. Lauderdale, *Preterm Babies, Fetal Patients, and Childbearing Choices*

Harris Wiseman, *The Myth of the Moral Brain*

Jason Schwartz and Arthur L. Caplan, eds., *Vaccine Ethics and Policy: An Introduction with Readings*

Holly Fernandez Lynch, Barbara E. Bierer, I. Glenn Cohen, and Suzanne M. Rivera, eds., *Specimen Science: Ethics and Policy Implications*